THE MOON

THE MOON

FROM IMAGINATION TO EXPLORATION

FOREWORD BY
ADRIAN WEST

CONTENTS

FOREWORD WRITER

Adrian West—known as *VirtualAstro* online—is an astronomer, presenter, and author with one of the largest independent astronomy followings on social media. Passionate about making stargazing accessible, he leads stargazing tours in Oxfordshire, UK, and has written for the BBC, Met Office, National Trust, and others. Adrian's best-selling book, *The Secret World of Stargazing*, and articles for online science magazines inspire readers worldwide. As host of *The Night Sky Show*, a sell-out theater experience, he shares the wonders of the universe in an entertaining way.

CONTRIBUTORS

Carolin Crawford (main consultant) held the position of Public Astronomer at the Institute of Astronomy in Cambridge for many years and was Professor of Astronomy at Gresham College from 2011 to 2015. Her research focused on using telescopes and satellite observations to investigate the environments of the most massive galaxies in the universe.

Philip Eales studied physics, planetary geology, and remote sensing at University College London. As well as writing about astronomy, Earth science, and climate, he runs a computer graphics company specializing in the visualization of astronomical and geographical data for clients including the European Space Agency.

Carolyn Kennett is an astronomer and author. She is the Chair of the Society for the History of Astronomy and a Fellow of the Royal Astronomical Society. She has written books about planetary science and Cornish prehistory and recently contributed to DK's *Cosmos*.

Hilary Lamb is a multi-award-winning science writer. With a background in physics, she has worked as a contributor and consultant on more than a dozen DK books, including *The Physics Book* and *Simply Quantum Physics*, and is nonfiction editor of the literary magazine *Tamarind*.

Giles Sparrow is a science and astronomy author, journalist, and consultant and a Fellow of the Royal Astronomical Society. He has contributed to many DK titles, including *Universe*, *Spaceflight*, and *Cosmos*. His other recent books include *The History of Our Universe in 21 Stars*.

Andrew Szudek is a writer and editor who studied philosophy at the University of Cambridge. He has worked on numerous nonfiction titles, including books on astronomy and the history of science.

FOREWORD

The Moon has been Earth's closest celestial companion for billions of years. Born from the debris created by a cataclysmic collision between Earth and a Mars-sized planet, it has since been an integral part of Earth's evolution. The Moon has helped create the perfect conditions for life in this quiet corner of space and has a profound influence on Earth's rhythms—stabilizing the seasons, controlling the tides, and affecting the development and behavior of living things.

The Moon was sacred to Neolithic humans and its phases were an integral part of their calendars. Early cultures and civilizations around the world continued to look to the Moon in their agriculture, ritual, and religion, leaving proof of its significance in the form of colossal stone monuments that precisely mark out the alignments of the Moon and the Sun. Many cultures today still use the Moon to mark time, planning religious ceremonies and holidays around its cycles. The brilliant, changing face of the Moon—shining with the Sun's reflected light—has fired humankind's imagination, inspiring art, folklore, and myth across the eons.

The Moon has also, from ancient times, been a subject of interest to astronomers. Even the novice stargazer can view the Moon through binoculars or a small telescope, and its alien landscape can be striking. The Moon's surface has thousands of craters, mountains, valleys, and maria—plains that resemble seas when seen from Earth. Some of these features date from the Moon's creation, while others are the result of billions of years of meteorite impacts.

Until very recently, Earth's natural satellite had always been out of reach, a mystery. But in July 1969, with a giant leap for humanity, Apollo astronauts Neil Armstrong and Edwin "Buzz" Aldrin were the first people to set foot on its surface, followed by several subsequent missions. The Apollo program ended in the mid-1970s, and those pioneering landings were consigned to history books. Now, over 50 years later, humans are planning a return to the lunar frontier, this time with a longer-lasting presence. The Moon is a potential springboard to Mars and beyond, marking an exciting new chapter in space exploration.

My journey in astronomy was heavily influenced by the Moon, and still is. This book will take you on an epic tour of our nearest heavenly neighbor, with arresting images, detailed illustrations, and easy-to-follow explanations that reveal the key role played by the Moon in shaping our daily lives, science, culture, and art. If you have ever looked up at the Moon in wonder, then this book is for you.

ADRIAN WEST
ASTRONOMER, PRESENTER,
SOCIAL MEDIA INFLUENCER, AND AUTHOR

EARTH'S COMPANION

When the solar system was young, the newly formed Earth is thought to have been struck by a planet-sized body known as Theia. In the resulting fireball, they both disintegrated and reformed into the Moon and Earth. They orbited each other—initially very closely— and the Moon was large enough to stabilize Earth's axis, which moderated its climate. This created an environment that made it favorable for life to develop on Earth.

VITAL STATISTICS

Earth's closest neighbor in space and the sole planetary body to have been visited by humans, the Moon is both a familiar sight and a strange, hostile world.

The Moon is the only celestial object with surface features that can be seen with the unaided eye. Early observers thought that each dark area was a sea (*mare* in Latin) and the bright areas between them were land (*terra* in Latin). Scientists now know that the Moon is almost totally dry, with a surface of heavily cratered solid rock topped by a thick blanket of dust. With no atmosphere and a mantle too cool to generate much tectonic activity, the Moon has remained almost unchanged for several billion years, making its surface older than the oldest geological features of Earth. It offers a window back in time, through which we can learn much about the early solar system and Earth's origins.

Composition

The Moon is large for a planetary satellite, but its gravity is too weak to retain any significant atmosphere. Like the terrestrial planets (Earth, Mars, Venus, and Mercury), the Moon has a crust and mantle of silicate rock around an iron core. Relative to its size, the Moon's core is much smaller than Earth's, and only a small part of it is liquid. Without a deep, circulating pool of liquid metal in its outer core, the Moon cannot generate a magnetic field like Earth's. As a result, its surface offers no protection from the harmful charged particles in the solar wind. Nevertheless, 12 men walked on the Moon during short visits between 1969 and 1972, and there are plans to return before 2030 and to establish a permanent lunar base.

Orange indicates lava rich in iron oxides

Young craters appear bright

Photo taken at full Moon was used to brighten shadowed part of lunar disk

6,972 craters on the Moon are 12.4 miles (20 km) or more across.

Near side
Subtle variations in color and shade across the Moon's surface are revealed in this composite of photographs taken from Merseyside, UK. For most of the photos, the Moon was 78 percent illuminated.

Blue indicates lava rich in titanium oxides

Dark lava plains cover 31 percent of Moon's near side

MOON PROFILE

Average distance from Earth
238,855 miles (384,400km)

Orbital period
27.3 Earth days

Rotation period
27.3 Earth days

Diameter
2,160 miles (3,476km)
= 0.27 x Earth

Length of a day on the Moon
29.5 Earth days

Volume
5.26 billion cubic miles
(21.9 billion cubic km)
= 0.02 x Earth

Mass
81 billion billion tons
(73.5 billion billion tonnes)
= 0.01 x Earth

Surface temperature
–387° to 253°F (–233° to 123°C)

Surface gravity
0.166 x Earth

Escape velocity
5,313 mph (8,550 km/h)
= 0.21 x Earth

Far side
Only visible from a spacecraft, the far side is more heavily cratered and has far fewer lava plains than the near side, as shown by this image from NASA's Lunar Reconnaissance Orbiter.

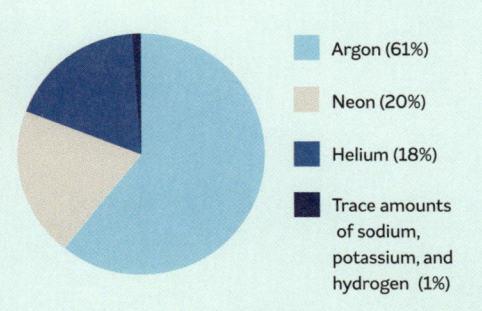

SIZE

The Moon is larger than the dwarf planet Pluto and almost as large as the planet Mercury. It is the solar system's largest moon relative to the size of its parent planet.

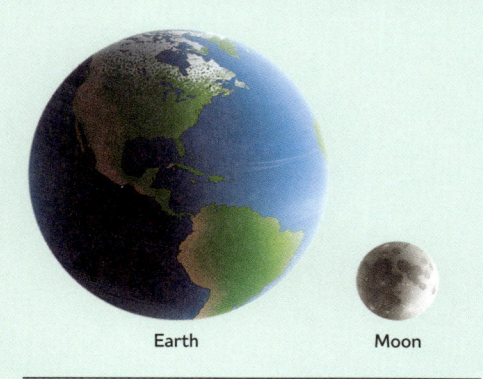

Earth   Moon

STRUCTURE

Like Earth, the Moon has a solid crust and mantle composed of rock and a core of iron. The core is much smaller than Earth's, with a solid inner core and a thin, liquid outer core.

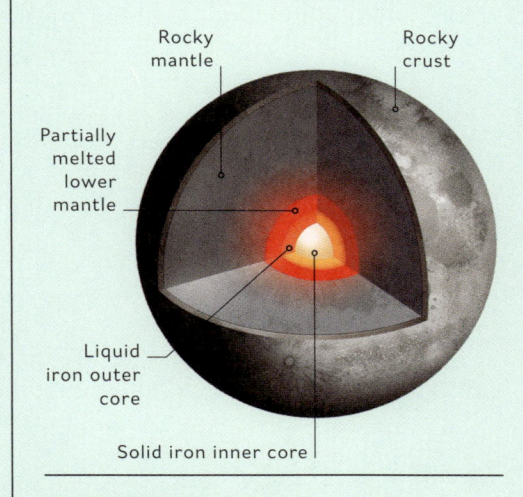

Rocky mantle

Rocky crust

Partially melted lower mantle

Liquid iron outer core

Solid iron inner core

ATMOSPHERE

The Moon has no breathable atmosphere but has gases (mainly argon, neon, and helium) in tiny quantities near its surface. These exert a pressure similar to that of Earth's exosphere (upper atmospheric layer).

Argon (61%)

Neon (20%)

Helium (18%)

Trace amounts of sodium, potassium, and hydrogen (1%)

1. Theia draws ever nearer to Earth, moving along a similar orbit.

2. Theia strikes Earth a glancing blow, unleashing a huge fireball.

Cataclysmic collision
The giant impact hypothesis proposes a violent crash between the early Earth and a smaller planet, Theia, named after the mother of the Moon goddess Selene in Greek myth (see pp.52–53).

A VIOLENT BIRTH

Among the four rocky planets of the inner solar system (Mercury, Venus, Earth, and Mars), Earth is alone in having a large moon. The origin of such a sizable natural satellite presents a puzzle for planetary scientists.

Analysis of lunar rock samples taken by the Apollo astronauts show that the composition of the Moon's crust is remarkably similar to Earth's mantle, so it is unlikely to have formed elsewhere in the solar system. However, the Moon has a far lower iron content than Earth: its core is less than 2 percent of its total mass, but Earth's iron core is 30 percent of its mass. The Moon's surface also has fewer volatiles, such as water. These differences mean that they cannot have formed at the same time from the same disk of material. The analyses also show that an ocean of magma covered the early Moon.

The giant impact hypothesis

The most widely accepted explanation of the Moon's origin proposes that a Mars-sized planet named Theia struck a glancing blow to Earth 4.5 billion years ago. The energy of the collision would have melted both Theia and

Earth's mantle, ejecting a spray of white-hot vaporized rock, which then cooled and coalesced to form the Moon.

Incomplete picture

The giant impact hypothesis could explain many characteristics of the Earth–Moon system, including Earth's large axial tilt, the Moon's small iron core, its early magma ocean, and the orientation and evolution of its orbit.

However, the hypothesis doesn't sit well with the similar composition of these two celestial bodies. If the Moon had formed largely from Theia, then we would detect a greater difference in its mineral content. A recently developed theory, known as the synestia hypothesis, might provide the final piece of the puzzle. It suggests conditions in which both bodies could have completely melted before forming their present structures.

6. Earth and the Moon as they are today.

3. Heat from the collision melts Theia and Earth's mantle.

4. Debris is ejected into Earth's orbit, where it quickly condenses to form the Moon.

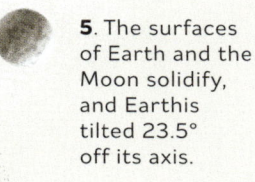

5. The surfaces of Earth and the Moon solidify, and Earthis tilted 23.5° off its axis.

THE SYNESTIA HYPOTHESIS

Recent computer modeling has revealed a new class of transient astronomical object named a synestia. It is a rotating disk of molten and vaporized rock formed during cataclysmic impacts. If the early Earth was rotating much faster than it does today and a larger Theia made a direct blow instead of a glancing blow, then a much more energetic collision could have completely melted early Earth and Theia. The resulting synestia could have then recondensed into Earth and the Moon.

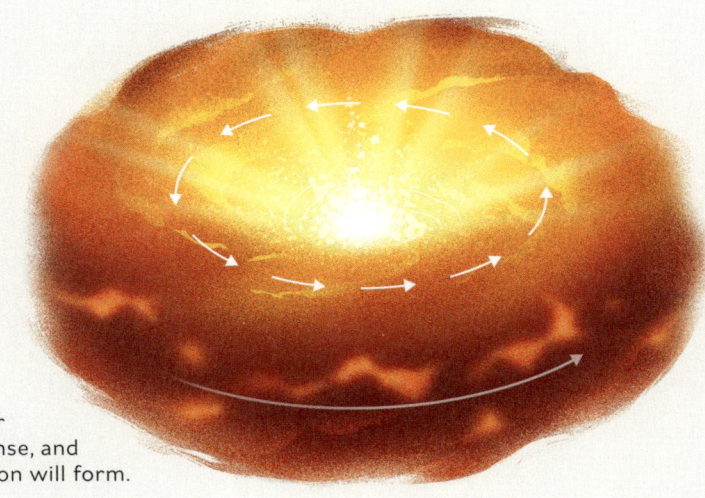

A new planetary object
A synestia has a slowly rotating inner part, from which a planet will condense, and a faster outer disk, from which a moon will form.

ANCIENT UPHEAVALS

Although the Moon appears to be unchanging today, it had a
turbulent early history—one that can inform our understanding
of the origins of Earth and the wider solar system.

The heat unleashed by Theia's impact with Earth (see pp. 12–13) left the Moon with a surface of magma (liquid rock), which was maintained for a while by heat from gravitational compression and the hard rain of molten debris. This magma ocean extended hundreds of miles deep and existed for more than a hundred million years—long enough for the Moon's chemical components to start separating, with lighter elements floating to the top and heavier ones sinking deeper into the interior (see pp. 198–199). This chemical zoning, which has been measured in rock samples brought back from the lunar highlands, was locked in place as the crust solidified 4.5 billion years ago.

Change from above and below

The Moon's new crust was born into a hostile environment of high-energy debris that was left over from the formation of the planets. Several asteroids crashed into the early Moon,

Molten lava

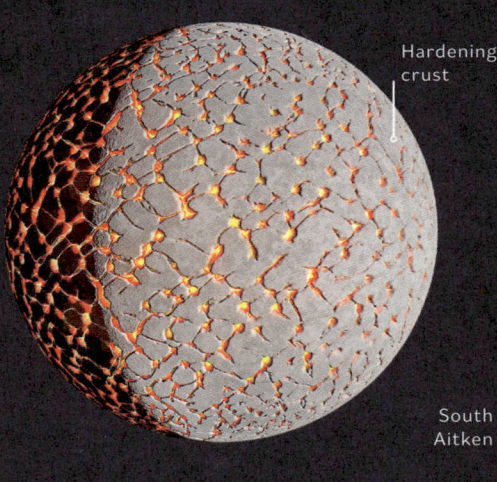
Hardening crust

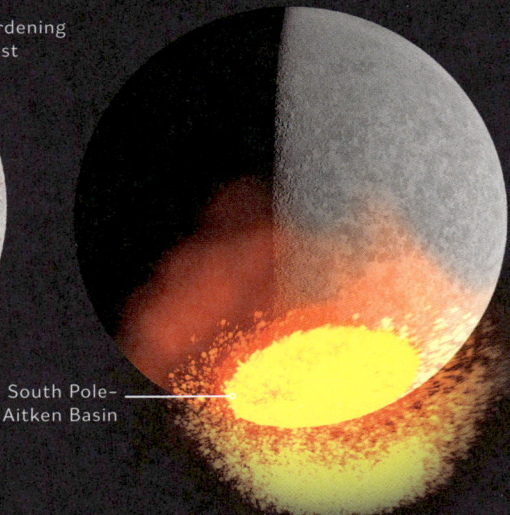
South Pole–Aitken Basin

Magma ocean
After Earth's collision with Theia
4.5 billion years ago (BYA), the Moon
may have coalesced from the resulting
debris within 100 years. Then, for
more than 100 million years, an ocean
of magma covered its entire surface.

Crust formation
As heat dissipated and the surface cooled,
a frothy crust formed that was rich in
aluminum and calcium but lacking heavier
metals, such as iron. This crust survives
in the lunar highlands and beneath the
younger volcanic mare plains.

Early basins
During the pre-Nectarian period, 33 large
asteroids left circular depressions in the
crust. Twelve more impacted during
the Nectarian, followed by three in the
early Imbrian. Each formed a basin, some
of which were later flooded by lava.

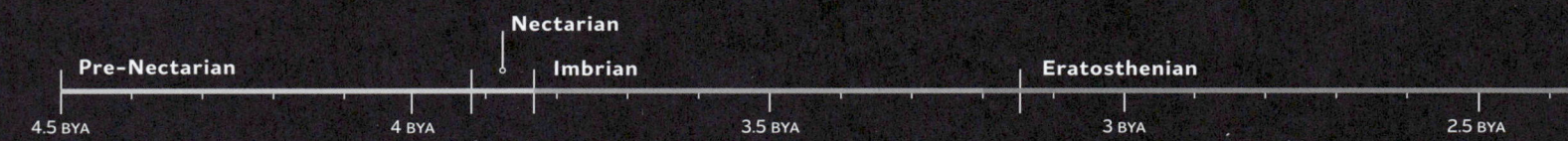

Nectarian

Pre-Nectarian   Imbrian   Eratosthenian

4.5 BYA   4 BYA   3.5 BYA   3 BYA   2.5 BYA

Geological periods
The Moon's geological periods are named after
two basins (Nectaris and Imbrium) and two
large craters (Eratosthenes and Copernicus).

fracturing its crust and forming large circular basins. Innumerable smaller impacts left lesser craters in the crust. This period of rapid change, known as the Late Heavy Bombardment, is thought to have been the last stage of the solar system's birth, in which material from its primordial, planet-forming disk was thrown into the inner solar system by the gravitational fields of the gas-giant planets, particularly Jupiter. The crater size distribution is similar across the Moon, Mercury, and Mars, so it seems likely that all the planets of the inner solar system, including Earth, experienced the Late Heavy Bombardment.

The crustal fractures left behind in the large-impact basins provided a route for later volcanic eruptions from beneath the crust. Highly fluid lava flowed vast distances across low-lying areas, including the basin floors, which were left with relatively smooth, dark surfaces. These maria (or "seas") are the darker patches that are visible on the face of the Moon today (see pp.118–119).

Asteroid impact

Lava flow

Meteor impact

Young craters

Mare

Late Heavy Bombardment
Throughout the pre–Nectarian and Nectarian periods, the Moon was subjected to an intense bombardment of asteroids and meteors. These impacts saturated the highland terrain—so much so that new craters completely erased the old.

Magma lakes
Throughout the Imbrian period and into the Eratosthenian, lava erupted from the Moon's interior and covered the floors of the impact basins and some large craters. The most recent volcanic events may have been 1 billion years ago.

Cratering
Most of the Moon's craters date from the Late Heavy Bombardment. Cratering continues today but with less frequent, and smaller, impacts. The youngest have bright "rays" of debris; the older ones have been eroded by later impacts.

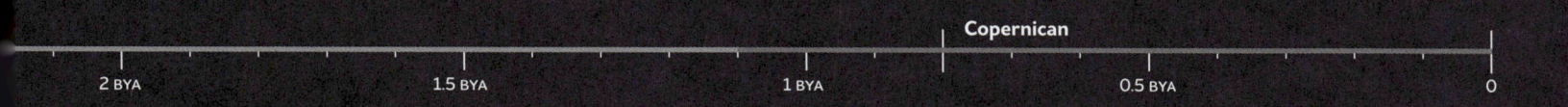

Copernican

2 BYA   1.5 BYA   1 BYA   0.5 BYA   0

THE EARTH-MOON SYSTEM

The Moon is the largest satellite in the solar system relative to its parent planet. Its presence has had a strong influence on Earth's unique geological and biological history.

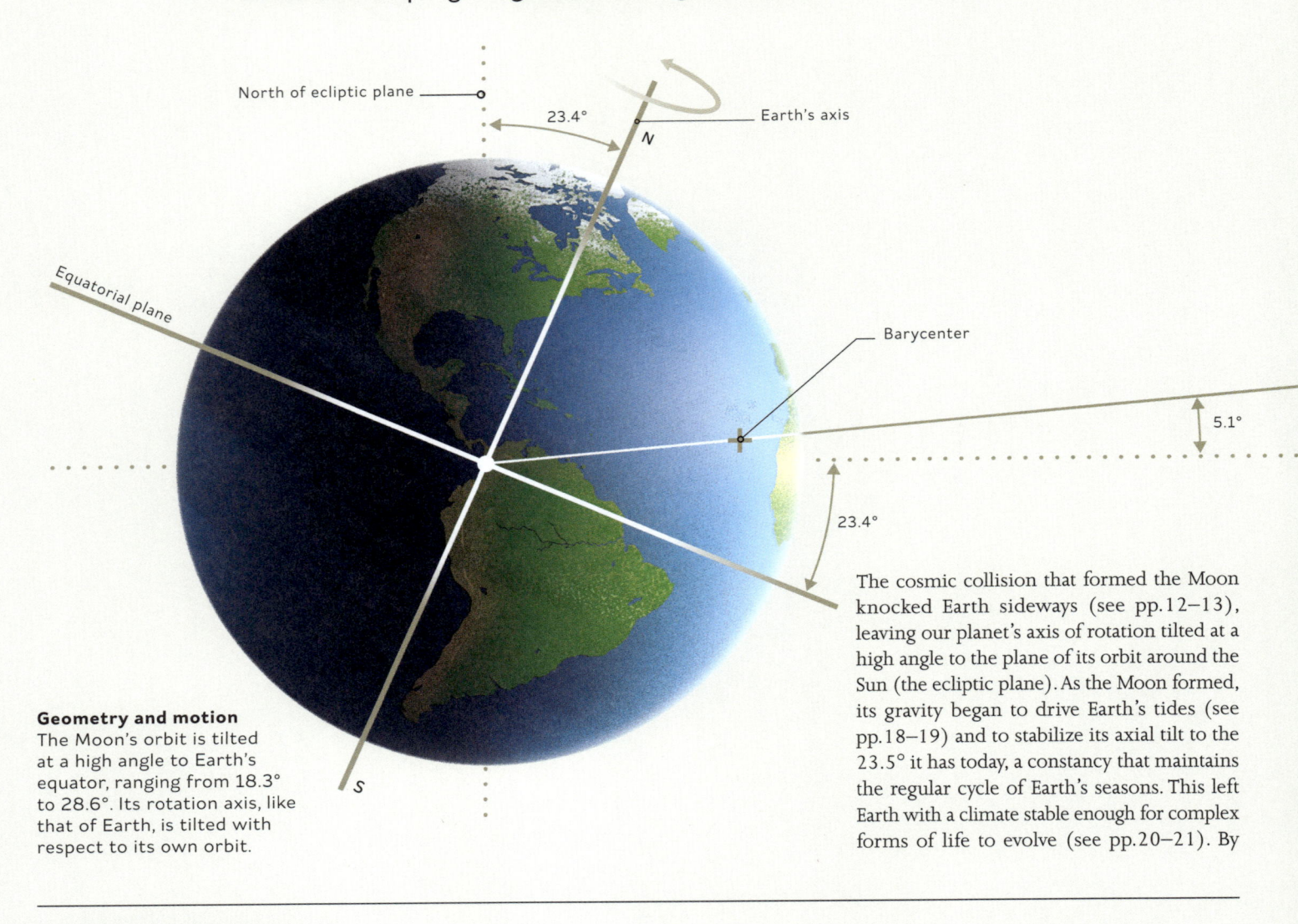

North of ecliptic plane

23.4°

Earth's axis

N

Equatorial plane

Barycenter

5.1°

23.4°

S

Geometry and motion
The Moon's orbit is tilted at a high angle to Earth's equator, ranging from 18.3° to 28.6°. Its rotation axis, like that of Earth, is tilted with respect to its own orbit.

The cosmic collision that formed the Moon knocked Earth sideways (see pp.12–13), leaving our planet's axis of rotation tilted at a high angle to the plane of its orbit around the Sun (the ecliptic plane). As the Moon formed, its gravity began to drive Earth's tides (see pp.18–19) and to stabilize its axial tilt to the 23.5° it has today, a constancy that maintains the regular cycle of Earth's seasons. This left Earth with a climate stable enough for complex forms of life to evolve (see pp.20–21). By

---

PERIGEE AND APOGEE

The Moon's orbit is not quite circular. Like all orbiting bodies, it follows an elliptical path around its parent body. The distance from Earth to the Moon varies by about 5 percent over the course of its orbit.

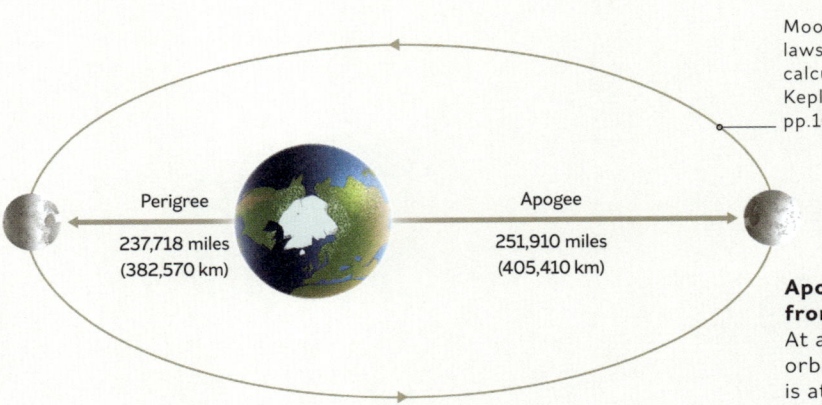

Moon's orbit follows laws of planetary motion calculated by Johannes Kepler in 1609 (see pp.100–101)

Perigree
237,718 miles
(382,570 km)

Apogee
251,910 miles
(405,410 km)

Perigee (closest to Earth)
A full Moon at perigee is known as a "supermoon." It appears 14 percent larger and 30 percent brighter than at apogee.

Apogee (farthest from Earth)
At apogee, the Moon's orbit is slower than it is at perigee.

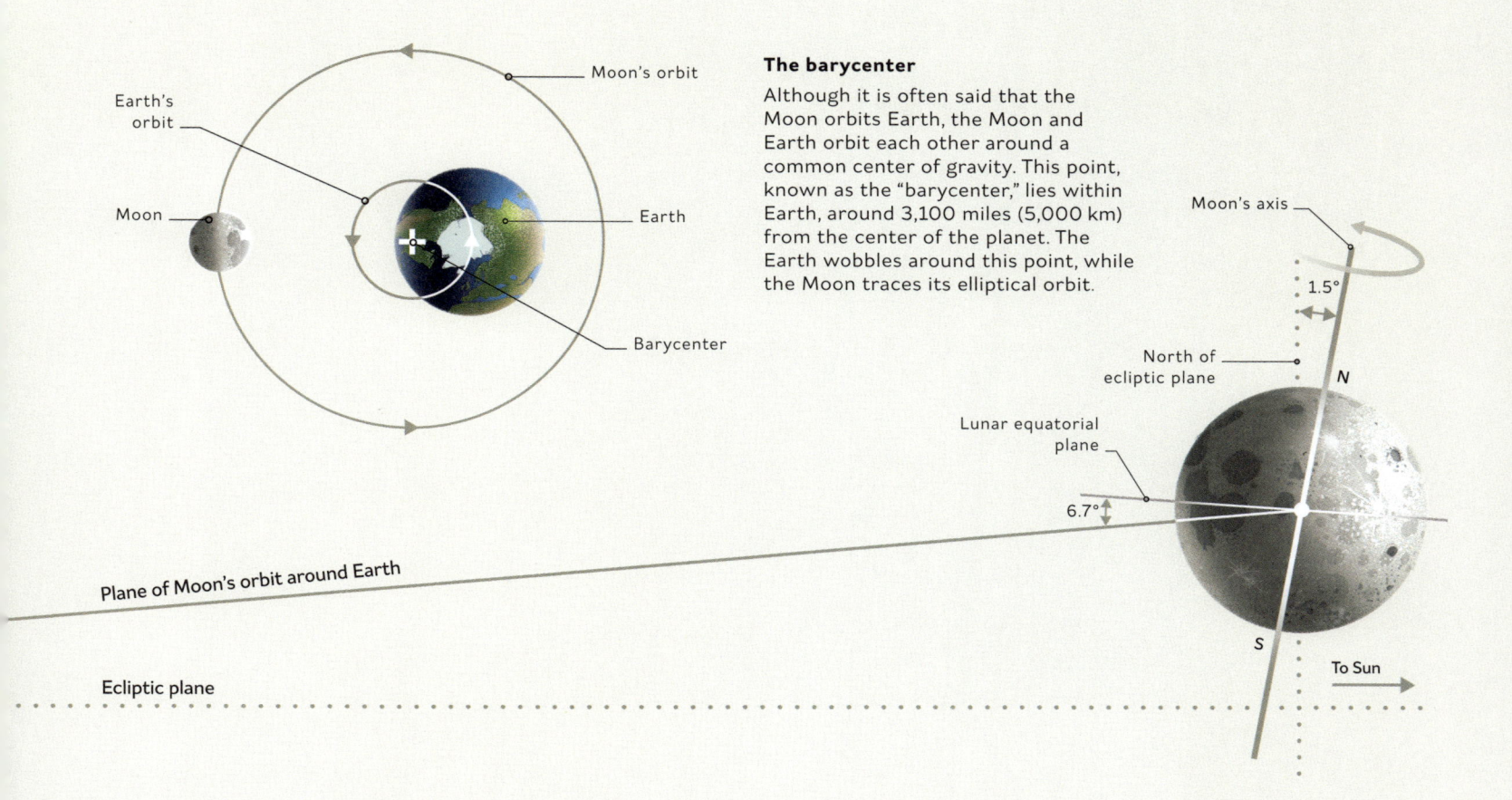

Earth's orbit

Moon's orbit

Moon

Earth

Barycenter

The barycenter

Although it is often said that the Moon orbits Earth, the Moon and Earth orbit each other around a common center of gravity. This point, known as the "barycenter," lies within Earth, around 3,100 miles (5,000 km) from the center of the planet. The Earth wobbles around this point, while the Moon traces its elliptical orbit.

Moon's axis

1.5°

North of ecliptic plane

Lunar equatorial plane

6.7°

N

S

To Sun

Plane of Moon's orbit around Earth

Ecliptic plane

contrast, the planet Mars does not have a stabilizing moon, so its axial tilt shifts, causing unpredictable climatic changes.

Cosmic coincidences

The other large moons of the solar system lie close to the equatorial plane of their parent planets. But the impact that made the Moon left its orbit at a high angle to Earth's equator and closer to the ecliptic plane. By coincidence, the Moon's own rotational axis is further tilted

from its orbit, making it almost perpendicular to the ecliptic plane. This leaves the Moon with no seasons; parts of its polar regions are always in shadow and may bear ice (see pp.196–197).

An even bigger coincidence is the fact that the Moon appears the same size in the sky as the Sun. This was not always so; the Moon was once far closer to Earth (see pp.20–21) and is receding a few centimeters each year. Their similar size, as seen from Earth, enables the spectacle of a total solar eclipse (see pp.38–39).

> "It is best to regard the **Earth–Moon system** as a **double planet**."
> Patrick Moore, amateur astronomer and broadcaster (1923–2012)

SYNCHRONOUS SPIN

The Moon rotates exactly once on its axis each time it orbits Earth. This means that only one side of the Moon is always visible from Earth. Although it is popularly known as the "dark side of the Moon," the far side sees as much Sun as the near side.

Tidal locking

Ever since the Moon's formation, its rotation has been slowed by Earth's gravity pulling at the rock of the lunar interior. This "tidal pull" keeps one side of the Moon facing Earth.

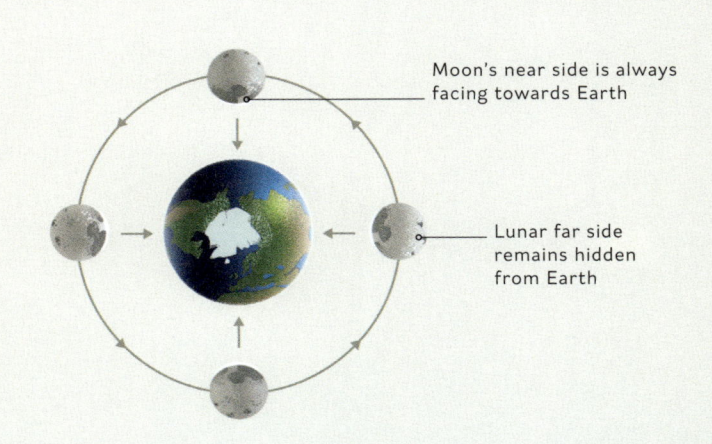

Moon's near side is always facing towards Earth

Lunar far side remains hidden from Earth

TIDAL CYCLES

The Moon's gravitational pull is largely responsible for the regular movement of the tides in Earth's oceans, but Earth's shape and motion and the gravitational pull of the Sun also play a role.

Earth's tides are best known as the regular rise and fall of the sea surface where it meets the land. High tide occurs when the water reaches its highest point inland. Low tide is when it has receded to its farthest point away from land. The tidal range is the height difference between the two extremes.

Ancient sailors used the gentle tides in the Mediterranean Sea as an aid to launch and beach their ships. When they ventured into the North Atlantic, they found the much greater tidal range to be more of a challenge, as it hides and then reveals dangerous reefs and generates powerful tidal currents at the entrance to bays

and narrow straits. Today, knowledge of tides remains crucial to commercial shipping, fishing, and leisure pursuits and unlocks a potential source of carbon-free energy.

While fixed tide gauges have long measured coastal tides, tidal ranges in the open ocean were hard to gauge until the advent of satellites with radar altimeters that can measure the sea's height to within a few centimeters. Although largely driven by the orbit of the Moon and Earth's daily rotation, tides are also modified by the Sun, landmasses, the shape of the seafloor, and the size of the sea. Some places have two high tides a day, some have only one.

HIGH AND LO

The sea level cha
twice-daily cycl
and low tide. The
around the worl
about 12° eastw
of Earth. Therefo
50 minutes later

Earth

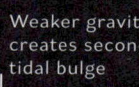

Strong gravity
The Moon's grav
water towards i
"bulge" as a high
planet that is ne

Weaker gravit
creates secon
tidal bulge

Weak gravity
The Moon's grav
slightly more str
ocean. This creat
on the opposite s

Low tid
sides of
water is

Two tides a day
The combined for
gravity and Earth
produce two tida
the orbiting Moo

Tidal range
The height of the sea's surface can change by up to 23 ft (7 m) between low and high tide at Porthcawl, south Wales, shown here at noon (top) and 8 p.m. (bottom).

MONTHLY CYCLES

The Sun contributes its own tidal pull on Earth's oceans. Although much more massive than the Moon, the Sun is so distant that it exerts less than half the tide-raising force of the Moon. Its influence is mostly seen in the monthly cycle of "spring" and "neap" tides, when the Sun either reinforces or counteracts the lunar tide. There is also a small, truly seasonal "declinational" tide due to the angle of the Sun relative to the equator, and a corresponding tidal effect caused by the angle of the Moon relative to the equator.

Spring tides

When Earth, the Moon, and the Sun line up at full Moon and new Moon, the Sun's pull reinforces that of the Moon, making high tides higher and low tides lower than average. The increased tidal range around these times is known as a spring tide.

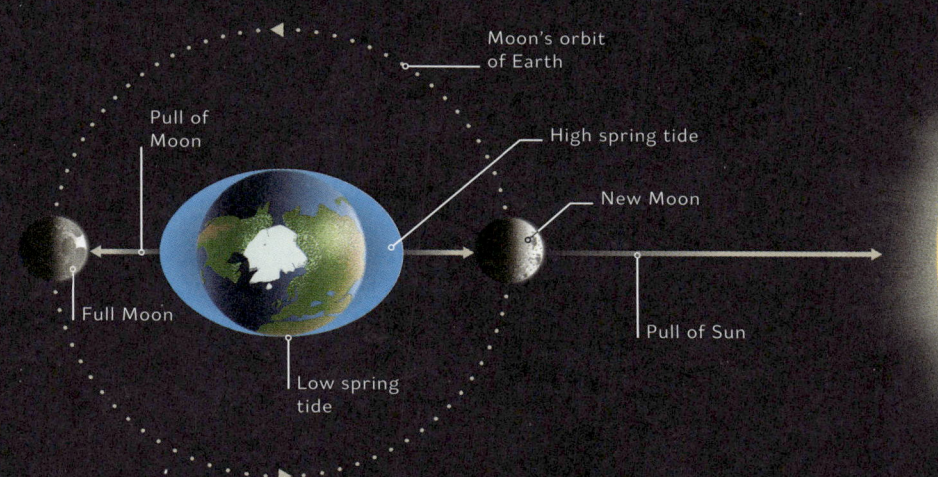

Moon's orbit of Earth

Pull of Moon

High spring tide

New Moon

Full Moon

Pull of Sun

Low spring tide

Neap tides

A week after spring tide, at first-quarter or last-quarter Moon, the Sun and the Moon are 90° apart, and the Sun's pull counteracts that of the Moon, causing high tides to be lower and low tides to be higher than normal. This reduced tidal range is called a neap tide.

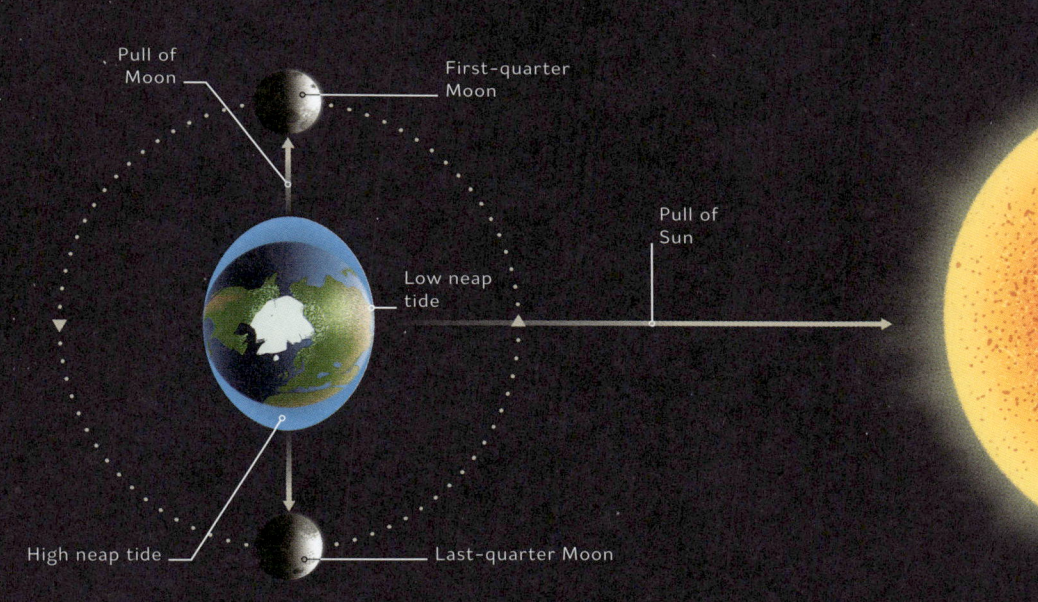

Pull of Moon

First-quarter Moon

Pull of Sun

Low neap tide

High neap tide

Last-quarter Moon

TIDES AROUND THE WORLD

Tidal range varies in response to the configuration of the coastline and the shape of the seafloor. The highest and lowest tides occur in coastal areas. In parts of the open ocean, known as amphidromic points, there are almost no tides (shown in dark blue on the map at right). In the northern hemisphere, the crest—or highest part—of the tide rotates counterclockwise around these areas. In the southern hemisphere, the rotation is clockwise.

Twenty-year lows

This map shows the lowest tides relative to mean sea level computed over 20 years. The lowest tidal ranges occur in small, enclosed seas and in parts of the open ocean.

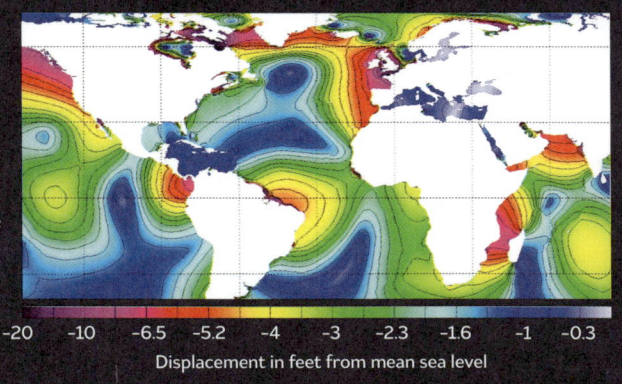

-20   -10   -6.5   -5.2   -4   -3   -2.3   -1.6   -1   -0.3

Displacement in feet from mean sea level

Strong lunar tides ebbing and flowing across vast distances may have provided the ideal conditions for life to develop on Earth.

THE ORIGINS OF LIFE

We may owe our lives to the Moon. It is possible our planet's satellite played a role in kick-starting life on Earth and likely helped animals move from the ocean to the land.

By 3.9 billion years ago (BYA), Earth had a well-established ocean. The Moon was then around three times closer than it is today, exerting nine times its current gravitational pull on Earth's ocean and creating much stronger tides. Earth was also rotating about twice as fast (see below), so high tides would have been only six hours apart. Stronger, more frequent tides helped distribute heat around the planet, stabilizing Earth's climate and increasing the chances of life taking hold.

First life

Evidence suggests life began about 3.8 BYA. The early ocean water was a "soup" of organic molecules, from which the first life may have formed around hydrothermal vents on the deep ocean floor or in hydrothermal springs in shallow water. At these vents and springs, hot, mineral-rich water gushed out from Earth's interior. But there is a third possible origin location: tidal pools. In some places, the strong tides amplified by the Moon's proximity would surge miles inland, filling rock pools and withdrawing again after a few hours. This rapid cycle of flooding and evaporation, maintained over a long period of time, may have created conditions in which strands of organic molecules could repeatedly assemble and disassemble, leading to the development of more complex molecules, including the nucleic acids DNA and RNA—the building blocks of all life on Earth.

The leap onto land

It is possible that, much later, the Moon also gave animals an evolutionary "nudge" to move from the ocean onto land—an event thought to have happened around 400 million years ago. At that time, with the Moon 10 percent closer to Earth, tides were still stronger than today. Fish species at regular risk of being stranded by powerful tides may gradually have evolved limblike fins that they used to flip themselves back into the water, and those often caught in oxygen-poor tidal pools developed primitive lungs that enabled them to take oxygen from the air. In time, with further evolutionary steps, air-breathing four-footed animals emerged and colonized the land.

Growing distance
The Moon is much farther away than it was 3.9 BYA, when it would have appeared 2.9 times larger in the sky. By bouncing laser beams off its surface, scientists have shown that it is still receding from Earth.

Moon's orbit 3.9 BYA: estimated distance 83,140 miles (133,800 km)

Earth

Moon's orbit today: 238,855 miles (384,400 km)

THE RECEDING MOON

The pull of the Moon's gravity makes the ocean bulge out as a high tide that follows the orbiting Moon (see pp. 18–19). Earth rotates much faster than the Moon orbits, which causes friction between the seabed and the tidal bulge. This friction drags the tidal bulge slightly ahead of the Moon. The Moon's gravity still pulls the tidal bulge toward itself, but the Moon is also accelerated along its orbit by the tidal bulge's own gravity. The combination of friction and gravity slows Earth's rotation, transferring energy from Earth to the Moon. It is this Earth-to-Moon energy transfer that is causing the Moon to move away from Earth, albeit by only a tiny amount—1.5 in (3.8 cm)—each year.

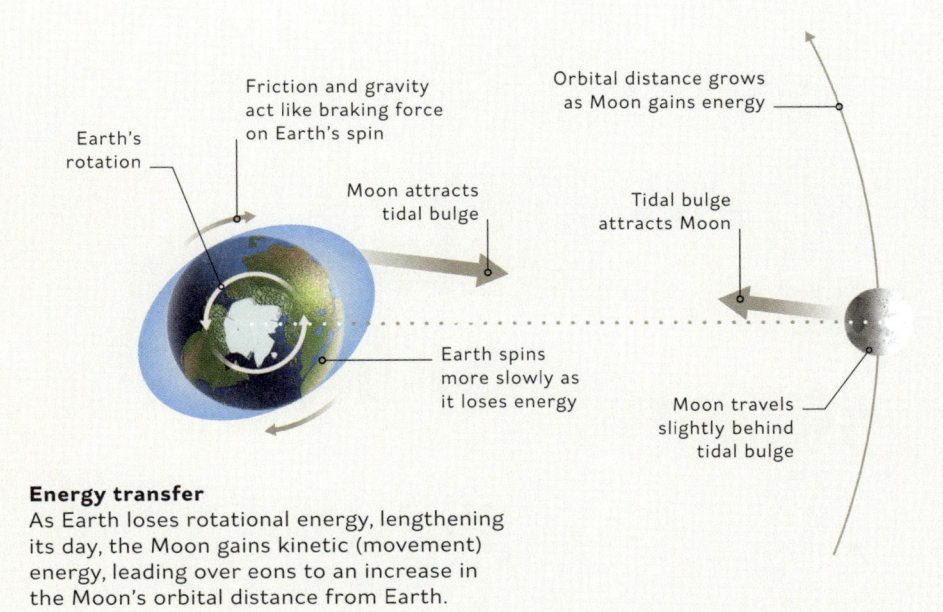

Earth's rotation

Friction and gravity act like braking force on Earth's spin

Moon attracts tidal bulge

Orbital distance grows as Moon gains energy

Tidal bulge attracts Moon

Earth spins more slowly as it loses energy

Moon travels slightly behind tidal bulge

Energy transfer
As Earth loses rotational energy, lengthening its day, the Moon gains kinetic (movement) energy, leading over eons to an increase in the Moon's orbital distance from Earth.

ANCIENT ASTRONOMY

Prehistoric communities tracked the Moon's phases to use as a kind of timekeeper—one that gathered days into months and months into years. The Moon's position in the sky was also important, indicating the best times for planting seeds or reaping harvests. By 700 BCE, Babylonian astronomers had discovered the 18-year lunar eclipse cycle, and by 400 BCE, the Greeks had observed that the Moon returns to the same position in the sky every 19 years.

THE FIRST CALENDARS

Prehistoric people used the Moon as a natural timekeeper. The predictable, ever-repeating pattern of the lunar phases provided an observable method of marking the passage of days and months.

During prehistory, people began to recognize the Moon's consistent waxing and waning as they observed the roughly 29.5-day lunar cycle from one new Moon to the next. Paleolithic hunter-gatherers used the regular changes in the lunar phases to set times for daily activities and cultural rituals. They probably found the easily observable pattern of the Moon simpler to follow than the longer solar year.

One way of tracking the lunar months was by using bones as tally sticks. The 42,000-year-old Lebombo bone, from the mountains between Eswatini and South Africa, has 29 notches and is believed to have been used as a lunar-phase counter. Similarly, the Ishango bone from the Democratic Republic of Congo, possibly dating to 20,000 BCE, features marks that may have been used to track the Moon's cycles. Discovered in France, the Blanchard bone (see opposite) has gouges instead of straight tally marks, which probably represent a more intricate phase count.

Bones were not the only medium used for tracking the Moon's phases. On the walls of the Lascaux caves in France, beneath a painted deer, is a row of 13 dots symbolizing half of the lunar cycle. This is followed by an empty space, possibly left to indicate a new Moon. Farther into the cave, another row of dots is placed under a horse; this row of 29 dots symbolizes a complete lunar cycle.

The nonstop lunar cycle was a reliable timekeeping tool until Neolithic times, when the emergence of farming made tracking the Sun essential for agricultural societies to predict and prepare for seasonal changes. »

Ishango bone
The 60 notches on this bone may have been made by someone observing the Moon. The lines may correspond to the days of two lunar cycles, and the lengths of the marks may indicate lunar phases.

Lascaux painting
Painted around 16,000 years ago, a row of 29 dots in France's Lascaux caves possibly stands for the lunar cycle. To prehistoric communities, the Moon's phases may have been part of a greater cosmic picture, in which animals represented constellations.

BLANCHARD BONE

Estimated to be around 30,000 years old, the Blanchard bone, from the Dordogne region of France, has 69 marks carved into its surface. While these marks, which form a curving line, vary in shape, they nevertheless can be grouped into categories: some are arched, some are hooked, and some are circular. The hooked symbols, which are finer, were most likely made with a different tool. These shapes and their serpentine pattern are thought to represent the Moon's celestial path through its cycle of phases.

> "Under the microscope the **gray bird bone** [the Blanchard bone] became a 'library'—which I sat 'reading' and **pondering** for many days."
>
> Alexander Marshack, American journalist and prehistorian, *The Roots of Civilization* (1972)

Notches along edges may have acted as coordinates for plotting Moon's position

Blanchard bone is made of antler and is 4 in (10 cm) in length

Hooked symbol, made with fine-pointed tool

Direction of arched symbols shows whether Moon is waxing or waning

Measuring time

The marks on the Blanchard bone are read by following the serpentine curve of the symbols, starting at center-left. The sequence marks several new Moons and ends at a full Moon. The arched symbols indicate the waxing and waning phases, while whole dots may represent new Moons and full or nearly full Moons. Not every day in the cycle is represented on the Blanchard bone.

Moon waxes as it heads towards full

Moon is recorded as full over course of four consecutive nights

In this part of sequence, arched symbols get smaller as Moon wanes

Sequence begins here; second symbol is probably a full Moon

The twisting path may represent the Moon's changing position in the sky

Nebra sky disk

Measuring approximately 12 in (30 cm) in diameter, the Nebra sky disk is a unique bronze-and-gold artifact that showcases the best of Bronze Age metalworking and astronomical knowledge. Found near Nebra, in Germany, the disk was made from locally sourced copper, as well as tin and gold from Cornwall, UK. Dating from around 1600 BCE, it is one of the earliest-known European representations of the cosmos.

The disk's design evolved over time, the gold elements being added in three stages. The largest gold circle represents either the Sun or a full Moon, possibly with a crescent Moon to its right. The dots represent stars, possibly including the groupings now known as the Big Dipper and Cassiopeia, while a cluster of seven stands for the Pleiades. Archaeologists have interpreted the two curved sections that were originally on either side as solstitial arcs, marking the extent of the rising and setting Sun on the summer and winter solstices. A third arc, at the bottom, may symbolize a solar boat (a mythological vessel that carried the Sun across the sky) or a partial eclipse.

Rivet holes around the disk's edges suggest the disk was mounted on a shaft, possibly to be held aloft during ceremonies. It may also have been used to inform people how to align their solar and lunar calendars for agricultural or religious purposes, thus serving both as an astronomical tool and a ritual object.

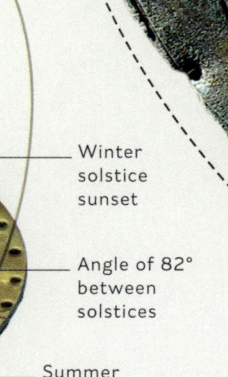

North

Second gold solstitial arc, now missing from edge of disk, may have fit into groove here

Large golden circle, probably representing Sun or full Moon

West

Shaft may have been attached by rivet holes here

1

South

1. The solstices

Originally, the disk had two gold arcs on its edges that marked the places at which the Sun rises and sets at Nebra, Germany. Since Nebra has a latitude of 82°, each of these arcs made an angle of 82° on the disk. The northern and southern ends of the arcs marked the sunrises and sunsets of Nebra's summer and winter solstices.

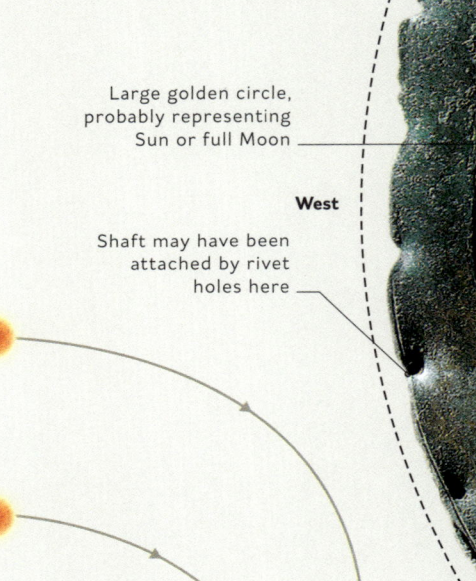

Winter solstice sunrise

South

Winter solstice sunset

Angle of 82° between solstices

Summer solstice sunrise

Summer solstice sunset

North

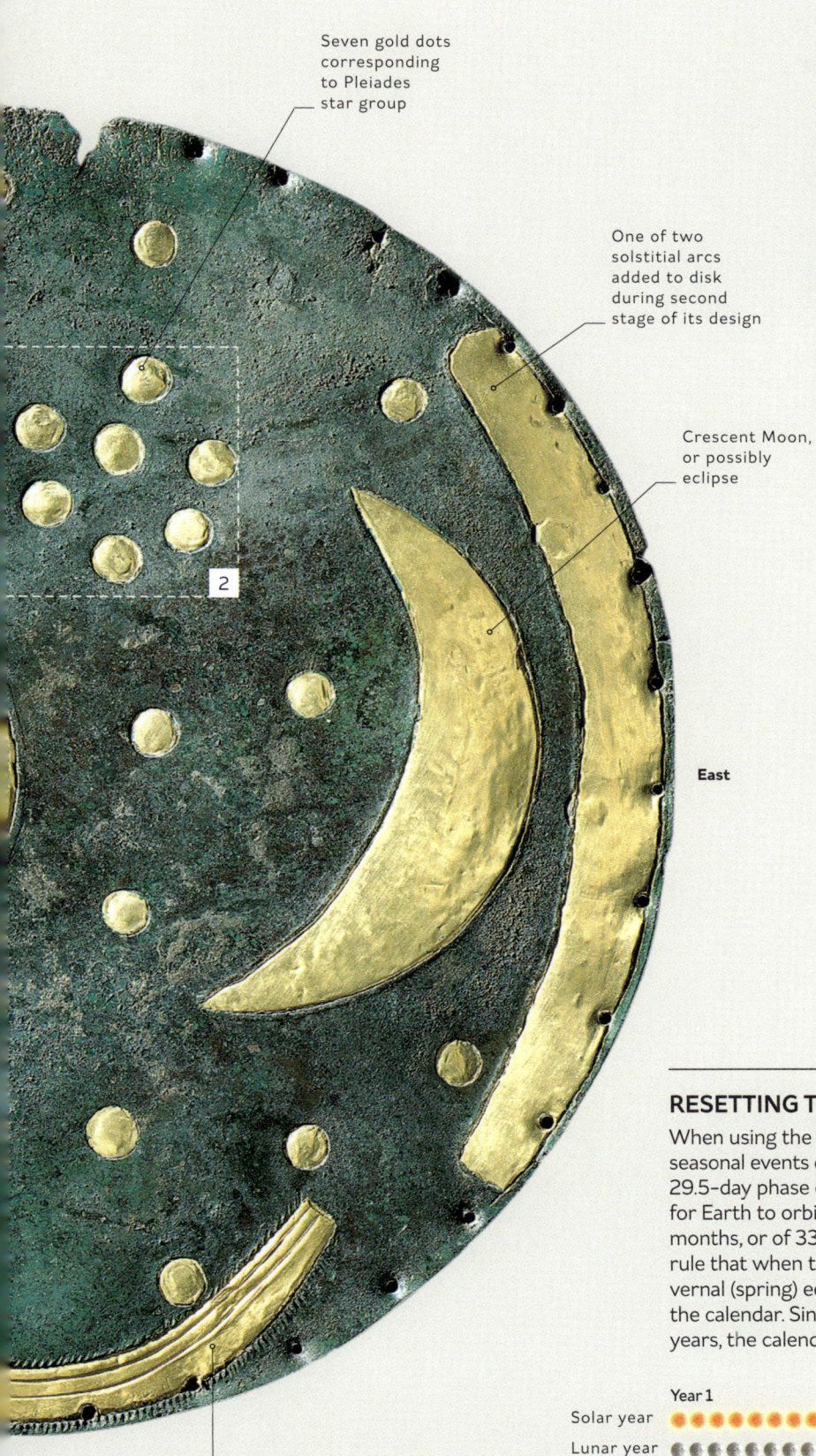

Seven gold dots corresponding to Pleiades star group

One of two solstitial arcs added to disk during second stage of its design

Crescent Moon, or possibly eclipse

2

East

Solar boat

2. The Pleiades
The brightest stars in the Pleiades cluster can be seen with the naked eye. Their blue hue and distinctive grouping make them a prominent evening feature in the southeastern sky at harvest time. Rising before the bright red star Aldebaran and Orion's belt, they were used as timekeepers by many cultures, including that of ancient Egypt.

> "The **astronomical rules** that are **depicted** wouldn't be imaginable without **decades** of **intensive observation**."
>
> Harald Meller, German archaeologist, *Archaeology Magazine* (May/June 2019)

RESETTING THE CLOCK BY THE PLEIADES

When using the lunar cycle as a timekeeper, the timings of annual and seasonal events drifts against the solar year. This is because the Moon's 29.5-day phase cycle is not directly divisible into the 365.25 days it takes for Earth to orbit the Sun—there is a shortfall of 11 days after 12 lunar months, or of 33 days after three years. The Nebra sky disk depicts the rule that when the Pleiades appears next to a crescent Moon near the vernal (spring) equinox, an extra lunar month must be added to realign the calendar. Since this conjunction occurs approximately every three years, the calendar is never more than a month out of alignment.

	Year 1	Year 2	Year 3
Solar year			
Lunar year			

Pleiades aligns with Moon at this point

13th Moon

LUNAR MONTHS

The ever-changing face of the Moon has long marked the passage of time beyond the daily motion of the Sun crossing the sky. It is this repeating cycle of lunar change that gives us the "month."

The Moon's orbit
When the Moon lies between Earth and the Sun, it is in full shadow (new Moon). Two weeks later, when its orbit has taken it to the opposite side of Earth from the Sun, it is fully sunlit (full Moon).

The Moon takes 29.5 days to change from new Moon to full Moon and back again. This lunar, or synodic, month has given a rhythm to the year and to human activity. As well as influencing tides and seasons (see pp.16–17 and pp.18–19), it has dictated the timing of farming activities such as planting and is the basis of most calendars. However, the lunar month does not synchronize exactly with the Moon's orbit around Earth or with Earth's orbit around the Sun. As a result, there is neither a round number of days in a lunar month nor a round number of lunar months in a year. For this reason, the modern calendar splits the year into 12 months, most of which are a day or two longer than a lunar month, with an extra day added every four years to keep the months and years aligned.

There are other lunar months of different duration, measured by observing the Moon against the stars (the sidereal month), the time between lunar equinoxes (the tropical month), the time the Moon takes to move from one perigee to the next (the anomalistic month), or to pass from one ascending node to another (the draconic month; see pp. 30–31).

Day 0

Day 4

Day 7

Day 11

Illuminated face
The Moon waxes and wanes (gets brighter and darker) over the course of the lunar month. Roughly half the Moon's face is illuminated at first and last quarter. In the crescent phase, less than half is lit, while the gibbous phase is more than half-illuminated.

New Moon

Waxing crescent

First quarter

Waxing gibbous

THE SYNODIC MONTH

It takes the Moon 27.3 days to complete one 360-degree orbit of Earth, but Earth is also traveling around the Sun at the same time and in the same direction. Because of this, the period between full Moons is a little longer, at 29.5 days, and the Moon moves 390° in its orbit. The duration of the Moon's orbit can also be measured against the backdrop of the stars, giving the sidereal month. The longer period between successive full Moons is the synodic month. Many calendars are based on the synodic month, including the Islamic and Jewish calendars.

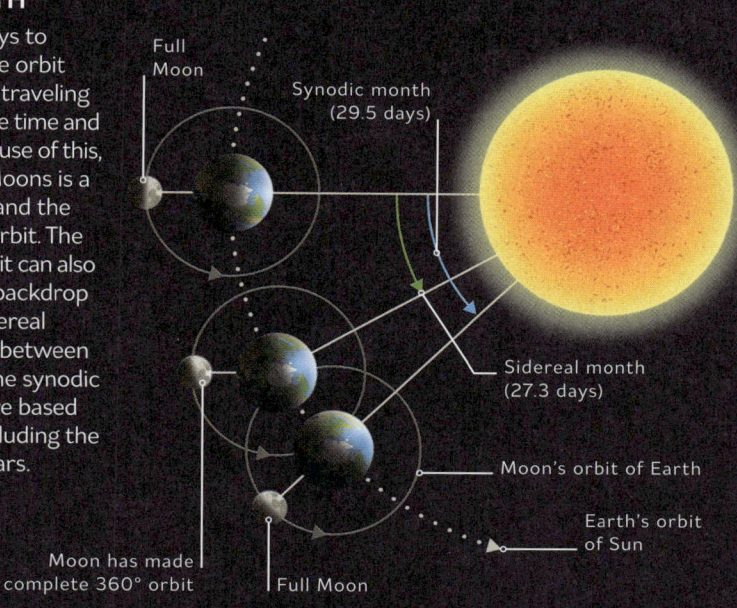

Full Moon

Synodic month (29.5 days)

Sidereal month (27.3 days)

Moon's orbit of Earth

Earth's orbit of Sun

Moon has made complete 360° orbit

Full Moon

PHASES OF THE MOON

The Moon's orbit around Earth causes the angle (or phase) of its illumination by the Sun to vary, drastically altering the Moon's appearance as viewed from Earth (see below). From a fully illuminated disk, the Moon shrinks to a crescent before darkening entirely, only to return to full illumination again after 29 days have passed.

Day 15

Day 18

Day 22

Day 26

Day 29

Full Moon

Waning gibbous

Last quarter

Waning crescent

New Moon

LUNAR PRECESSIONS

The Moon's motion, like that of any orbiting body, is not fixed in space or time, but has been evolving ever since the Moon's birth, under the gravitational influence of Earth and the Sun.

Like a spinning top, the Moon moves in three-dimensional space, orbiting Earth and rotating on its axis. But while a spinning top eventually loses stability and wobbles increasingly as time passes, the motion of an astronomical body such as the Moon gradually evolves into a stable state through gravitational interaction with its parent planet and the Sun. The Moon is now tidally locked to Earth (see pp.16–17),

and its motion is characterized by repeating cycles of change to its rotation and orbit. A cyclical change in the orientation of a rotating body is known as a precession.

Multiple precessions

The Moon's motion displays three precessions, one affecting the tilt of its rotation axis and two that affect its orbit. The precession of

APSIDAL PRECESSION

The Moon's elliptical orbit around Earth is disturbed by the gravitational pull of the Sun, which causes the major (long) axis of the orbit to precess eastward (or counterclockwise when viewed from above Earth's north pole). The orbit's perigee and apogee (see pp.16–17), which lie at the ends of the major axis, are known as the apsides of the orbit, so this is the apsidal precession. One complete precession of 360° takes 8.85 years.

AXIAL PRECESSION

The Moon's rotation axis is tilted at 6.7° to its orbit, and Earth's gravity tries to pull it back upright. This causes the rotation axis to precess 360° westward every 18.6 years. This rate of change—about 19° per year—mirrors that of the nodal precession (see opposite page), which keeps the Moon's geographic north pole, its orbital north pole, and the ecliptic north pole aligned and the Moon's rotation axis tilted at a constant angle of 1.5° to the ecliptic north pole.

Orbit precesses eastward
(by about 1° every 9 days)

Major axis/line
of apsides

Minor
axis

Apogee
(apside)

Perigee
(apside)

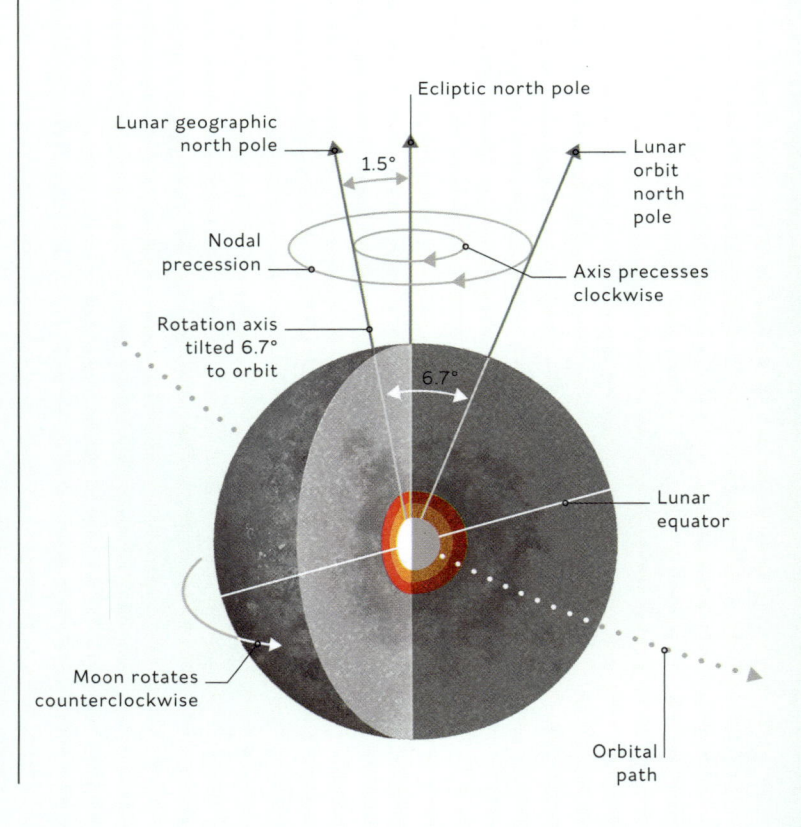

Ecliptic north pole

Lunar geographic
north pole

1.5°

Lunar
orbit
north
pole

Nodal
precession

Axis precesses
clockwise

Rotation axis
tilted 6.7°
to orbit

6.7°

Lunar
equator

Moon rotates
counterclockwise

Orbital
path

its axis (axial precession) proceeds at the same rate as cyclical changes to the orientation of its orbit (nodal precession), but in the opposite direction to the orbit. This is a feature of any tidally locked moon, and it results in the Moon's rotation axis remaining at a fixed angle relative to the ecliptic plane—the plane containing Earth's orbit around the Sun (see below and pp.16–17).

The nodal precession accounts for the difference between the length of the sidereal month and the draconic month (see below and pp.28–29) and long-period changes in the height of Earth's ocean tides, such as spring, neap, and seasonal tides (see pp.18–19). Artificial satellites orbiting Earth display nodal precession, too, as do moons orbiting other planets.

The Moon's orbit around Earth is elliptical, and the long, or "major," axis of the ellipse (known as the line of apsides) precesses, too. This is known as apsidal precession. Earth also undergoes axial and apsidal precession, but on much longer timescales: 25,771 years for its axial precession and 112,000 years for its apsidal precession. Earth's precessions affect its long-term climate by causing changes in the distribution of solar heat over the planet's surface.

Describing precession
Italian astronomer Giovanni Domenico Cassini (1625–1712) devised three laws that describe the motion of the Moon, including its axial and nodal precession. Cassini published his laws in 1693.

NODAL PRECESSION

The Moon's orbit is tilted at 5.1° to the ecliptic plane, and the points where the orbit crosses the ecliptic plane are called nodes. Each orbital cycle has two nodes—the ascending node, where the Moon rises above the ecliptic plane, and the descending node, where the Moon drops below the ecliptic plane. On each orbit, the Moon receives a small gravitational pull from the Sun, which causes the line of nodes to rotate (precess) around Earth.

Year one
The Moon passes through the same node every 27.2 days—a time period called the draconic month (named after mythical dragons that "eat" the Moon during an eclipse). A full moon or a new moon around this time means the Sun, Moon, and Earth are aligned and an eclipse may occur (see pp.38–41).

Year two
A year later, the line of nodes has drifted 19.4° westward. A complete precession of 360° is thus completed in 18.6 years; during this time, the angle of the Moon's orbit relative to Earth's equator varies between 18.5° and 28.5°, which influences the long-period tidal range in Earth's oceans.

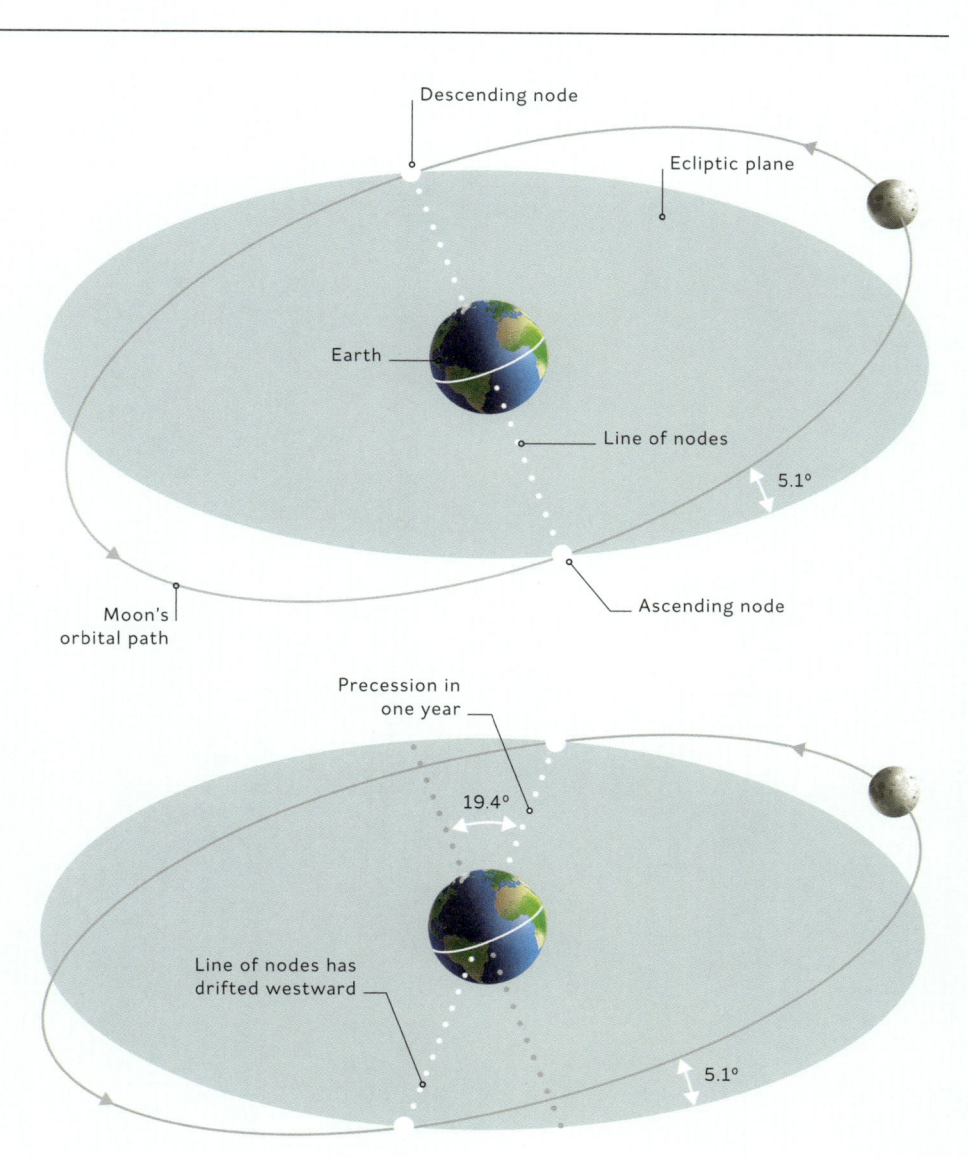

Descending node

Ecliptic plane

Earth

Line of nodes

5.1°

Moon's orbital path

Ascending node

Precession in one year

19.4°

Line of nodes has drifted westward

5.1°

PREHISTORIC RITUALS

Archaeological evidence suggests that the Moon played a key role in our ancestors' understanding of the seasons, and more broadly of life, death, and fertility.

Venus sculpture
Carved out of limestone and painted with red ocher, the Laussel Venus holds a bison horn etched with 13 lines.

Without written records to guide us, we only know the prehistoric world indirectly, via the artifacts left by ancient civilizations. However, these relics, which range from exquisite figurines to imposing stone monuments, clearly show that Neolithic humans (10,000–2200 BCE) were keen observers of the night sky and took particular interest in the Moon. Evidence reveals that even before this time, people recognized that the Moon's 29.5-day phase cycle matches the menstrual cycle and that the average pregnancy lasts nine lunar months (see pp. 28–29). It is easy, then, to see how the Moon became linked to women and childbirth, and possibly—since the Moon disappears for three days during the lunar cycle—to mythological ideas about death and rebirth.

One of the most striking images that associates women with the Moon is the Laussel Venus—one of the many so-called "Venus sculptures" that were carved in Europe

Tracking the Moon
The stones of Callanish I, on the Outer Hebridean island of Lewis in Scotland, form a circle at the center of a larger cross.

between 40,000 and 10,000 BCE. Discovered in France, the sculpture depicts a woman reaching down to her swollen, possibly pregnant belly, while in her other hand she clasps a horn etched with 13 lines. These lines may mark the waxing days of the Moon's phase cycle or the 13 potential full Moons in the solar year. The horn is distinctly Moon-like, and the woman may represent fertility.

Stone monuments

Archaeological remains also bear witness to our ancestors' interest in the Moon. Its image is carved in the templelike complex of Göbekli Tepe, which was built by the first Neolithic farmers some time after 10,000 BCE in what

is today southern Türkiye. The Moon is depicted full and within a crescent—an image that the Babylonians and classical Greeks later repeated. By the late Neolithic period, people were raising huge stone pillars in parts of the Atlantic seaboard, the alignments of which offer clues about the beliefs of their builders. The stone ring of Callanish I, for example, is carefully placed in the Scottish landscape. To the south, the hills resemble a reclining woman, over whose body, every 18.6 years, the Moon traces its lowest arc in the sky (see box). A similar understanding of the Moon's long-term cycle can be seen at Stonehenge in England, but at Callanish, it is profoundly linked to Earth and the female form.

Early temple carving
A pillar at Göbekli Tepe bears an etching of a circle within a crescent, possibly a crescent moon cradling a full moon or an eclipse in progress.

ANCIENT ALIGNMENTS

Due to a peculiarity of the Moon's orbital tilt (see pp.34–35), it takes the Moon 18.6 years to return to exactly the same point in the sky. During that time, it rises both higher and lower than the Sun rises at the summer and winter solstices. Archaeologists have discovered that many Neolithic sites are arranged in such a way that their stones are aligned with these key astronomical moments—including the time when, in the northern hemisphere, the Moon is at its lowest point in the skies.

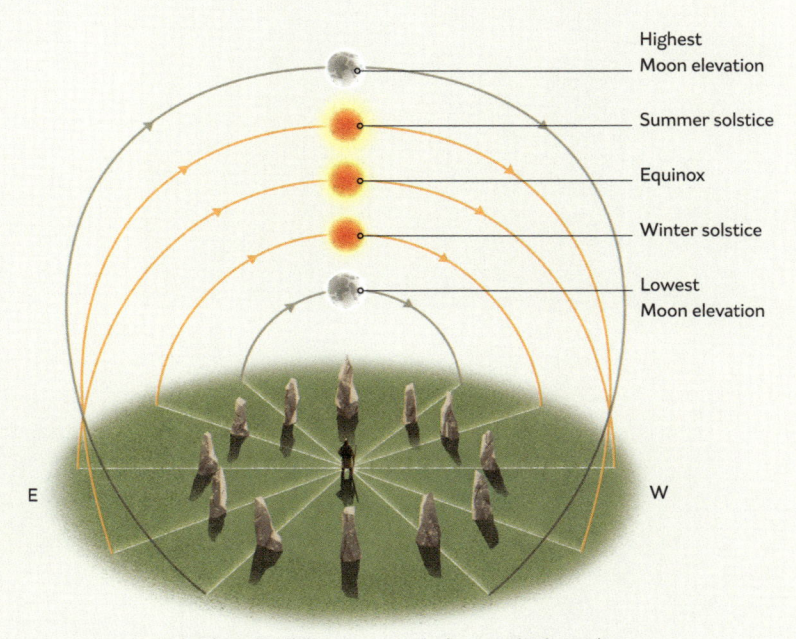

Highest Moon elevation

Summer solstice

Equinox

Winter solstice

Lowest Moon elevation

E — W

A stone circle, aligned with key events in the lunar and solar cycles

LUNAR AND SOLAR CYCLES

The Moon's movement across the sky is much more complicated than the simple rotation of the stars around the North Star, having more in common with the daily and seasonal path of the Sun.

Earth's axis of rotation is tilted at 23.4° to the plane of its orbit around the Sun (the ecliptic plane), giving rise to the seasons. When one of the hemispheres is tilted toward the Sun, the Sun climbs high in the sky and more solar heat is absorbed by the ground over longer days—that hemisphere has summer. Six months later, the same hemisphere is tilted away from the Sun, which stays closer to the horizon, giving less heat over shorter days—it is winter.

Standstills

For a few days of the year, around its highest and lowest points in the sky, the midday elevation of the Sun appears to stand still. These times are called solstices, from the Latin *sol* (meaning "Sun") and *sistere* (to "stand still"). The Moon undergoes a similar pattern of rising and falling in the sky, but on a much shorter monthly cycle that corresponds to its orbit around Earth. The high and low lunar standstills are called lunistices.

Lunar complexity

The Moon's track across the sky is complicated by the cyclical change, or precession, of its orbit (see pp.30–31), which is tilted at 5.1° to the ecliptic plane. Sometimes, this brings the Moon's orbit closer to Earth's equatorial plane; sometimes, it tilts it farther away.

The combination of the Moon's orbital tilt and the 23.4° tilt of Earth's axis—and also, therefore, of its equator—determines how high in the sky the Moon will reach on a given date. Over the period of a month, the Moon can range 28.5° above and below the celestial equator if its orbit is tilted beyond the ecliptic plane (23.4 + 5.1 = 28.5), or 18.3° above and below if it lies closer than the ecliptic plane (23.4 − 5.1 = 18.3). When the Moon's orbit lies closer to the equatorial plane, Earth's oceans experience a stronger gravitational pull away from Earth's rotation axis, thereby increasing both the height and range of the tides (see pp.18–19).

Celestial monument
At Stonehenge, a Bronze Age monument in Wiltshire, UK, the major standing stones are aligned with the summer solstice sunrise and the winter solstice sunset. Some minor stones appear to be aligned with the southernmost moonrise during a major standstill of the Moon.

SOLSTICES

A solstice is a day when the Sun reaches its declination (highest or lowest angle) above the horizon. In the northern hemisphere, the summer solstice occurs around June 21, and the winter solstice around December 21. The summer solstice sees the most northerly sunrise and sunset and the most daylight hours. The winter solstice brings the most southerly sunrise and sunset and the fewest daylight hours. At the spring and fall equinoxes, the Sun rises directly in the east and sets directly in the west; day and night are approximately equal length.

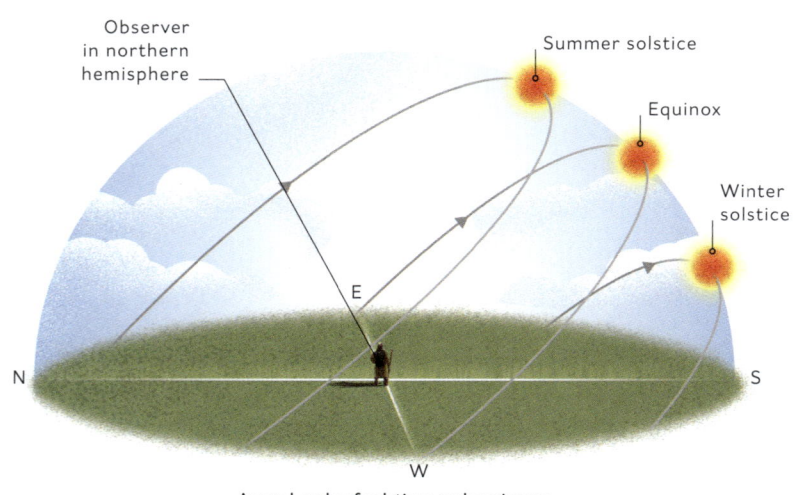

Annual cycle of solstices and equinoxes

"The moving Moon went up the sky, and nowhere did abide. Softly she was going up, and a star or two beside."

Samuel Taylor Coleridge, British poet, *The Rime of the Ancient Mariner* (1798)

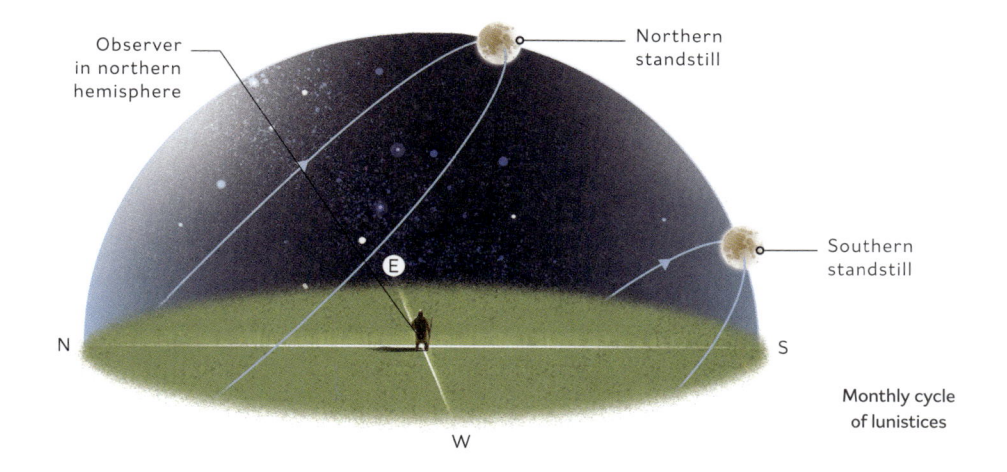

Lunistices
A lunistice, or lunar standstill, occurs twice a month as the Moon appears to pause upon reaching its highest and lowest angle (declination) above the celestial equator. The direction of moonrise and moonset shifts northward and returns southward over the same time period.

Monthly cycle of lunistices

MAJOR AND MINOR STANDSTILLS

Due to the precession of the Moon's orbit, the amplitude (angular range) of the lunar standstills' monthly cycle also changes—but over a much longer cycle of 18.6 years. The maximum declinations of 28.5° north and south of the celestial equator (known as major standstills) combine to give a range of 57° over this period. The combined minimum declinations of 18.3° north and south of the celestial equator (minor standstills) give a range of 36.6° over the same timescale.

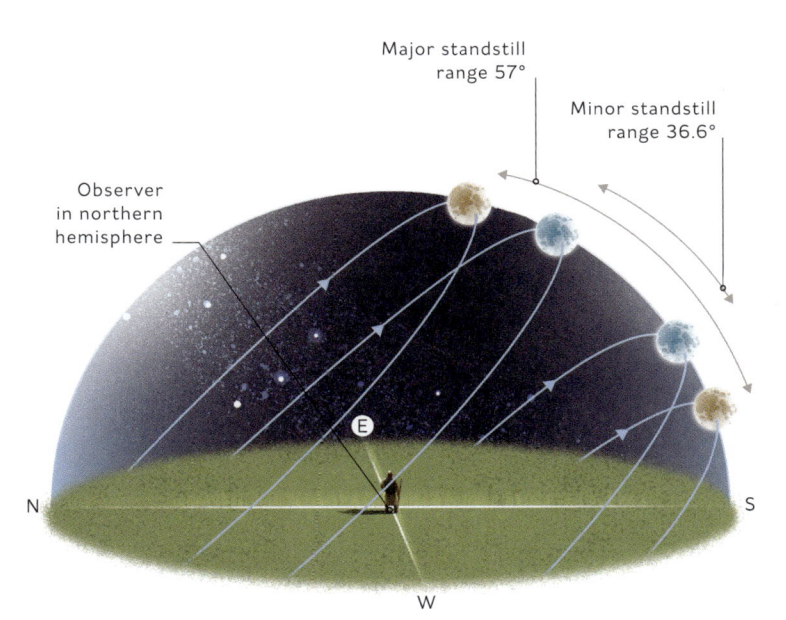

Ranges of lunar standstills (lunistices) over 18.6 years

Major standstills
At a time when the Moon's orbit tilts farther from Earth's equator than the ecliptic plane does, the Moon can reach a maximum declination of 28.5° north and south of the celestial equator.

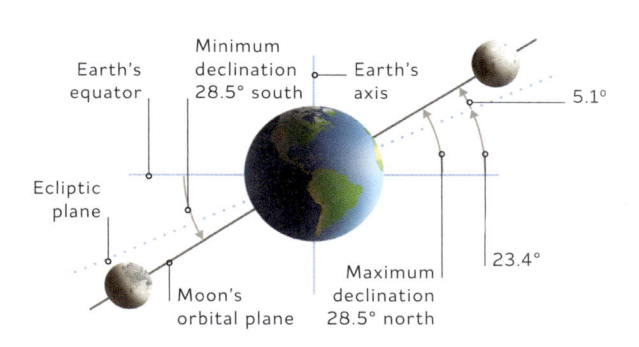

Minor standstills
When the Moon's orbit tilts less from Earth's equator than the ecliptic plane does, the Moon's declination is limited to a maximum declination of 18.3° north and south of the celestial equator.

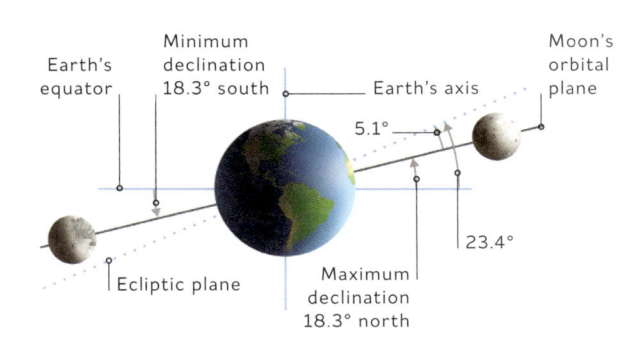

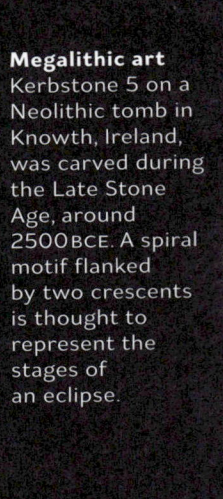

Megalithic art
Kerbstone 5 on a Neolithic tomb in Knowth, Ireland, was carved during the Late Stone Age, around 2500 BCE. A spiral motif flanked by two crescents is thought to represent the stages of an eclipse.

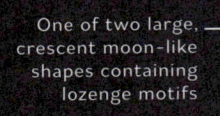

One of two large, crescent moon-like shapes containing lozenge motifs

The central spiral has six layers moving outward in a clockwise direction

THE EARLIEST RECORDED ECLIPSES

The alignment of the Earth, Moon, and Sun creates a striking cosmic spectacle called an eclipse. People have long been captivated by the Moon turning blood-red or the Sun disappearing during the day.

Oracle bone
The ancient Chinese text inscribed on this shoulder bone of a sheep or cow records a total lunar eclipse that occurred during the Shang dynasty in December 1192 BCE.

Before the advent of writing, people marked rocks or other items to record the eclipses they witnessed. Written accounts of eclipses date back more than 3,000 years to Babylonia and ancient China, where eclipses were considered to be omens from the gods that could influence everything from agriculture to war. Early records from these places reveal that people were aware of the cyclic nature of eclipses (see pp.38–41) and

that they used this to predict future occurrences and ward off misfortune by appeasing the gods.

Averting disaster in Mesopotamia
Astronomer priests in Mesopotamia kept meticulous records of celestial events on clay tablets. Seven of the 12 *Enuma Anu Enlil* astrological tablets (1595–1197 BCE) are dedicated to lunar eclipses. Driven by a

The Battle of Gaugamela

This 15th-century manuscript depicts Alexander the Great consulting his astrologers about a lunar eclipse in 331 BCE. The eclipse occurred as Alexander's Macedonian army prepared to confront the forces of Persian king Darius III at Gaugamela (in modern-day Iraq). Despite being outnumbered, the Macedonian troops believed it foretold Darius's defeat, boosting their spirits before their victory.

Six circles adorn the top-right side of the curbstone, possibly representing six months

> "If an **eclipse** (of the Moon) takes place and **Jupiter is present** ... the **King is safe**; a noble **dignitary will die** in his stead."
>
> Tablet 20 of the *Enuma Anu Enlil* (1595–1197 BCE)

desire to influence their future by mitigating danger, the elite developed sophisticated methods to predict eclipses. When an eclipse was due, the king would be replaced with a substitute to face the full wrath of the gods instead. The surrogate was then sacrificed so the true king could return to his position unharmed.

One legend tells of King Erra-Imitti (d. 1861 BCE), who appointed his gardener in his place. However, he died during the eclipse, leaving the gardener to claim his throne.

Harmony in ancient China

From around 1200 BCE, people in ancient China recorded the dates of eclipses on oracle bones – turtle shells or animal shoulder bones carved with text and used for divination.

They believed that solar eclipses indicated the emperor's loss of favor with the gods and that corrective action was needed. Records from the Zhou dynasty (1046–256 BCE) show a systematic approach to predicting eclipses, which was considered essential for maintaining the harmony between heaven and Earth.

Historical records
Written in cuneiform script, this clay tablet lists lunar eclipses from 609 to 447 BCE in Babylon in modern-day Iraq. It also notes the murder of the Persian king Xerxes I by his son in 465 BCE.

ECLIPSES

When the Moon, Earth, and Sun align, the shadows they create cause eclipses. Among the most spectacular of natural sights, eclipses have inspired myths and legends in different cultures across the ages.

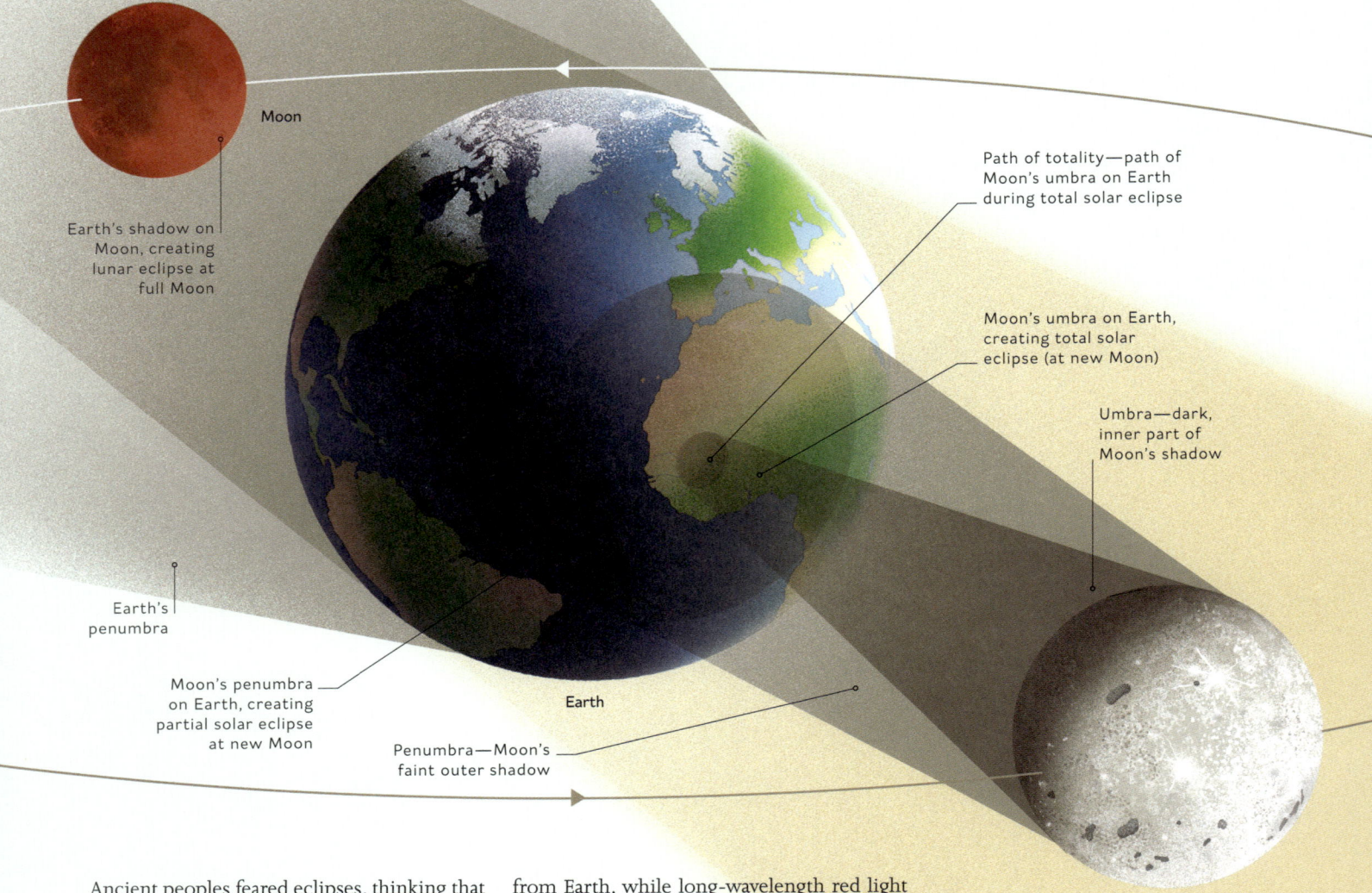

Moon

Earth's shadow on Moon, creating lunar eclipse at full Moon

Path of totality—path of Moon's umbra on Earth during total solar eclipse

Moon's umbra on Earth, creating total solar eclipse (at new Moon)

Umbra—dark, inner part of Moon's shadow

Earth's penumbra

Moon's penumbra on Earth, creating partial solar eclipse at new Moon

Earth

Penumbra—Moon's faint outer shadow

Moon

Ancient peoples feared eclipses, thinking that they were caused by monstrous creatures devouring the Sun or Moon. Modern astronomers welcome them, however, as a chance to learn more about the Sun's atmosphere and the Moon's surface.

Lunar eclipses

When the Moon lies exactly on the opposite side of Earth from the Sun, it falls into Earth's shadow and a lunar eclipse occurs. If skies are clear, the Moon often looks red during a total lunar eclipse (see opposite page). The way sunlight behaves as it passes through Earth's atmosphere depends on the light's wavelength: short-wavelength blue light is scattered away

from Earth, while long-wavelength red light is refracted (bent) toward the Moon, producing a "blood Moon" (see pp. 80–81).

Using telescopes fitted with heat-sensitive cameras, astronomers can investigate the thermal properties of the Moon's rocks as a lunar eclipse plunges its surface into shadow.

Solar eclipses

When the Moon lies exactly between Earth and the Sun, the Moon obscures the Sun, casting a shadow over part of Earth's surface and causing a solar eclipse. The Sun is 400 times larger than the Moon and also 400 times farther away from Earth. Consequently, the Moon appears the same size as the Sun in the

sky and, during a total solar eclipse, the Moon covers the Sun, briefly cutting off all sunlight, turning day into night, causing the temperature to drop and winds to weaken and shift. With the Sun's dazzling disk obscured, details such as flares and eruptions rising from the Sun's surface can be seen, and the corona (the Sun's faint outer atmosphere) is revealed. Observers should always view solar eclipses using specially designed eclipse safety glasses, never with the unaided eye (see pp. 244–45). »

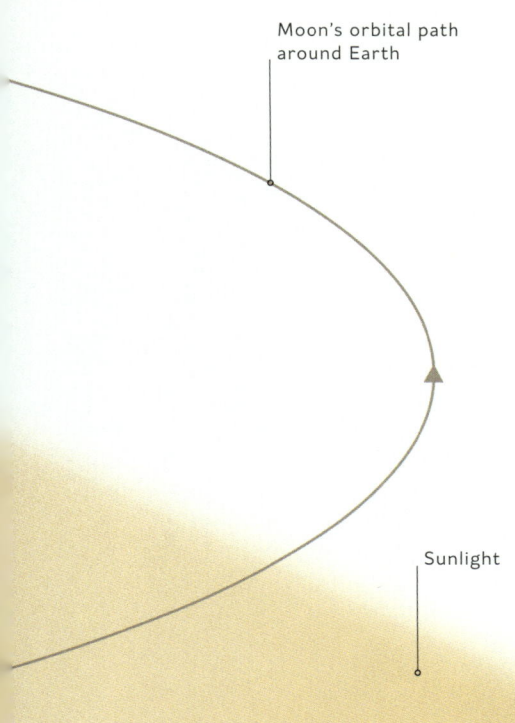

Lunar and solar eclipses
The Moon's orbit around Earth sometimes takes it through Earth's shadow, creating a lunar eclipse. At other times during its orbit, the Moon casts its own shadow on to Earth's surface, causing a solar eclipse.

Moon's orbital path around Earth

Sunlight

TYPES OF LUNAR ECLIPSE

During a total lunar eclipse, the Moon is fully shadowed for up to 100 minutes. It does not always completely disappear and often has a red glow from light refracted by Earth's atmosphere. In penumbral lunar eclipses, the Moon passes through Earth's faint outer shadow (or penumbra). It dims only by about 10 percent, so such eclipses often go unnoticed. However, they are easy to see when the Moon passes into Earth's full shadow (the umbra), even if only part of the lunar disk is obscured, giving a partial eclipse.

Total lunar eclipse

Penumbral lunar eclipse

Partial lunar eclipse

TYPES OF SOLAR ECLIPSE

During a total solar eclipse, which can last up to 7 minutes, the Moon entirely blocks out the Sun, and the Sun's corona becomes visible. During a partial solar eclipse, only part of the Sun's disk is obscured and there may be little difference beyond a general drop in the brightness of the day. If the Moon is near the most distant part of its orbit, it may cover 90 percent of the Sun's disk, leaving a ring of sunlight (an annular eclipse); the Sun's corona is not visible.

Total solar eclipse

Partial solar eclipse

Annular solar eclipse

"… the **Sun has perished** out of heaven, and **an evil mist hovers** over all."
Homer, ancient Greek poet, *The Odyssey*
(thought to refer to a total solar eclipse in 1178 BCE)

Sun

The eclipse cycle

If the Moon orbited Earth in the same plane as Earth orbits the Sun, we would have a solar eclipse and a lunar eclipse every month, at new Moon and full Moon respectively. But the Moon's orbit is tilted at 5° to Earth's orbit, so eclipses are far less common events. Eclipses can only occur when the Moon crosses the plane of Earth's orbit around the Sun—known as the "ecliptic plane". This occurs twice on each lunar orbit, at points called nodes; only when both nodes align with the Sun can an eclipse occur (see diagram below).

The Moon's orbit around Earth is also slightly elliptical, which causes the apparent size of the Moon in the sky to vary between 90.2 and 106 percent of the Sun's diameter. A total solar eclipse occurs only when the Moon appears as large as the Sun, or even larger; if it is smaller, part of the Sun remains visible, and there is an annular eclipse.

Frequency of eclipses

Globally, there can be up to five total or annular solar eclipses in a year, but on average, one occurs about every 18 months. Total lunar eclipses are less common: there can be up to three per year, although in some years there are none. A total lunar eclipse is visible to the population of an entire hemisphere, whereas a total solar eclipse can be viewed only from a small area of Earth's surface.

> "... like ... a **string of brilliants** disappearing like **snow** under a **white heat**."
>
> Mabel Loomis Todd, on Baily's beads, *Total Eclipses of the Sun* (1894)

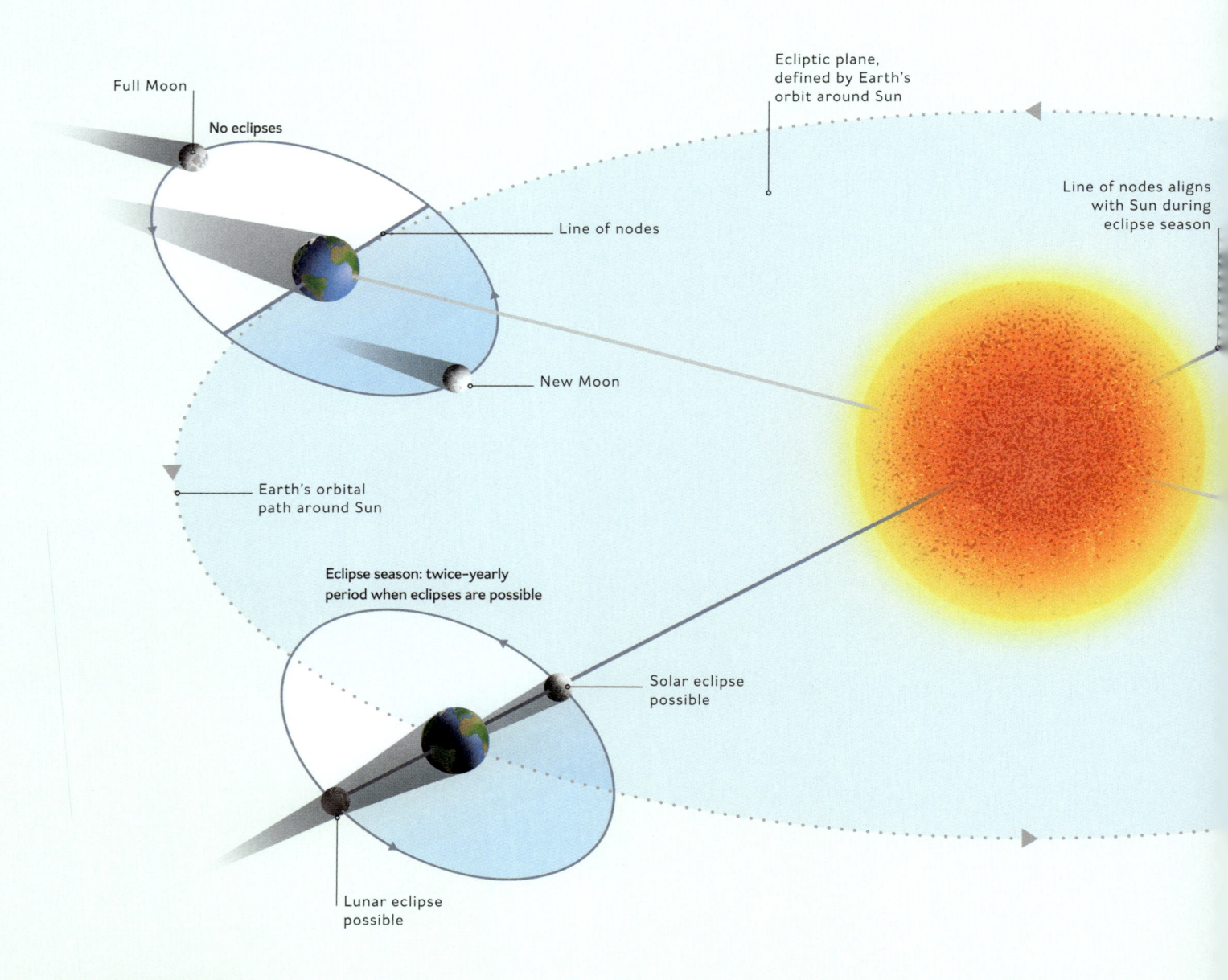

Full Moon

No eclipses

Ecliptic plane, defined by Earth's orbit around Sun

Line of nodes aligns with Sun during eclipse season

Line of nodes

New Moon

Earth's orbital path around Sun

Eclipse season: twice-yearly period when eclipses are possible

Solar eclipse possible

Lunar eclipse possible

Pinkish hue from Sun's hydrogen gas

Baily's beads

Eclipse effects
Just before a total solar eclipse, bright spots of light may be visible around the Moon's circumference. Called "Baily's beads," they are caused by the last few vestiges of sunlight breaking through valleys on the uneven lunar horizon. When only one is left on the Sun's halolike corona, it creates an effect known as a "diamond ring." The process repeats when the Moon starts to move away from the Sun, marking the end of totality. The composite of 35 photographs above records the 5 minutes of totality during the 2019 solar eclipse. It was taken by Chile's Cerro Tololo Inter-American Observatory.

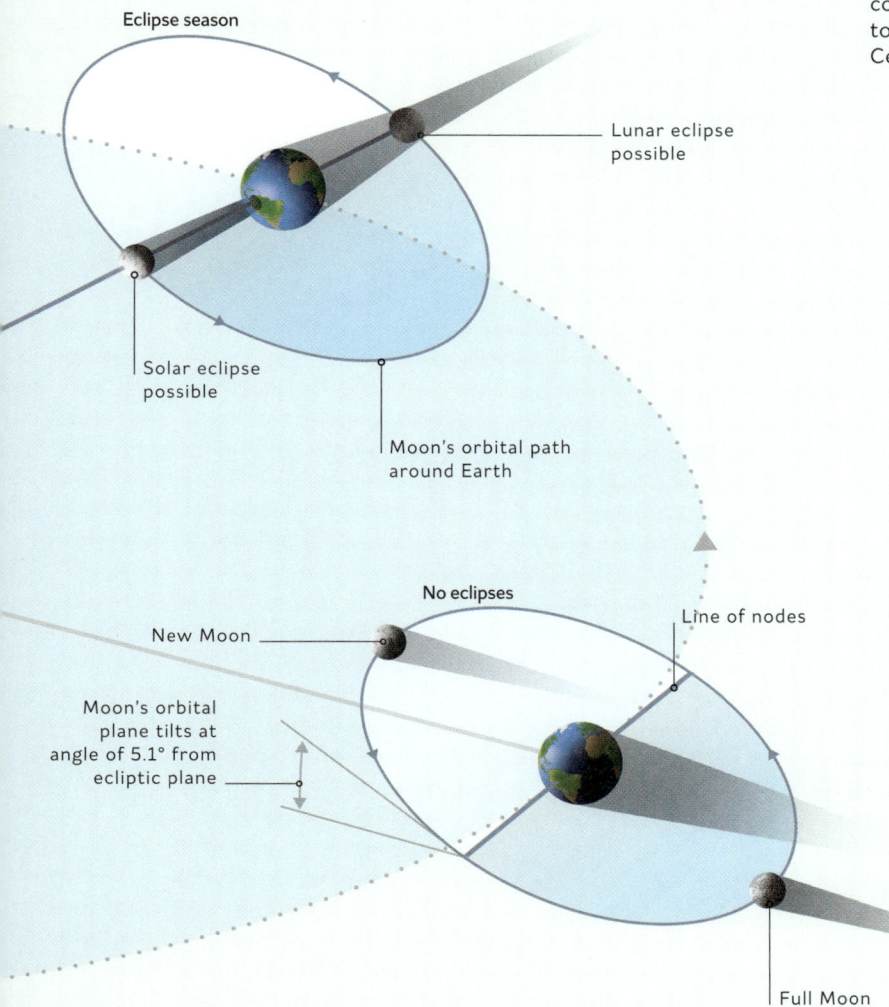

Eclipse season

Lunar eclipse possible

Solar eclipse possible

Moon's orbital path around Earth

No eclipses

New Moon

Line of nodes

Moon's orbital plane tilts at angle of 5.1° from ecliptic plane

Full Moon

Solar eclipse shadow
The area of total shadow (the umbra) during the solar eclipse of April 8, 2024, was 115 miles (185 km) across. This image of the Moon's shadow was captured by the EPIC camera on NASA's high-orbiting DISCOVR satellite.

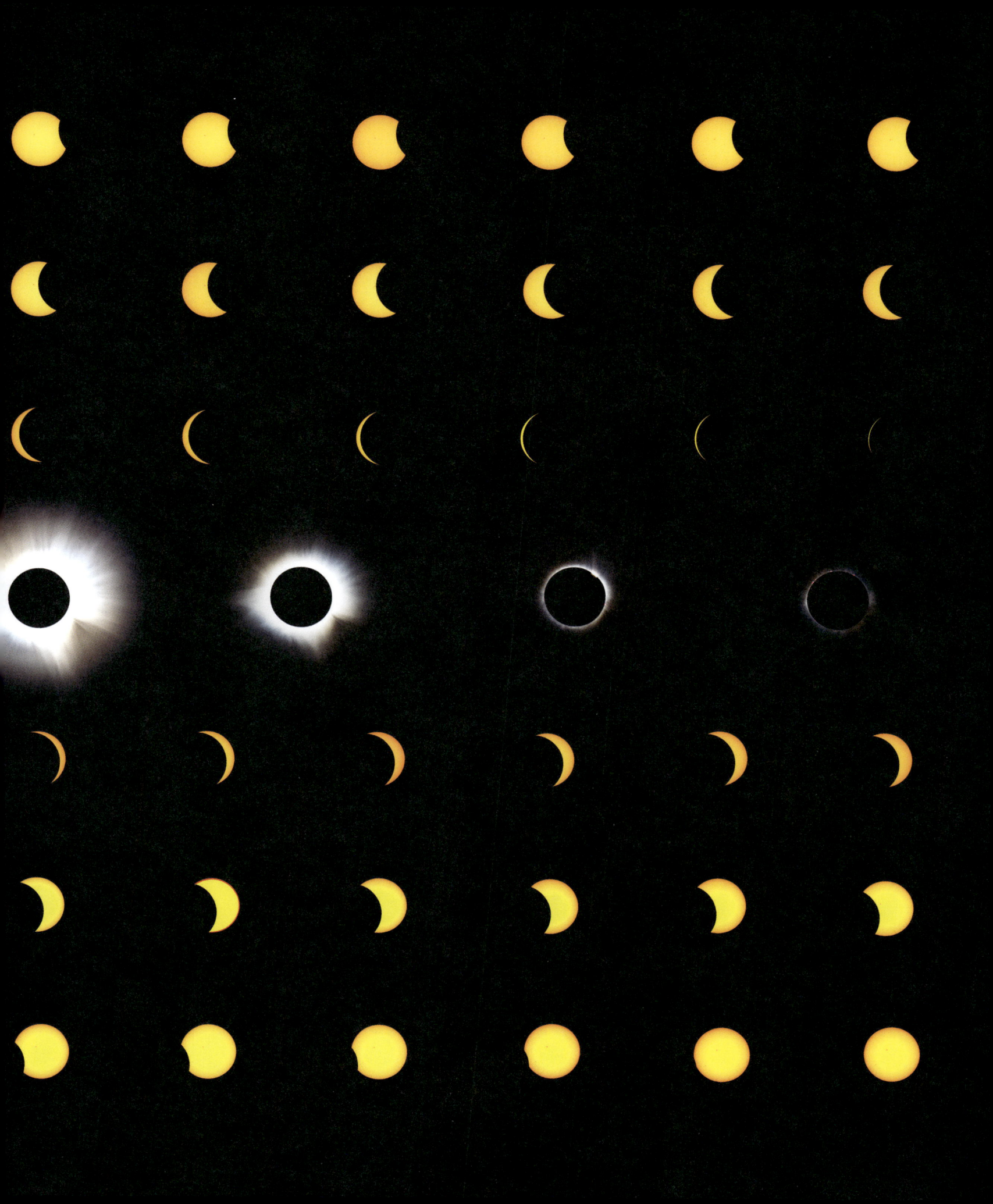

THE SACRED MOON

The earliest written records by Mesopotamian civilizations describe the Moon as a male god called Nanna, or Sin. Nanna was depicted as having humanlike qualities and living among people on Earth.

By 6000 BCE, farming was well established along the Fertile Crescent, which covered parts of modern-day Türkiye, Iraq, Iran, and Syria. As populations embraced more settled lifestyles, they organized themselves into village communities in which religious ideologies thrived and astronomical observations were shared. Such societal changes were supported by the development of writing. Texts inscribed on clay tablets in a script called cuneiform record the Moon's motions and details of the worship deities such as Nanna, who was symbolized by the crescent Moon.

Disk of priestess Enheduanna
Enheduanna, high priestess of the Moon god Nanna, is shown at left pouring liquid in a libation ritual. A priest and two priestesses stand behind her.

The cult of Nanna

As Nanna's influence spread, he became synonymous with Sin, or Suen, another lunar deity, known as the "shining cow," who governed the night's light and cattle. The cult of Nanna-Sin centered upon the Sumerian city-state of Ur, in what is now southern Iraq. Ur was also the residence of the lunar cult's high priestesses. The most famous of these was Enheduanna, appointed by King Sargon the Great (r. 2334–2279 BCE), her father.

Great Ziggurat of Ur
Constructed around 4,000 years ago, this step-pyramid temple dedicated to the Moon god Nanna is thought to have originally risen to more than 98ft (30m) in height. Today, only its foundation layer remains, standing 36ft (11m) tall. The facade of the ziggurat was rebuilt in the late 20th century.

Regularly spaced, flat-faced buttresses stabilize high vertical walls

Upper layers of mud-brick temple are in ruinous state

"Luminous brilliance that fills the **holy heavens** ... your **divine powers** are holy!"

Enheduanna, Sumerian high priestess, *A Hymn to Nanna* (c. 2250 BCE)

As leader of the cult, she oversaw ceremonies called libations, where liquids such as wine were offered to the gods. She also ensured the accuracy of the lunar-based calendar, which extended her influence into political and economic governance.

Monumental temple

King Ur-Nammu (r. 2112–2095 BCE) began the construction of the Great Ziggurat at Ur. Dedicated to Nanna-Sin, this vast temple in the form of a step pyramid was the cult's focal point. People gathered to pray to the Moon, seeking prosperity for the year ahead. »

Crescent Moon (god Sin) is positioned above all other sky gods

Scorpion represents stellar constellation traveling across face of stone

Transaction record
The crescent Moon sits between Venus and the Sun on this boundary stone, which witnesses a transaction (recorded at the base) granting more land to a man named Gula-eresh (1125–1100 BCE).

How it once looked
The original step pyramid had ramps leading to the central staircase, which accessed an upper temple.

Wide central staircase gives access to ziggurat's higher levels

Number dashes are calculation of length of diagonals in square

The early Mesopotamian civilizations developed an administrative calendar based on the Moon's phases, alternating between months of 29 and 30 days. They also had a civil or day-to-day calendar consisting of 12 lunar months, with an extra month added periodically in order to maintain alignment with the solar year. This adjustment was based on the lunar phases and the first sightings or heliacal (near-dawn) risings of bodies such as the Pleiades, Arcturus, or Sirius. Each month began at the new Moon's appearance. This was recorded by religious leaders, such as the high priestess Enheduanna (see previous page), who made any necessary amendments to ensure the calendar remained accurate.

Sexagesimal system
The cuneiform number dashes on this clay tablet from the 19th or 18th century BCE show a calculation using the sexagesimal (base–60) number system. Mesopotamians used this system to measure time.

Unified timekeeping
Since the calendar relied on astronomical observations made at temples in various locations, different places often had slightly misaligned calendars. The city of Nippur began the standardization of the calendar around the

VISUAL TOUR

1. Venus
The planet Venus is a star and known as the goddess Ishtar.

2. Sin
Sin is a horizontal crescent Moon and takes a central position.

3. The Sun (Shamash)
The Sun—a star in a circle—is Shamash, god of justice and light.

4. Goddess Nanaya
Nanaya sits with hands raised, ready to receive the young girl.

5. King Melishipak
Melishipak raises one hand to Nanaya and holds his daughter's hand with the other.

Overseen by Sin
On this stele from the 12th century BCE, Sin (the crescent Moon) oversees King Melishipak (1186–1172 BCE) as he presents his daughter to the goddess Nanaya.

3rd millennium BCE. This calendar was widely adopted in southern Mesopotamia, unifying many city-states under a common timekeeping system. The calendar began in the spring, around March–April, with the first year coinciding with a full Moon on the spring equinox. The new year was important to farmers, being a time of planting and renewal.

Calendar refinements

Over time, the calendar, which remained fully observational, was refined so that each month began with the waxing crescent Moon. The extra month was inserted roughly every two to three years to ensure seasonal alignment.

These practices are well documented, with guidelines preserved in the MUL.APIN tablet of around 1000 BCE. This artifact contains advanced astronomical and mathematical knowledge detailing the Moon's rising and setting times, its path across the sky, and its position relative to other celestial bodies.

The Mesopotamians also invented the sexagesimal number system, which uses a base of 60 (just as the decimal system utilizes a base of 10). The sexagesimal system is still in limited use today, with 60 seconds in each minute and 60 minutes in each hour. Angles and geographical coordinates also use the sexagesimal system.

> ## "I, **Nanna–Suen**, have caused six hundred cows to give birth to calves for the **house of Enlil**."
>
> From a translation of the Sumerian poem *Nanna-Suen's Journey to Nibru*, by A. J. Ferrara (1973)

Moon movements
This section of an ivory prism, inscribed in red, features calculations indicating the Moon's position and phases, the timing of tides, and the length of the day. Dating from 800 BCE, it comes from modern-day Iraq.

MINOAN ASTRONOMY

The ancient Minoans of Crete built grand palace-temples, such as Knossos, to honor their deities. At one site, a libation bowl bears the name Mena, a word etymologically linked to the Moon. The temples were adorned with bull's horns, symbolic of the curving crescent Moon. Key sites, such as the palace-temple at Zakros, were oriented to the extremes of the lunar cycle, and the Minoan lunar calendar used the heliacal rising of Arcturus to maintain alignment with the solar year.

Flowers held aloft by bare-chested priestess

Casting mold
A priestess is shown worshipping at a sun disk on this Minoan mold from Palaikastro, Crete, dating from 1370–1200 BCE. A crescent Moon appears on the reverse; the casting from this mold may have been part of a geared mechanism for predicting eclipses.

Geared wheel containing cruciform Sun symbol

Small disk containing cross on bell-shaped base

CREATION STORIES

Beliefs and myths about the Moon's creation not only explain its origin, but also serve as guides to how people should interact with each other and with the spirit realm. Such stories often emphasize cosmic balance, renewal, and harmony. A common theme is the Moon and Sun being born together: the duality of the event speaks of the need for balance in the universe, with the Sun governing by day, the Moon by night. One Japanese myth tells how the eyes of the god Izanagi gave birth to the Moon and Sun deities, while in the Cook Islands, the Sun and Moon are seen as the eyes of the primordial mother, who can only open one eye at a time. In West Africa, Mawu and Lisa, the Moon and Sun, are twin deities whose union maintains cosmic order.

Divine origins are also a recurring theme. Christian scriptures tell of God creating the Moon and Sun to provide light to the world. In Finnish mythology, the Moon is born from an egg nurtured by the goddess Ilmatar, symbolizing the Moon's connection to nature and its cyclical rhythms.

1. From the eye of a god
Returning from visiting his wife, Izanami, in the underworld, Japanese deity Izanagi bathes to purify himself. As he wipes his eyes, Amaterasu, the Sun goddess, is born from one eye; from the other, the Moon god, Tsukuyomi, is born (seen at back left in this 1870 image by an unknown artist).

2. Ilmatar gives birth to the Moon
The virgin goddess Ilmatar lies back on the sea in this illustration by Joseph Alanen of a scene from the Finnish saga *Kalevala*. As the wind builds around her, a duck flies across the sky overhead and lays an egg that falls onto Ilmatar's lap. From this egg, the goddess gives birth to the Moon.

3. Vari-ma-te-takere and Avatea
In a creation story from the Cook Islands, a primordial woman, Vari-ma-te-takere, plucks off part of her body. It becomes Avatea, the first human (shown as half-man, half-fish in this basalt relief). Avatea opens his eyes, and they shine brightly: the right eye as the Sun, the left as the Moon.

5

6

> "**God** made ... the **lesser light** (the Moon) to **rule the night**."
>
> Genesis 1:16–18, the *Amplified Bible* (AMP)

4. Consort of the Sun

Mawu-Lisa, or Mahu-Lisa, of West African tradition—shown here in this painting by Michael Constantine—are twin deities who made the world. As the Moon and the Sun, their lives are intricately entwined. The Fon people see the Moon, Mawu, as an older female consort of Lisa, the male Sun.

5. Biblical story

According to the Christian Bible, God made the Sun and the Moon on the fourth day of creation. In this medieval manuscript illustration from Vincent de Beauvais's *Speculum Historiale*, God holds the smiling Moon aloft—one of the two celestial lights that would illuminate the world below.

6. Babylonian god of creation

Dating from c. 1156–1025 BCE, this limestone fragment depicts a dragon-like *mushhushshu* and a spade symbol. Both were associated with Marduk, the chief Babylonian deity, who brought forth the Moon in the ancient epic creation story *Enuma Elish*.

THE NIGHT TRAVELER

In Egyptian mythology, the god of the Moon was Khonsu, the "night traveler." However, the Moon was also associated with Thoth, a god who played a key role in maintaining the natural order.

The Egyptians depicted Thoth as a baboon or a man with the head of an ibis. He was believed to control time and was credited with creating the lunar calendar of 12 phases, totaling 360 days. The Moon's movements were interpreted as journeys through Duat (the underworld), where Thoth's wisdom and magical abilities were essential to navigate the dark, chaotic reams. Meanwhile, the Moon's transition from darkness to light reflected Thoth's ability to restore balance and order.

Khonsu stele
Thought to aid walkers at night, the Moon god Khonsu is depicted here as a young man seated in front of offerings, accompanied by lunar symbols. The name Khonsu means "to traverse the night."

Osiris, the god of the afterlife, was also associated with the Moon. The repeated lunar phases reflected the eternal nature of life and the possibility of rebirth. The waning and waxing of the Moon represented Osiris's death and regeneration—in other words, the soul's journey through death to new life.

The Eye of Horus
Also known as the Wadjet, the Eye of Horus commonly appeared on Egyptian coffins. It brought protection and good health after death, particularly to royalty. It is linked to the Moon through the myth of Thoth restoring Horus's eye after it was damaged in a fight with his rival, Set—a restoration similar to the Moon's rejuvenation each month.

The waxing Moon
Dating from the 1st century CE, the astronomical ceiling at the Temple of Hathor, Dendera, features the waxing moon as 14 steps ending with the restored "Eye of Horus" (the full Moon). At the far right stands Thoth, who healed Horus's eye after battling Set, god of chaos and storms.

Thoth raises his arms to heal Horus's eye

Eye of Horus is shown restored after Thoth's battle with Set

THOTH'S GAMBLE WITH THE MOON

The Sun god Ra cursed the sky goddess Nut, forbidding her to give birth on any day of the year. Thoth helped by gambling and winning extra light from the Moon. This extended the calendar from 360 days to 365 days, allowing Nut to have her children.

Relief of Thoth wearing a lunar-disk headdress

THE TWELVE MONTHS

The ceiling of the Tomb of Senenmut (c. 1479–1458 BCE) at Deir el-Bahri, Thebes, features 12 circles representing the lunar months, making up 360 days. Each circle is divided into 24 sections—the hours of the day. Circumpolar constellations are also depicted: at the top is Ursa Major as a bull. He faces the falcon-headed god Anu.

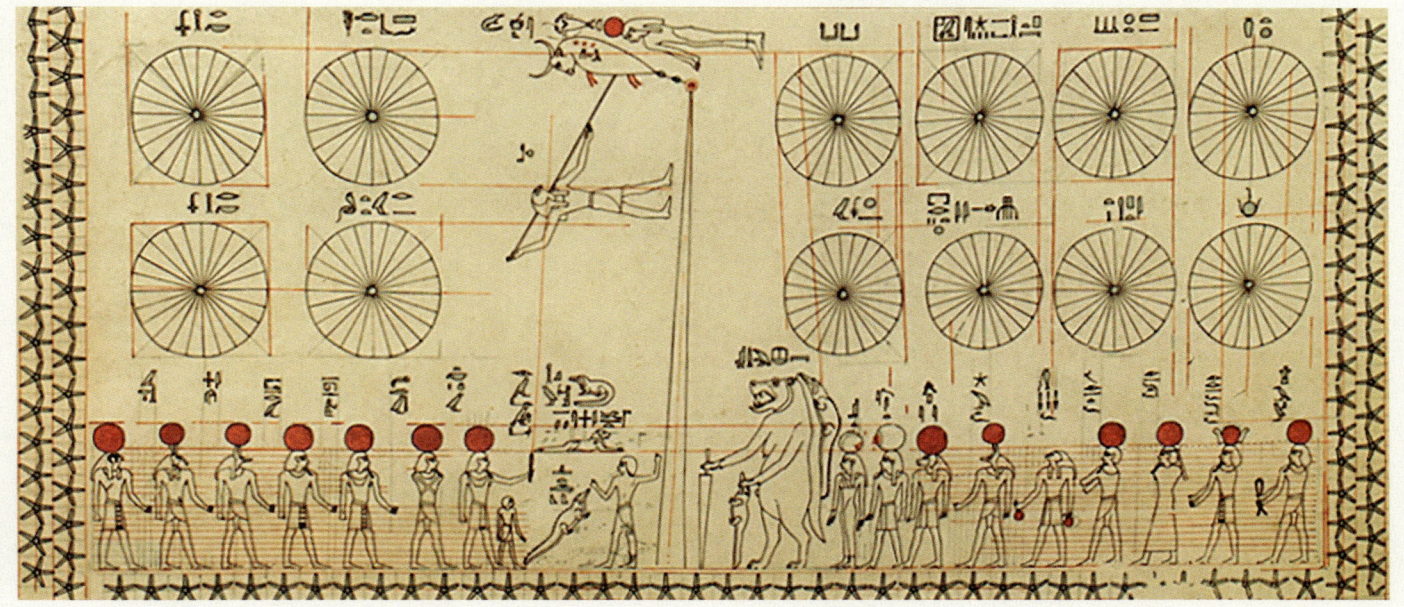

Facsimile of Senenmut's astronomical ceiling

CLASSICAL CALCULATIONS

In ancient Greece, many thinkers sought to understand the Moon's nature and behavior. They studied the Moon through observation and reasoned inquiry, actively laying the foundations of science.

The earliest ventures into astronomy in Greece owed much to Babylonian scholars, whose knowledge of celestial motion was passed on through cultural exchanges. One of the first Greek philosophers to contemplate the Moon was the polymath Pythagoras (c.570–495 BCE). He postulated that the Moon is a spherical body situated in a cosmos of order and harmony.

Pythagoras's work is now lost, but books about his teachings written by later students have preserved many of his discoveries, although some became obscured by myth. One legend describes him listening to a blacksmith strike an anvil and noting that the rhythm of the hammer was similar to the harmonic movement of the heavens. The Pythagorean idea of musical harmony and symmetry in the cosmos would remain influential for many centuries, even enduring into the Renaissance period.

Understanding the Moon

Anaxagoras (c.500–428 BCE) deviated from the popular Greek idea that the Moon was a divine entity, the goddess Selene. Instead, he perceived it as a physical object composed of rock. This radical idea meant that the Moon

Terminator line
Anaxagoras correctly proposed that the terminator, the line dividing the sunlit part of the Moon from the part in shadow, changes position as different parts of the lunar surface appear illuminated from Earth's perspective.

Fresco of Anaxagoras
Anaxagoras is shown pointing at a globe in this 19th-century painting. The shadow of his hand falls on its surface similar to how the shadow of the Moon falls across Earth in a solar eclipse.

Earth-centered universe
Most Greek philosophers—including Aristotle and Plato—considered Earth to be at the center of the universe, with the Moon being the closest celestial body orbiting around it.

1. Earth
2. Moon
3. Mercury
4. Venus
5. Sun
6. Mars
7. Jupiter
8. Saturn

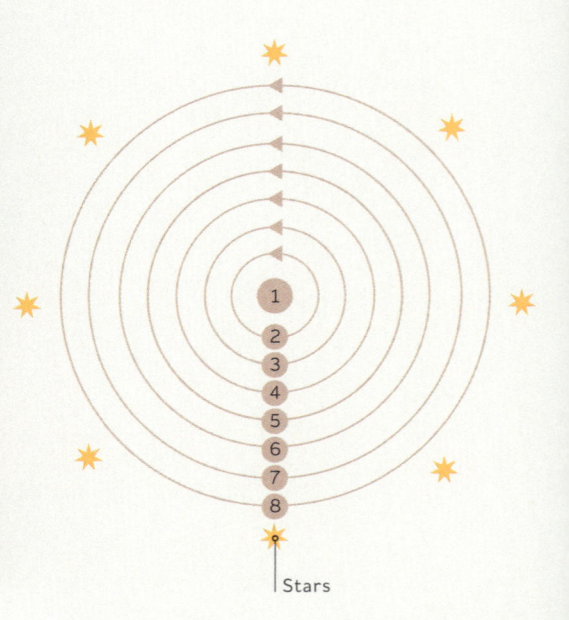

Stars

was subject to natural laws and could be understood through observation and rational thought instead of via mythological or supernatural explanations. Anaxagoras recognized that the Moon was illuminated by reflected light from the "red-hot stone"—the Sun. He also gave early explanations of the phases of the Moon and the cause of eclipses.

Modeling the cosmos

Instead of focusing on scientific details, Plato (c.428–348 BCE) explored the Moon symbolically in his philosophical dialogues. He argued that celestial bodies, including the Moon, moved in elegant, uniform circles. He also devised a simple geocentric (Earth-centered) model of the solar system. Plato's acknowledgment of the Moon's role in the cosmos reflected the growing complexity of Greek thought regarding celestial phenomena.

Aristarchus of Samos (c.310–230 BCE) viewed the cosmos differently. He proposed a heliocentric solar system, suggesting that Earth and the planets revolve around the Sun, but his model was not widely accepted. »

Moon goddess
The ancient Greek deity Selene was the personification of the Moon. As in this mosaic, she was often depicted with a crescent Moon on her head.

VISUAL TOUR

1. Horned goddess
Crescent Moon "horns" protrude from Selene's head.

2. Lunar chariot
Selene crosses the heavens in her chariot, bringing light to night sky.

3. Selene's steeds
Two horses pull Selene's chariot; four pulled the chariot of the Sun god Helios.

4. Endymion asleep
Selene visits her mortal lover Endymion while he sleeps.

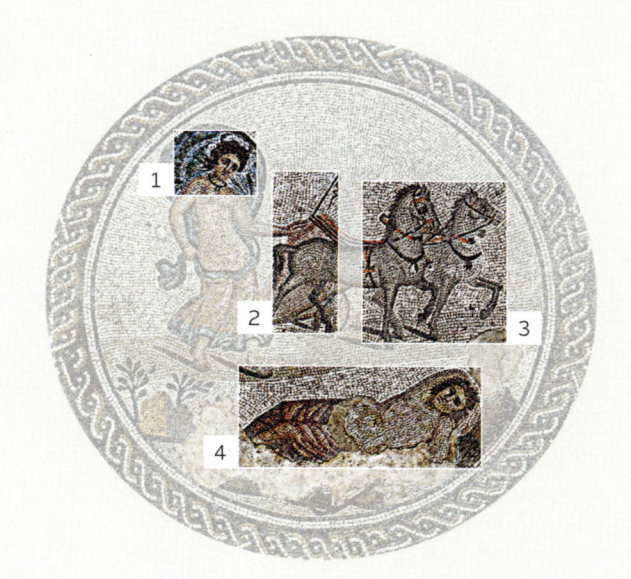

Aristotelian influence

The philosopher Aristotle (384–322 BCE) advocated a geocentric solar system, with the Moon, Sun, and planets orbiting Earth. He viewed the Moon and other celestial objects as occupying a place of perfection; visible features on the lunar surface must therefore be due to corruption from our realm. Despite Aristotle's influence, some thinkers were skeptical of his ideas. Plutarch (46–120 CE) suggested that the Moon's dark patches were shadows and rivers and that the surface was not perfect, but had mountains and chasms.

Measuring the Moon

Formative ideas concerning the Moon's shape and composition prompted fresh questions, such as "How big is the Moon?" and "How far away is it?" Aristarchus (see pp.52–53) made pioneering estimations of the sizes of Earth, the Moon, and the Sun and the distances between them, while another astronomer, Hipparchus (c.190–120 BCE), measured the Earth–Moon distance using lunar parallax (the apparent shift in the Moon's position when viewed from two different vantage points). Hipparchus also developed a model to explain the Moon's orbit and phases.

Hipparchus
One of the foremost astronomers of antiquity, Hipparchus is credited with making the first known star catalog.

COSMIC GEOMETRY

Aristarchus realized that Earth and the Moon make a right triangle with the Sun when the Moon is at first- or third-quarter phase. At just such a time, he measured the angles between the Sun and the Moon in the sky. Then, without knowing the distances between them (see opposite), he used Pythagoras's theorem to calculate the geometry of all three celestial bodies.

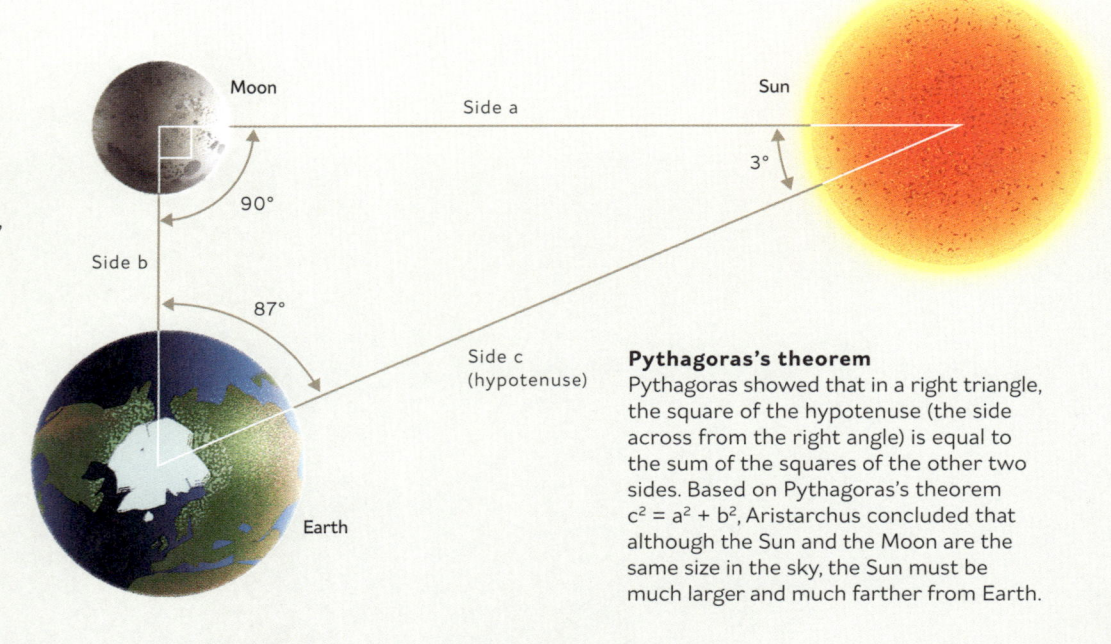

Moon

Side a

Sun

3°

90°

Side b

87°

Side c (hypotenuse)

Earth

Pythagoras's theorem
Pythagoras showed that in a right triangle, the square of the hypotenuse (the side across from the right angle) is equal to the sum of the squares of the other two sides. Based on Pythagoras's theorem $c^2 = a^2 + b^2$, Aristarchus concluded that although the Sun and the Moon are the same size in the sky, the Sun must be much larger and much farther from Earth.

THE SIZE OF THE MOON

Aristarchus saw that during the longest lunar eclipses, the Moon passes through the center of the shadow cast by Earth (see pp.38–39). He also realized that the position of the Moon at the start and the end of the eclipse indicates the size of Earth's shadow. He concluded that by calculating how many Moons fit into this shadow, he could determine a size ratio between Earth and the Moon.

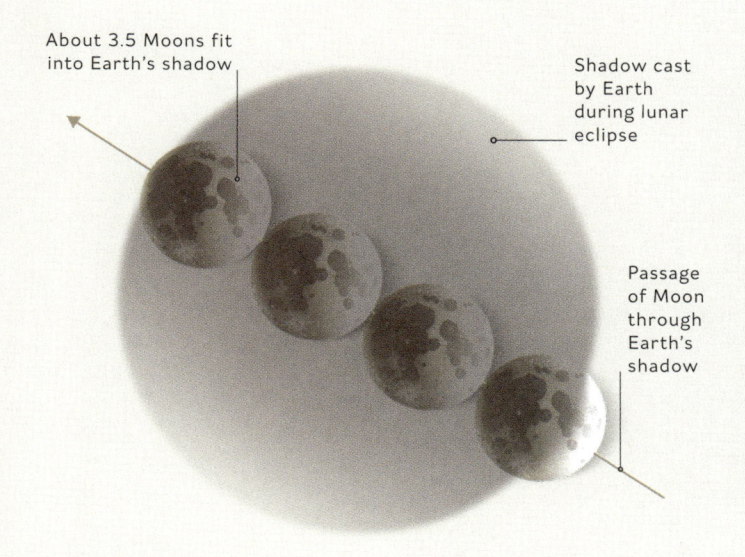

About 3.5 Moons fit into Earth's shadow

Shadow cast by Earth during lunar eclipse

Passage of Moon through Earth's shadow

Aristarchus's estimate

After a lunar eclipse, Aristarchus estimated Earth to be 3.5 times larger than the Moon—close to current measurements of the two bodies' equatorial diameters, which give a ratio of 3.67. After Eratosthenes (see right) had deduced Earth's diameter from its circumference, the Earth–Moon ratio could be used to calculate the Moon's actual diameter.

EARTH'S CIRCUMFERENCE

Eratosthenes (c.276–194 BCE) knew that when a stick is placed in the ground, its shadow forms an angle with the stick—and that sticks placed at different latitudes at the same time have different shadow angles. He concluded that if he knew the distance between two such sticks and the difference between their shadow angles, he could calculate the circumference of Earth.

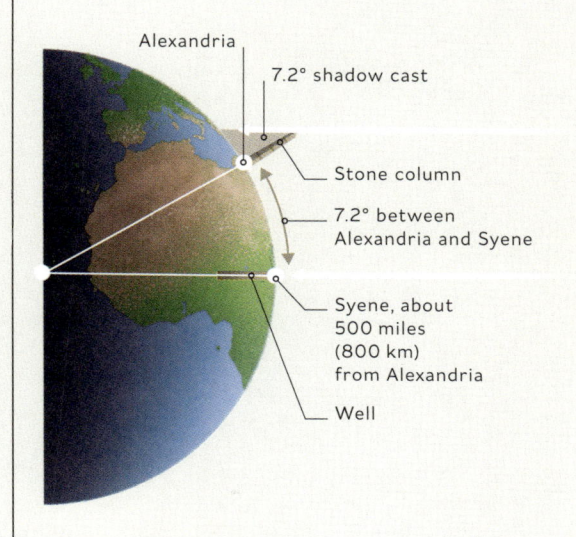

Alexandria

7.2° shadow cast

Stone column

7.2° between Alexandria and Syene

Syene, about 500 miles (800 km) from Alexandria

Well

Eratosthenes's experiment

At the summer solstice (see pp.34–35), the midday Sun shone directly down a well at Syene, while in Alexandria, it cast a shadow of 7.2°. From this, Eratosthenes calculated Earth's circumference to be 24,855 miles (40,000 km), which is almost correct. He also estimated that the Moon is 2.5 times smaller than Earth—only a slight miscalculation.

EARTH–MOON DISTANCE

Hipparchus measured the Earth–Moon distance during a solar eclipse, using observations made at the Hellespont (where the eclipse was total) and Alexandria (where one-fifth of the Sun was visible). Because he already knew the distance between the Hellespont and Alexandria, all he had to do was measure the difference between the Sun's angle at the same time at the two locations— the rest was simple trigonometry.

Measuring a solar eclipse

Hipparchus made a right (90°) triangle between the two locations and the edges of the Sun and the Moon. Measuring the Moon's angle at Alexandria then enabled him to calculate the final angle and the other two distances.

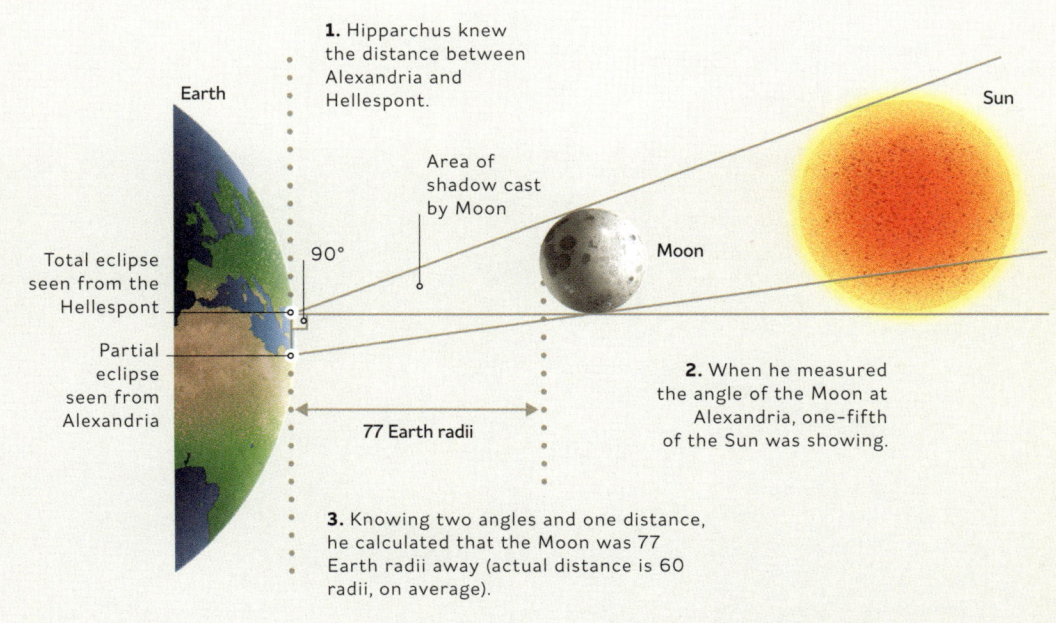

1. Hipparchus knew the distance between Alexandria and Hellespont.

Earth

Area of shadow cast by Moon

Sun

90°

Moon

Total eclipse seen from the Hellespont

Partial eclipse seen from Alexandria

77 Earth radii

2. When he measured the angle of the Moon at Alexandria, one-fifth of the Sun was showing.

3. Knowing two angles and one distance, he calculated that the Moon was 77 Earth radii away (actual distance is 60 radii, on average).

A MODEL OF THE COSMOS

Often described as the world's oldest computer, the Antikythera mechanism is a device that ancient Greek astronomers used to represent the movements of astronomical bodies.

In 1901, divers recovered a bronze object from the wreckage of a sunken ship off the Greek island of Antikythera. It turned out to be a fused mass of several objects, one of which the politician Spyridon Stais (1859–1932) correctly identified as a gear. Under further investigation, more fragments separated and began to reveal their secrets. Together they formed a device, which has been dated to c.205–c.70 BCE. Subsequent dives brought the total number of fragments to 82.

Tracking the heavens

Research by figures such as German philologist Albert Rehm (in 1905–06) and scholar Derek de Solla Price (from the 1950s) provided key insights into the fragments. In the 1970s, scientists used first X-ray and then CT scans to produce high-resolution images of them. These images enabled engineers to reconstruct the pieces and to see how they fit together.

The complete structure had a hand crank at its side and a series of dials on its front, on which representations of the Sun, the Moon, and the planets turned within zones denoting the constellations of the zodiac. This zodiac ring lay within a larger circle, which analysts now believe to be an ancient Egyptian calendar with the names of the months in Greek. Two dials on the back of the box enabled the user to monitor various dates and astronomical cycles. These included the Moon's 19-year Metonic cycle (see pp.88–89) and its 76-year Callippic cycle.

Salvaged remains
This bronze fragment, containing a large, central gear, was one of the first pieces of the mechanism to be recovered.

Detailed reconstruction
This computer-generated model shows the front and back plates of the Antikythera mechanism. The front plate (left) contains the zodiac and calendar dials, while the back plate indicates long-term solar and lunar cycles. The original device was made from wood and bronze and measured around 7 x 4 x 13 in (18 x 10 x 33 cm).

USING THE ANTIKYTHERA MECHANISM

Turning the handle on the side of the machine sets the date and time on its front-facing dial. This also sets the locations of the Sun, the Moon, and the planets in the zodiac constellations on that particular day. The ball representing the Moon rotates on its axis, mimicking the Moon's phase cycle, and revolves at the same rate as the synodic month (see pp. 28–29). At the same time, turning the handle moves two dials on the back of the box. These keep track of important astronomical and cultural events and enable the user to synchronize their lunar and solar calendars.

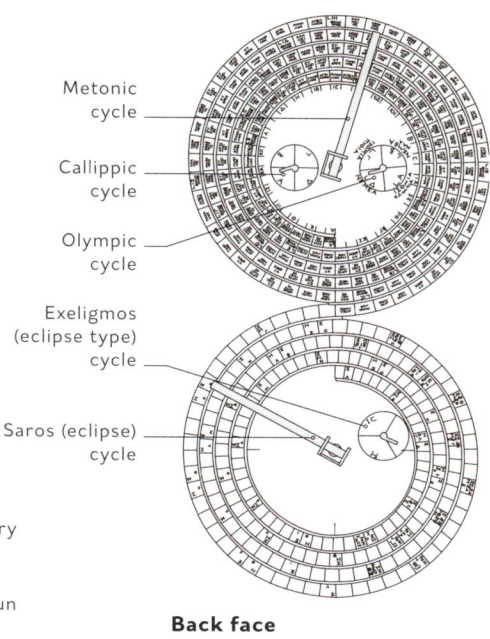

- Metonic cycle
- Callippic cycle
- Olympic cycle
- Exeligmos (eclipse type) cycle
- Saros (eclipse) cycle

Back face
The back dials keep track of eclipses, eclipse types, the dates of important sports festivals, and the Metonic and Callippic cycles.

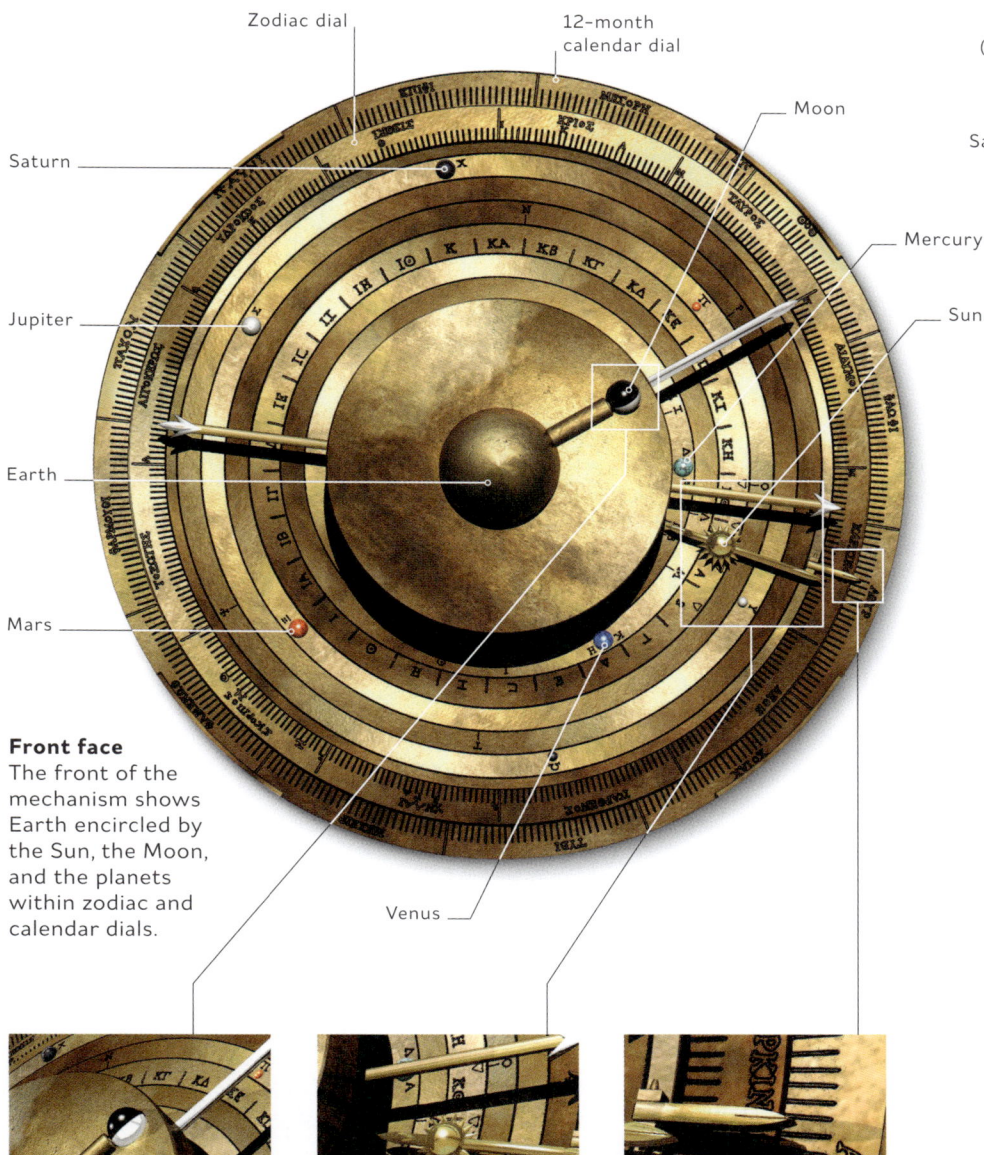

- Zodiac dial
- 12-month calendar dial
- Saturn
- Moon
- Jupiter
- Mercury
- Sun
- Earth
- Mars
- Venus

Front face
The front of the mechanism shows Earth encircled by the Sun, the Moon, and the planets within zodiac and calendar dials.

Lunar phases
The black-and-white ball rotates relative to the Sun's position on the dial, demonstrating the phase of the Moon at any given time.

Lunar nodes
No supporting evidence survives, but scientists have suggested that the Sun dial's triangles mark lunar nodes (see pp. 30–31), indicating eclipses.

Date pointer
A small pointer indicates the date in the Egyptian calendar on the mechanism's outer dial. The months are written in Greek letters.

BYZANTINE COMPUTER

Dating from c.520 CE, this bronze gear mechanism was part of a Byzantine portable sundial-calendar. Far less sophisticated than the Antikythera mechanism, it is the second-oldest-known device of its kind.

The gears of a 6th-century computer

DREAMTIME STORIES

The indigenous peoples of Australia hold diverse and profound beliefs about the Moon. Their cultural and spiritual ideas, passed on via oral tradition and artwork, are called Dreamtime stories.

Many Aboriginal groups perceive the Moon to be male and the Sun female. The Eulayhi call the Moon Bahloo and believe he travels with three pet snakes. In one Dreamtime story, Bahloo descends to Earth and asks the people for help crossing a river. The people agree but demand that he leaves his snakes behind, as they are scared of them. Angered, Bahloo carries the snakes across the river himself, declaring that, unlike him, the people will not be reborn after death. This story, which explains mortality and fear of snakes, is a good example of how Aboriginal cultures connect natural phenomena, such as the Moon, with cultural beliefs, values, and practices.

Immortal Ngalindi

A story from the Yolngu tells how the Moon was once a plump, lazy man named Ngalindi. In punishment, his wives chopped pieces off him, as illustrated by the waning moon. Ngalindi climbed a tree to escape but died from his wounds, which is symbolized by the new Moon. After three days, he revived, like a waxing Moon, only for the cycle to repeat monthly. Beings on Earth were immortal before Ngalindi's death, but his curse made death final for everyone else, allowing only Ngalindi to rise again.

Mityan's sorrow

The Boorong have a tale of Mityan, a man who tries to lure away one of the two wives of the ancestral spirit Unurgunite. A battle ensues, and lovelorn Mityan fails in his attempt. He is placed in the sky as the Moon, with battle scars on his face. His changing phases reflect his ongoing sorrow. As well as being a creation story, the tale is a moral warning of the dangers of jealousy. In recognition of the Boorong people's rich star lore, Unurgunite was adopted by the International Astronomical Union as the official name of the star Sigma Canis Majoris. Unurgunite's two "wives" appear as stars nearby; the Moon travels closer to one than the other as it moves through the sky.

Bark was stripped from a eucalyptus tree in wet season, cured over fire, and flattened under weights

Crescent engraving
Two figures reach towards a crescent in this rock engraving at Ku-ring-gai, north of Sydney. The pointed horns of the crescent suggest it represents the Moon rather than a boomerang, and the motif is interpreted as symbolizing an eclipse, with Sun-woman obscured by Moon-man.

The Moon Man, Alinda
This 1948 bark painting from Yirrkala is an artistic depiction of the male Moon tale. Dreamtime stories in Aboriginal art carry significant meaning, encoding historical, geographical, and spiritual information.

"... you would not do **what I, Bahloo, asked you** to do ... you have **lost the chance** of rising again **after you die**."

K. Langloh Parker, *Australian Legendary Tales* (1896)

MOON HALO

High-altitude cirrostratus clouds, which often come before bad weather, can cause a halo to appear around the Moon as light is dispersed by ice crystals in the clouds. Moon halos were used by Aboriginal Australians to predict rain. This knowledge, alongside animal behavior and other natural signs, helped communities plan activities like hunting and gathering.

Sunlight reflected by Moon is scattered by ice crystals in atmosphere

Halo forms about 22° out from Moon

A Moon halo may indicate approaching rain.

Round object could be Sun or full Moon

Pigment is made from ocher

Crescent Moon is boxed between leaf shapes

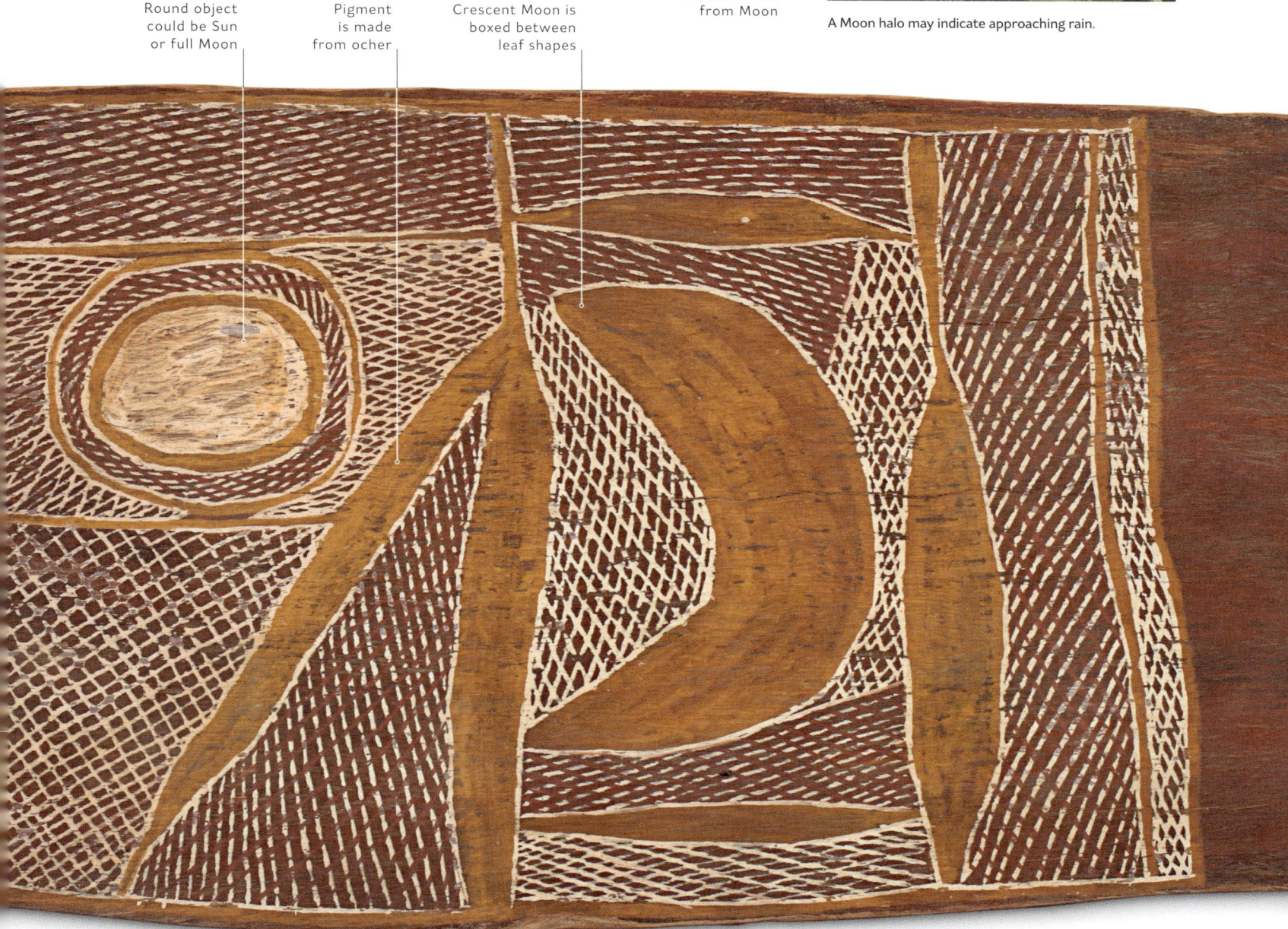

Chandra, the Hindu god of the Moon, rides across the sky on the auspicious *Purnima* (full Moon) in this 18th-century illustration of *Shakunavali* (Book of Dreams and Omens).

INDIAN LUNAR TRADITIONS

In India, people revere the Moon for its spiritual and astronomical significance. Ancient astronomers such as Aryabhata (476–550 CE) pioneered its study, and its symbolism continues to inform religious traditions today.

Around 499 CE, the mathematician and astronomer Aryabhata, from Kusumapura (modern-day Patna), wrote his 118-verse treatise *Aryabhatiya*. His poetic work advanced Indian understanding of the motions of celestial bodies. It described a geocentric model in which the Moon orbited Earth in epicycles (see pp. 100–101). Aryabhata correctly explained lunar eclipses but mistakenly believed the universe was filled with a substance known as "ether." Before there was any knowledge of gravity, ether was thought to fill space and aid celestial movements. Aryabhata's work placed the Moon in a new cosmological framework that included both the divine and the material worlds.

The Moon and Hinduism

The months of the Hindu calendar are primarily lunar. To keep them synchronized with the solar year, an additional month, known in Sanskrit as "Adhik Maas," is inserted roughly every three years. The new Moon (*Amavasya*) is auspicious and often associated with new beginnings and religious rituals.

Chandra the Moon god is revered in Hinduism as the father of Buddha. His influence on crops links the phases of the Moon with agriculture, and he traverses the night sky in a chariot drawn by 10 white horses or an antelope. As the keeper of the sacred Soma plant, Chandra holds the elixir that grants the gods immortality.

Chandra's favoritism toward Rohini, one of his 27 wives, led to a curse that caused him to wane. The deity Lord Shiva moderated the curse, allowing Chandra to wax as well as wane, much like the Moon, and Chandra came to symbolize the rhythm of life and time.

Many Hindu deities wear a crescent Moon, including Lord Shiva, who embodies the cyclical nature of time and existence. His crescent Moon also represents Shiva's role in maintaining balance and order in the universe.

Jain cosmology

By the 12th century, the Indian religion of Jainism had developed a complex model of the universe in which the Moon played a key role. Souls liberated from rebirth go to a place of perfection called *Siddhaśilā*, which is symbolized by a crescent Moon on the head of the Cosmic Man.

Movement of the Jain Moon
The verses of the *Sangrahani Sutra* contain the cosmological Jain worldview. Here, the four animals between the spheres help the Moon travel through its phases.

REACHING FOR THE MOON

This painting depicts one version of the story of the Moon and a young Lord Krishna, the god of protection and love. Held by Maa Yashoda, his foster mother, he points skyward to the crescent Moon and, with childlike curiosity, asks her to bring it to him. Not wanting to disappoint, she reaches out to the Moon's reflection in a plate of water. In Hinduism, Krishna is an incarnation of Vishnu, one of the principal gods, and is identified as the Sun during the day and as the Moon at night.

Krishna Reaching for the Moon (c.1820), in the *Parahi*, or Hills, style of Indian painting.

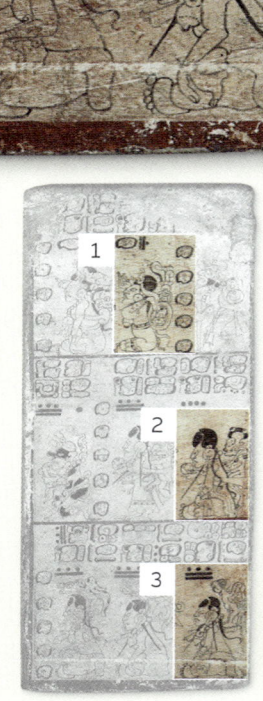

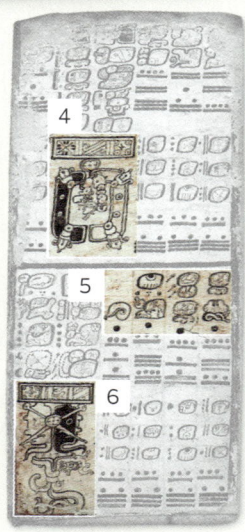

VISUAL TOUR

1. Divine intervention
The young Moon goddess looks up to the sky, reaching out to heal the sick.

2. Self-sacrifice
The burden of disease is carried on the back of the young Moon goddess.

3. The nature of disease
The disease reveals itself as a bird sitting on the goddess's shoulder.

4. Facing darkness
A tattooed face looks into the darkness of an eclipse, with the light left behind it.

5. Lunar charts
Glyphs give the dates of predicted eclipses—the codex lists 857 lunar-phase cycles.

6. Disappearing light
A toothed plant or creature eats the light during an eclipse. The rectangle above is the sky.

THE MAYAN TIMEKEEPER

The ancient Mesoamerican civilization of the Maya began around 2000 BCE and reached its peak between 250 and 900 CE. It is known for its advanced intellectual, architectural, and astronomical achievements.

Living in what is now southeastern Mexico, Guatemala, Belize, and parts of Honduras and El Salvador, the Maya were expert observers of the sky and accrued extensive knowledge about the movements of the Moon, the planets, and stars. Astronomical events were interwoven with their beliefs and informed their rituals and calendar.

By the 10th century CE, the Maya had built observatories such as El Caracol at Chichen Itza, and they had developed sophisticated methods to observe the Moon's cycles and predict eclipses. They recorded these on codices—long strips of bark-paper that folded up like an accordion. One example that survives today is the Dresden Codex, which includes the cycles of Venus, Mars, and Mercury; the Sun's zeniths and equinoxes; and solar and lunar eclipses.

Among many counting systems in the Maya calendar were the *Tzolk'in*, the *Haab*, and the Long Count. The *Tzolk'in* had two interlocking cycles that combined the numbers 1 to 13 with 20 named days. After 260 days, when all the combinations had played out (13 × 20), the cycle restarted. The *Haab* was a 365-day period with 18 cycles of 20 days, plus five "unlucky" days when the gods were appeased. Used for longer periods, the Long Count measured from a fixed point and repeated every 5,128 years.

Information as power

The Maya considered eclipses to be disastrous, linked to malevolent gods and destructive events like flooding. The ability to predict eclipses was reserved for the rulers, and it was a powerful tool for showcasing their divine connection as the descendants of sky gods. Rulers also used their astronomical knowledge to dictate the timing of rituals and the planting of crops, waiting for the first full Moon after the rainy season began. This shared belief system created unity among the population.

The goddess and the rabbit

In this late-Maya ceramic, Ixchel the Moon goddess embraces a rabbit. She represented fertility and love and gave birth to the rabbit, who, it was believed, could be seen on the face of the Moon.

ANCESTRAL PUEBLOANS

Contemporaries of the Maya, the Ancestral Puebloan culture in southwestern North America believed that the mother Moon, P'áh-hlee-oh, kept them safe. In the San Juan Mountains of Colorado, the tallest natural sandstone pillars are Chimney Rock and Companion Rock. Every 18.6 years, the Moon rises between them. The Ancestral Puebloans had strong cultural links to these rocks and built ritual chambers and platforms for watching the event.

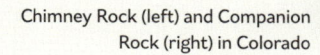
Chimney Rock (left) and Companion Rock (right) in Colorado

OBSERVATION AND INTERPRETATION

The Moon remained an important timekeeper long after humans established civilizations—the Arabic lunar calendar is still being used to set the dates of Islamic holidays today. However, the Moon was also an object of mystery and devotion, its monthly darkening and brightening embodying a cosmic demonstration of death and rebirth. For that reason, the Moon became a prominent symbol— and was even considered divine—in the religions and folklore of many cultures.

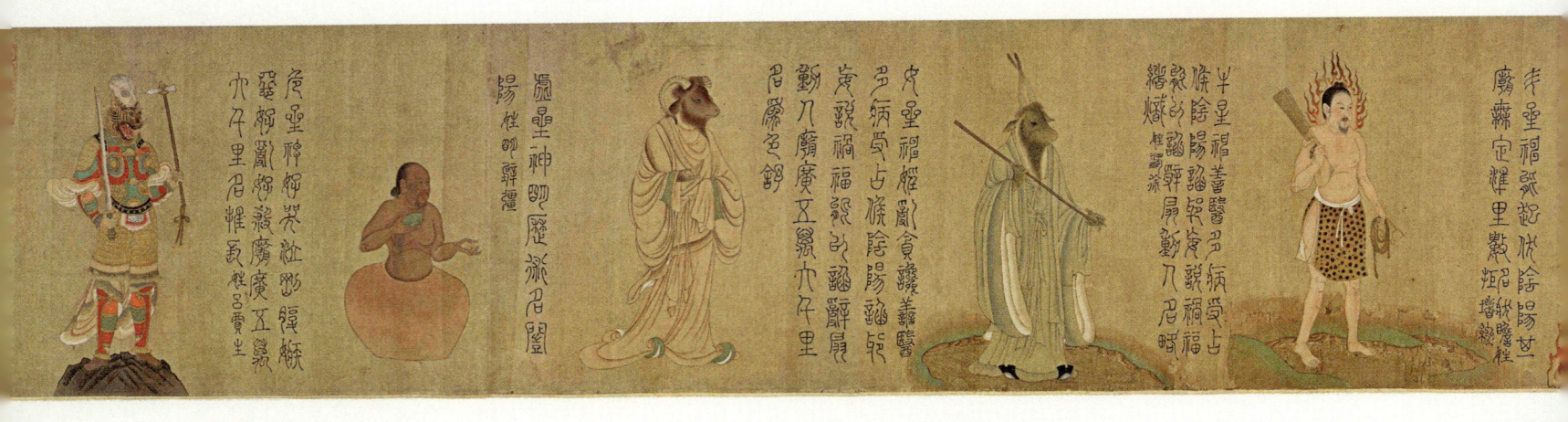

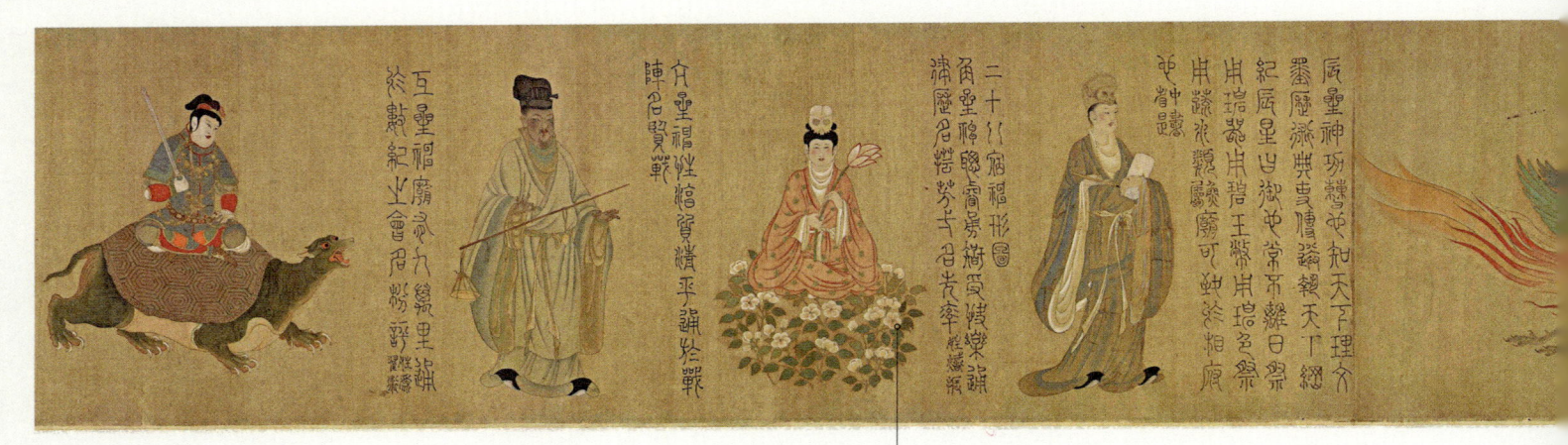

Bed of lotus flowers supports this deity

THE CELESTIAL EMPIRE

The Moon has long been central to Chinese culture and spirituality. Astronomy, fortune-telling, and belief systems have been closely aligned throughout China's history.

In Chinese culture, the Moon symbolizes the feminine Yin, in contrast to the masculine Yang of the Sun—Yin and Yang being the two complementary yet opposing forces that govern all aspects of life. It is the Moon, with her close links to water, that brings abundance at harvest time. This is joyously celebrated at the Mid-Autumn Festival, when families come together to honor Chang'e, a beautiful young woman who became the Moon goddess after drinking a potion of immortality. Exiled, she ascended to the Moon, where she now resides with her faithful companion Yutu, the Moon Rabbit (see pp.74–75).

At the Mid-Autumn Festival, candles and lanterns are lit and traditional mooncakes shared. This celebration, often led by women, is a time of reflection, gratitude, and hope for future prosperity. The reverence for Chang'e reaches beyond China's borders: in Malaysia, devotees visit the Thean Hou Temple, where a statue of the Moon goddess stands, to have their clothes and hair blessed in her name.

Lunar mansions

As the Moon moves across the sky, it appears to pass through patterns of stars known in Chinese astrology as the Twenty-Eight Lunar Mansions. The mansions correspond to the Moon's position during its 27.3-day orbit of Earth. Each mansion has a patron deity and belongs to one of four houses, which are ruled by animals: dragon, tortoise, tiger, and bird. The Moon's movement through the mansions is a key component in Chinese horoscopes, divination, and fortune-telling.

On the handscroll *The Five Stars and Twenty-Eight Mansions*, attributed to Zhang Sengyou (active c.490–540 CE)—but possibly by a later artist—the patron deities are shown walking or riding (and even seated on lotus flowers) as they move across the sky.

Palm raised, this
deity—representing
Saturn—rides a bull

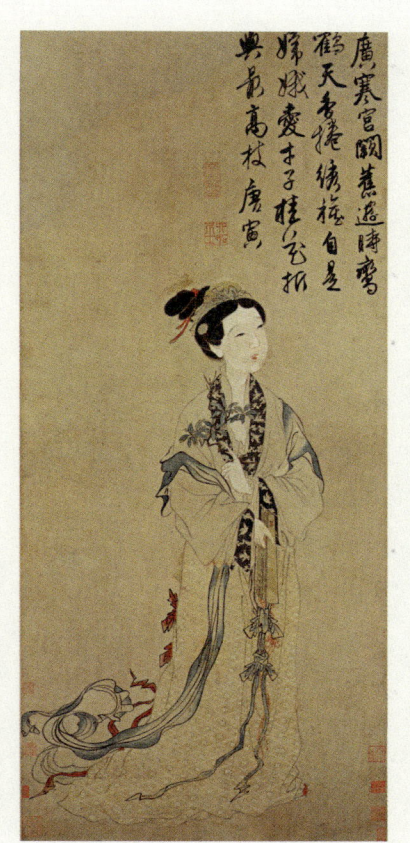

Lonely goddess
In this scroll painting,
dated 1498, Chang'e
stands in solitude wearing
traditional long, flowing
robes as she looks
skyward to the Moon.
Her loneliness will end
when she travels there
with the rabbit Yutu,
which will become her
companion in exile.

BUDDHIST THOUGHT

According to tradition, the Buddha
was born, attained enlightenment, and
passed away on days marked by a full
Moon. For Buddhists, full Moon days are
reserved for spiritual practices and
quiet contemplation. In this statue,
Bodhisattva Avalokiteshvara sits in
his Water Moon pose, gazing
at the Moon's reflection,
which symbolizes
a path to greater
understanding.

> "The Sun is like fire and **the Moon like water**. The fire gives out light and the **water reflects it**."

From *Ling Xian* (*Sublime Model*), by Zhang Heng (120 CE), translated by Joseph Needham (1986)

Astronomical innovations

The ancient Chinese developed advanced cosmological ideas about the Moon, supported by robust observational astronomy. By the 4th century BCE, astronomers Gan De and Shi Shen were independently observing the Moon and Sun, forecasting eclipses with impressive accuracy. Recorded on oracle bones used for divination, eclipses were seen as bad omens. Reliable eclipse predictions allowed for rituals to be conducted to avert disastrous events.

In the early Imperial era, astronomer Liu Hong (129–210 CE) developed a lunar theory that detailed the Moon's motion and provided a way of calculating the syzygy (straight-line configuration) of Earth, Sun, and Moon,

enabling more precise predictions of eclipses. By then, China had developed calendar systems for recording dynastic periods and celestial events; many were lunar-based, but over time, the calendar evolved into a uniform lunisolar system, refined by observations of the Moon. As mathematics advanced alongside astronomy, Zhu Chongzhi (429–500 CE) calculated the Moon's sidereal month (see pp. 28–29) to be 27.21223 days—close to the modern value.

Shen Kuo (1031–1095), a Song Dynasty polymath, gave explanations for the Moon's shape and phases. He argued that the Moon reflected the Sun's light, suggesting that if a sphere were half covered in white powder and viewed from the side, it would have a crescent

Gaocheng Observatory
A bar, or gnomon, extended between the two rooms at the top of the platform, casting a shadow onto a 102-ft (31-m) horizontal scale in front of the building. Together, these measured the Sun's position at noon throughout the year.

shape. Shen Kuo also improved the accuracy of astronomical equipment in his observatory and developed a mechanism called a water clock—similar in principle to the clock tower built at K'ai-feng (see below)—which enabled better timekeeping of lunar motion.

The observatory at Gaocheng, built in 1276, could track the positions of the Moon, Sun, and stars with high precision. Later, the Imperial Observatory in Beijing, founded in 1442, became a center for advanced study. Equipped with tools like the moondial (see right), Chinese scholars combined their observations with Arabic and classical Western knowledge to deepen their understanding of the Moon's properties and behavior.

Moondial
This moondial from Beijing's Imperial Observatory is similar to a sundial except that it tells the time by the position of the Moon in the sky instead of that of the Sun.

Hand points to direction of Moon; scale on outer disk then reveals time

Armillary sphere consisted of series of rings rotated by clockwork mechanism

Astronomical clock tower
This waterwheel-driven clock tower, built by the polymath Su Song, was in use at K'ai-feng from 1090. As well as telling the time, it had a device at the top, called an armillary sphere, that replicated the movement of the Moon and other celestial objects.

WESTERN INFLUENCE

This page from a 1646 reprint of *General Introduction to Calendrical Astronomy* by Chinese scholar Wang Yingming (d. 1614) shows a lunar eclipse. Wang's work was the first Chinese astronomical treatise to display the influence of classical Western ideas. It followed the Gregorian calendar and made no mention of omens and fortunes.

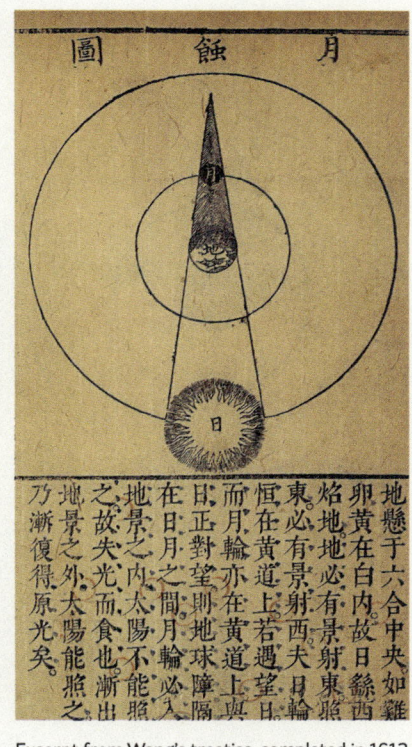

Excerpt from Wang's treatise, completed in 1612

LUNAR FESTIVALS

Around the world, the timings of a number of religious and cultural festivals share a common feature: they all depend on tracking the phases of the Moon instead of setting a date on the Western, or Gregorian, calendar, which is based on the solar year. The celebrations begin during different parts of the Moon's cycle, but those around the new Moon and the full Moon are the most popular. Most of the celebrations fall in the same season each year because they follow lunisolar calendars, which factor in both the Sun and the Moon's movements. However, Ramadan is based on a lunar calendar, so as the years pass, the festival moves through the seasons.

As religions and cultures have spread, many of these festivals have transcended their countries of origin and are now observed worldwide.

1. Ramadan
Timing varies · New Moon
Muslims spend the month of Ramadan fasting during the day and meeting to pray at mosques like the Sultan Ahmed in Istanbul, Türkiye. The ninth month of the Islamic calendar, Ramadan ends with the festival of Eid al-Fitr.

2. Diwali
Fall · New Moon
Here at the Golden Temple in Amritsar, India, Sikh women light candles to mark Diwali. Hindus, Sikhs, and Jains begin the five-day Festival of Lights two days before the new Moon of the Hindu month of Kartik, in October or November.

3. Rosh Hashanah
Fall · New Moon
The *shofar*, a ram's horn, is sounded for Rosh Hashanah. The two-day festival marking the Jewish New Year begins on the first day of Tishrei in the Hebrew calendar, which is typically in September or October.

4. Hobiyee
Spring · Waxing crescent Moon
Nisga'a performers in Vancouver, Canada, celebrate spring and the Nisga'a New Year. The two-day Hobiyee festival begins with the first crescent Moon of the year. If a star appears opposite the bowl of the crescent, the year ahead will be prosperous.

MATARIKI

The Māori festival of Matariki is not only timed to a lunar phase, but is based on the movement of a group of stars. In the southern hemisphere, the Matariki star cluster (also called the Pleiades, the Seven Sisters, or M45) disappears from sight for around a month every year. The first new Moon following its return, in midwinter, is the prompt for the start of the Māori New Year and its accompanying celebrations.

A Māori dancer performs to celebrate Matariki.

5. Chinese New Year
Spring · New Moon
Chinese dragon dancers perform at a fair in Beijing, China, in 2017. Also called the Spring Festival, Chinese New Year begins on the first day of the first lunar month of the year and ends with the Lantern Festival on the 15th day.

6. Holi
Spring · Full Moon
Hindus celebrate the arrival of spring with colorful powders and water. Timed for the last full Moon of the Hindu month of Phalguna, normally in February or March, the Holi festival brings winter to an end and wards off evil.

7. Easter
Spring · Full Moon
A procession in Spain during Semana Santa (Holy Week) reenacts Christ's journey to his crucifixion. The Easter period begins on the first Sunday after the full Moon that occurs on or after the spring equinox in March or April.

8. Reed Dance
Spring · Full Moon
In Africa, a Zulu woman leads her maidens in an eight-day rite of passage. The Reed Dance is a Zulu and Swazi tradition set around the full Moon that falls in August or early September—spring in the southern hemisphere.

THE MAN IN THE MOON

When the Moon is full, people around the world have perceived familiar shapes and figures in the patterns formed by its surface features, the most common being the face of the "Man in the Moon."

The Moon's dark-gray plains, known as "maria" (seas), stand out starkly against its lighter-colored highlands (see pp. 118–119), creating patterns that never change because the same side of the Moon always faces Earth (see pp. 16–17). In different cultures, these patterns have been interpreted in various ways, including as images of people, faces, and animal figures. This phenomenon—whereby the human brain creates familiar, meaningful images out of random jumbles of shapes—is known as pareidolia.

The Man in the Moon featured in European folklore, and in medieval times the Moon was often depicted with a face. In some tales, he is a woodcutter carrying sticks, sent to the Moon as punishment for wrongdoing. In Latvian tradition, it is an exiled woman; the "sticks" are yokes for carrying water buckets. She thinks herself more beautiful than the Moon, so the Moon takes her to himself.

Other cultures also imagined water carriers on the Moon: Scandinavians saw the brother and sister Hjuki and Bil, who were stolen by

Stained-glass depiction
This brilliant-yellow medieval image of a moon with a human face forms part of a church window in Burnham Deepdale, Norfolk, UK.

the Moon while drawing water from a well. The Māori of New Zealand see a woman with a pail of water. The view of her changes as the Moon moves (see pp.116–117), and as the pail appears to tip, it is said to forecast rain.

The rabbit in the Moon

In China, the rabbit companion of the goddess Chang'e is said to be visible making a potion for her (or medicines for mortals). The tale in Japan and Korea is slightly different, with the rabbit grinding ingredients for rice cakes. In one Buddhist story, a rabbit offers himself as a meal to a starving man in an act of virtue. Revealing himself to be the deity Śakra, the man draws the rabbit on the Moon's face as a symbol of selflessness, with the gray image surrounded by the smoke of the cooking fire. The Aztecs of Mexico also saw a Moon rabbit: it had been thrown there to stop the Sun from being outshone by the Moon.

"The **rabbit** in the **Moon** pounds the **medicine** in **vain**."

From *The Old Dust* by Li Bai, Chinese poet (701–762 CE)

Moon face

Looking up at the full Moon, a face appears to be staring back, with the two eyes, nose, and mouth all being outlined by dark lunar maria.

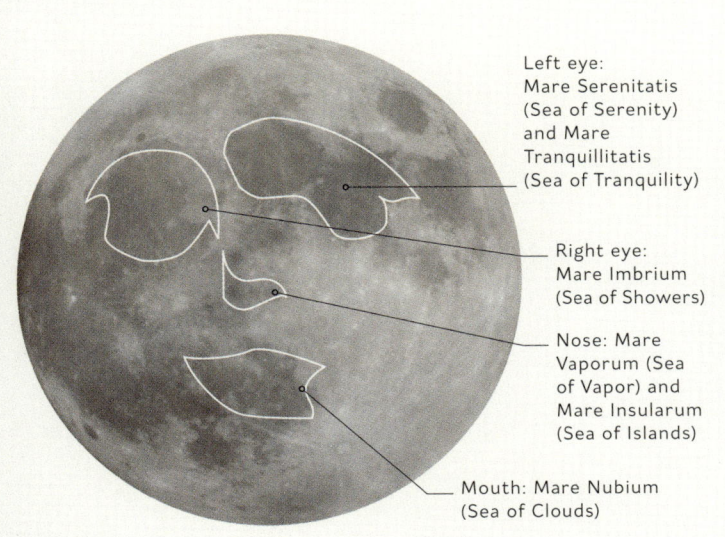

Left eye: Mare Serenitatis (Sea of Serenity) and Mare Tranquillitatis (Sea of Tranquility)

Right eye: Mare Imbrium (Sea of Showers)

Nose: Mare Vaporum (Sea of Vapor) and Mare Insularum (Sea of Islands)

Mouth: Mare Nubium (Sea of Clouds)

19th-century illustration from book of German folk tales

Banished woodcutter

A woodcutter carrying sticks can be "seen" on the Moon. According to German folklore, he was banished there as punishment for cutting firewood on the Sabbath, the day of rest, and his visible presence acts as a warning to others.

Woodcut showing Jade Rabbit mixing immortality potion

Lunar rabbit

In Chinese mythology, the goddess Chang'e fled to the Moon after consuming an immortality elixir, accompanied by a jade rabbit. The rabbit can be seen in the Moon's markings, pounding the potion for her with a pestle and mortar.

Toad in the Moon

The Salish people of the Pacific Northwest coast of America see a toad in the Moon. It is said that the toad made a mighty leap onto the Moon to escape the advances of a wolf that had fallen in love with it.

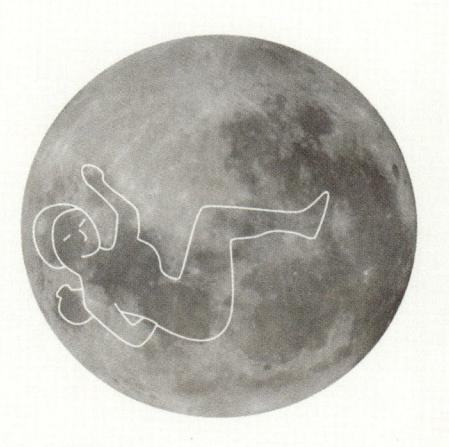

Water carrier

In Māori lore, Rona, a woman fetching water at night, fell when the Moon went behind a cloud. She blamed the Moon, which grabbed both her and her pail. She stayed on the Moon and was given control of Earth's tides.

FLIGHTS OF FANTASY

Long before space travel was possible, people envisioned visiting the Moon. Writers explored concepts of lunar travel, including ladders, boats, birds, flying machines, and later rockets. Nearly 2,000 years ago, the ancient Greek writer Lucian (c.120–180 CE) penned the earliest-known lunar tale, *The True History*.

With scientific advances, popular culture embraced new discoveries. In 1608, German astronomer Johannes Kepler wrote *Somnium*. In this dream narrative, humans breathe using damp sponges in their nostrils and are sedated due to the cold on their long journey.

Early science-fiction pioneers, such as the French novelist Jules Verne (1828–1905) and the British author H. G. Wells (1866–1946), presented imaginative tales of the Moon. Their stories expressed human curiosity and the desire to explore beyond known boundaries.

1. *The True History* (2nd century CE)
In ancient Greece, the satirist Lucian of Samosata wrote about a conventional sea voyage that soon takes an unexpected turn. A powerful storm lifts his ship into the sky and all the way to the Moon, where he encounters an array of bizarre and fantastical beings.

2. *The Bamboo Cutter* (c.10th century)
In this Japanese tale, a bamboo cutter finds a princess named Kaguya-hime inside a bamboo stalk. Her beauty attracts many suitors, to whom she gives impossible tasks. Eventually, Kaguya-hime reveals that she is from the Moon and must return to her celestial home, never to be seen again.

3. *Wan Hu* (late 14th century)
In a tale from China, Wan Hu is a 14th-century inventive dreamer who builds a chair fitted with 47 firework rockets. His ambitious gunpowder-fueled plan sees him propelled into the air, but instead of delivering him to the Moon, the rockets explode.

4. *The Man in the Moone* (1638)
Stranded on an island, Domingo Gonsales builds a wooden machine. Powered by 25 giant swans and a sail, it takes him to the Moon and back. The story, by the English bishop Francis Godwin, inspired many parodies, including one by the French satirist Cyrano de Bergerac.

5. *Baron Munchausen* (1785)
Among the many adventures of the fictional Baron Munchausen by German writer Rudolf Erich Raspe are trips to the Moon. He climbs a tall beanstalk and makes the return journey using a rope made of straw. For his second trip, he is propelled to the Moon in a ship during a hurricane.

6. *From the Earth to the Moon* (1865)
Author Jules Verne brought lunar travel into the 19th century with a story of three people from a gun club in Baltimore building a space gun and blasting off to the Moon. After a successful launch, their journey continued in a second novel, *Round the Moon* (1870).

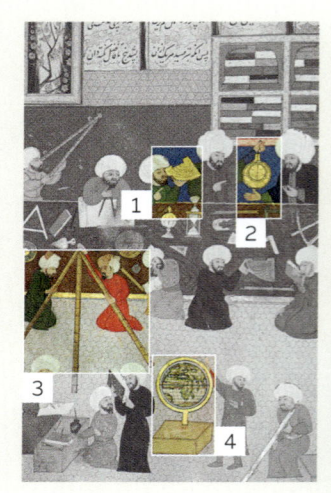

VISUAL TOUR

1. Islamic quadrant
An instrument, calibrated for the observatory's latitude, is used to tell the time.

2. Astrolabe
Taqī ad-Dīn calculates the phase of the Moon and its position relative to the stars.

3. Triquetrum
Also called a pyramid ruler, the device determines the altitude of celestial bodies.

4. Terrestrial globe
Possibly prophetic, it includes a southern landmass, which we now know as Antarctica.

Astronomers at the Galata Observatory
In the 16th-century *Şehinşahname* (*Book of the King of Kings*), Ottoman astronomers, including Taqī ad-Dīn, make observations and record their findings in logbooks. Built in 1577 in what is now Istanbul in Türkiye, the Galata Observatory burned down after only a few years of operation.

AN AGE OF DISCOVERY

Between the 8th and 14th centuries, the Islamic world made significant advancements in a range of fields, including science, maths, and astronomy.

From the 8th century, a golden age of learning flourished across North Africa, Southwest Asia, and Central Asia. Islamic scholars refined ideas from ancient Greece, India, and China and made new scientific discoveries. The 13th-century observatory at Maragheh, in modern-day Iran, had cutting-edge instruments and educated astronomers. One of their tasks was to time the start of each month in the Islamic calendar, based on the crescent Moon.

New lunar discoveries

Since ancient times, it had been believed that the Moon simply reflected sunlight, just like a polished mirror. However, around 1000 CE, the mathematician and astronomer Ibn al-Haytham (c.965–c.1040) from Basra, today in Iraq, challenged this idea and redefined the laws of reflection. He developed a new theory of optics, and his experiments were replicated by Galileo Galilei (1564–1642) nearly six centuries later. Ibn al-Haytham also studied the Moon illusion—the apparent enlargement of the Moon near the horizon—correctly identifying it as a perceptual illusion, rather than an atmospheric effect.

Born in modern-day Uzbekistan, Al-Bīrūnī (973–c.1052) made major contributions to the understanding of lunar phases and the Moon's size. His data on eclipses was later used by 18th-century astronomers to refine the speed of the Moon's orbit. Al-Bīrūnī also wrote about how the astrolabe device could measure and predict the position of celestial bodies such as the Moon.

The astronomer Ibn al-Shāṭir (c.1305–1375), from Damascus, in Syria, was concerned that the Greeks' model of the solar system (see pp.52–53) couldn't accurately predict the Moon's position. He added an extra lunar orbit called an epicycle—a circle whose center moves around the circumference of a larger circular path (see pp.100–101). With this adjustment, the model was able to make more accurate predictions of the Moon's position.

Golden astrolabe
A complex instrument refined by Arabic astronomers, the astrolabe is a hand-held map of the heavens that shows the paths of celestial bodies.

ISLAM AND THE CRESCENT MOON

Followers of Islam observe the Hijri calendar, which consists of 12 lunar months, each lasting 29 or 30 days. Every month begins with the sighting of the crescent Moon. The ninth month of the Islamic year, Ramadan, holds particular significance and is a time of fasting, charity, and other observances. The arrival of the next crescent Moon brings the end of Ramadan and the start of the festival Eid al-Fitr. Although the crescent Moon is widely used in the Islamic world, sometimes alongside stars, this iconography probably has its roots in much earlier civilizations. Many modern Muslims reject it as a symbol of their faith.

A crescent Moon tops the dome of the St. Petersburg Mosque in Russia (1913).

MOONLIGHT

Although the Moon emits thermal radiation, it produces no visible illumination of its own—only reflected light from the Sun and Earth.

For most of human history, the Moon has played a vital role, both as a timekeeper for tracking the months and seasons (see pp.24–27) and as a dependably changing light. A "hunter's Moon," for example, was the first full Moon of October, when hunters tracked animals that had fattened over the summer. Similarly, in September or October, a "harvest Moon" enabled farmers to work late into the night to bring in their crops. To both hunter-gatherer and farming communities, therefore, the Moon was associated with brightness, health, and fertility, as well as with death and renewal—a cycle that repeated each month as the Moon's light faded, died, and returned.

The Moon's brightening and darkening is due to its phases (see pp.28–29) and by the fact that moonlight fluctuates in intensity. At full Moon, sunlight reflects off the Moon's surface and reaches Earth at an angle of almost 180°—which gives it a greater intensity than when it reflects at almost 90° during its first and third quarters. Full Moons also vary in intensity, the brightest being supermoons, which occur when the Moon is closest to Earth (see pp.16–17).

Relative brightness

Weather permitting, a full Moon can seem extremely bright and cast sharp shadows on the ground. However, this apparent brightness is due entirely to the Moon's closeness to Earth—on average, it is only 238,855 miles (384,400 km) away. As demonstrated by an image taken by the Deep Space Climate Observatory (DSCOVR) satellite (see opposite), the Moon is in fact one of the darkest objects in the night sky. It is much more like an asteroid when compared to a moon such as Saturn's Enceladus, which, being covered with ice, is one of the brightest objects in the solar system. Nor does all of the Moon's light come directly from the Sun—a tiny portion of it is Earth's reflected sunlight, which has traveled 2.52 seconds out to the Moon and back (see below).

Earthlight
Not all of the Moon's light is directly reflected sunlight. On occasion, light reflected from Earth dimly illuminates the otherwise darkened portion on the Moon. Best seen at dawn or dusk, this phenomenon is known as "Earthlight."

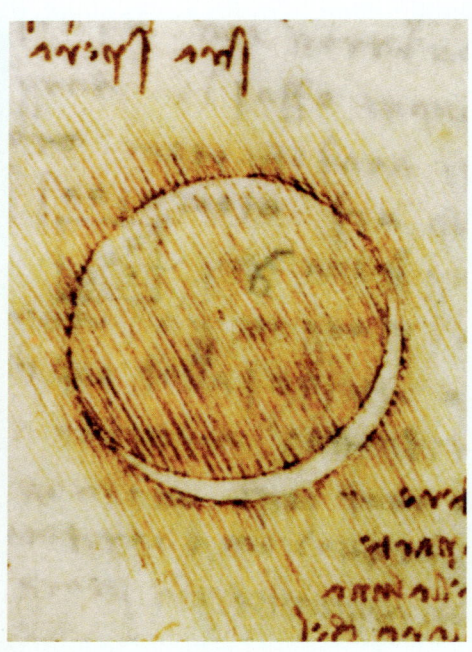

The "Da Vinci glow"
Florentine polymath Leonardo da Vinci (1452–1519) first suggested that it is Earth that illuminates the part of the Moon experiencing lunar night. In 1510, he sketched the phenomenon, which was later named after him.

ALBEDO

The amount of sunlight that an astronomical body reflects is known as its "albedo." If the object is 100 percent reflective, then it has an albedo of 1; if it is completely nonreflective, however, it has an albedo of 0. Most of the objects in the solar system have albedos that are somewhere in between: Earth's is approximately 0.30 (30 percent), Enceladus's (the moon of Saturn) is 0.81 (81 percent), and the Moon's is only 0.12 (12 percent). The Moon's low albedo means that, although it has brighter and darker areas, it absorbs most of the sunlight that it receives.

Earth's reflectivity
Taken in 2015 by the DSCOVR satellite from a distance of 1 million miles (1.6 million km), this image demonstrates that Earth reflects much more light than the Moon.

Light absorption
The molecular structure of an object determines how much light it will absorb and reflect. An object with a high albedo is highly reflective.

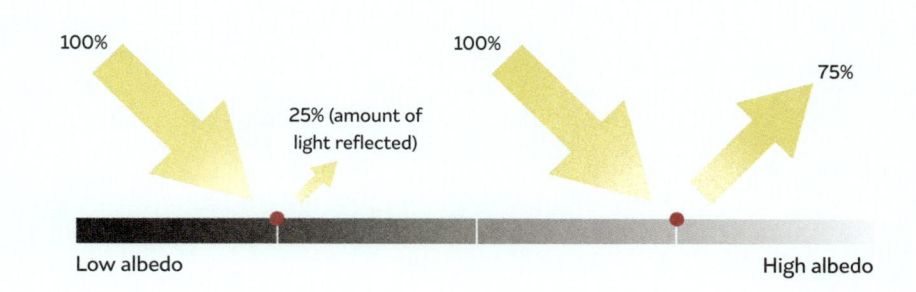

100%   100%   75%

25% (amount of light reflected)

Low albedo   High albedo

BLOOD MOON

The Moon turns red during a total lunar eclipse—a phenomenon known as a "blood Moon" (see pp.38–41). This happens because of the way the Sun's light interacts with Earth's atmosphere. At the moment of eclipse, as Earth hangs between the Sun and the Moon, its atmosphere bends the longer wavelengths of sunlight (at the red end of the spectrum) around Earth to reach the Moon's surface. At the same time, the shorter wavelengths of sunlight (at the blue end of the spectrum) are deflected by Earth's atmosphere in a process known as "Rayleigh scattering."

Sunset red
Blood Moons are caused by the same process that makes the Sun appear red at dawn and dusk. On the horizon, the Sun's light has to pass through a greater amount of atmosphere, and so is subject to a greater amount of scattering.

Refracted light
During a total lunar eclipse, Earth's atmosphere refracts the Sun's light so that only red wavelengths reach the Moon. Without this effect, the Moon would be completely dark.

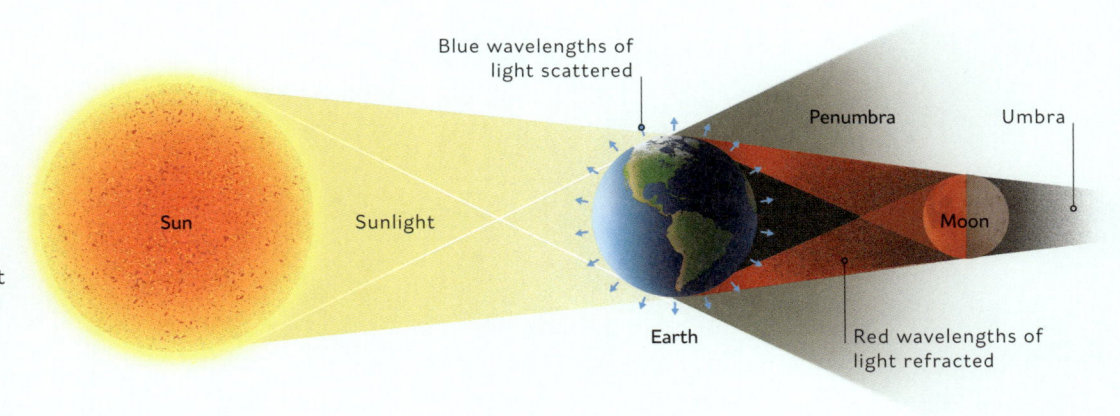

Blue wavelengths of light scattered

Penumbra   Umbra

Sun   Sunlight   Moon

Earth

Red wavelengths of light refracted

MEDIEVAL ASTROLOGY

In medieval Western and Arabic traditions, as in Indian and Chinese lore (see pp.60–61, 66–67), the Moon played a role in predicting human fate and was thought to shape people's health, fortune, and character.

To determine horoscopes, astrologers studied lunar phases, eclipse timings, and the Moon's daily position in the 12 zodiac constellations. These were used to analyze past events, advise on the present, and predict the future.

Astrological aids

During the Islamic Golden Age (see pp.78–79), scholars evolved sophisticated tools to study the movement of the Moon. Instruments such as the astrolabe allowed them to track its position with great precision. Astrologers consulted ephemerides—tables listing the Moon's expected positions—noted predictions on "dust boards" (sand-covered tablets), and studied the Moon's phase and location in the zodiac constellations. They also referred to texts such as the *Kitāb al-Bulhān* (*Book of Wonders*). Collated by Hassan Esfahani and completed in the 15th century, it combined astronomy and astrology and featured astronomical tables known as *zījes*. Artwork in such books often depicted the Moon with a human face (see pp.74–75) and alongside the zodiac sign Cancer (see below).

Astrology was deeply ingrained in Islamic society. Rulers often recruited prominent Islamic scholars such as Albumasar (Abū Ma'shar, 787–886 CE) and Al-Bīrūnī (973–c.1050) to provide accurate predictions. When founding Baghdad in 762 CE, Caliph Al-Mansur (c.714–775 CE) employed astrologers Nobakht Ahvazi and Mashallah ibn Athari to choose the ideal start time and date. Al-Bīrūnī wrote astrological guides but remained skeptical about some aspects of astrology and possibly saw it as a means of securing patronage for

From *Introductorium in Astronomiam* (1513), by Albumasar
This 9th-century work was first translated into Latin in the 12th century and became a great influence on astrology in medieval Europe. Here, an astrologer compares the size of a globe with the Moon and Sun.

From the *Kitāb al-Bulhān*
A crab—symbolizing Cancer—holds a winged, angry Moon. Cancerians were thought to be ruled by the Moon, which was believed to make them more emotional, caring, and sensitive.

more scientific endeavors. Astrologers were not exclusive to elite circles, however. They also often cast horoscopes in bazaars for a modest fee.

Like their contemporaries in the Islamic world, Western astrologers traced the Moon's position to cast horoscopes. Medieval lunaries, or "moonbooks," used the Moon's position to predict auspicious times for crop planting, religious rituals, and medical practices. (A rise in plague deaths in 1593 was attributed to the Moon's position in the sky, opposite Mars.)

Tarot cards originated in 15th-century Italy, the heart of the Catholic world, and were also used as tools to tell fortunes. The Moon was seen as a key card, indicating illusion.

Major advances in observational astronomy were rooted in astrological study of the Moon and other planetary bodies. In fact, leading Western astronomers such as Tycho Brahe and Johannes Kepler (see pp.100–101) both practiced astrology. As the foundations for modern astronomy were laid, however, the two fields diverged.

English folding almanac
Written in Latin on vellum, this document dates from c.1415–1420. It includes astrological tables, diagrams, a calendar, lists of solar and lunar eclipses, and details of religious feasts. A medical practitioner or patient would have consulted the almanac before a medical procedure.

"A holy man continues in **wisdom** as **the Sun**: but a **fool** is changed as **the Moon**."
Ecclesiasticus 27:12, Bible

La Lune
This highly decorated tarot card from 15th-century Italy shows two astrologers making astronomical observations of the Moon.

A page from Ars Magna Lucis et Umbrae (*The Great Art of Light and Shadow*), published in 1646 by Jesuit scholar Athanasius Kircher. Engraved by Pierre Miotte, it shows the 28 phases of the Moon.

Lunacy, a 19th-century Japanese woodblock print by Tsukioka Yoshitoshi—one of a series relating to the Moon—shows a maid mad with grief after her lover's death. His love-letter scroll streams up toward the cloud-shrouded crescent Moon.

THE MIND AND MEDICINE

The belief that the Moon affects human well-being has persisted for millennia. The term "lunacy," once used to describe mental-health disorders, originates from the Latin word *luna*, meaning "Moon."

The Greek philosopher Aristotle (see pp. 54–55) noted the high water content of the human brain and, knowing the connection between the Moon and tides, hypothesized that the full Moon might have an impact on humans. He speculated that the Moon's gravitational pull could influence water in the brain, potentially leading to symptoms ranging from mild forgetfulness to more severe issues such as epilepsy and episodes of mental illness.

Influence on women

In medieval times, it was widely accepted that women were susceptible to the Moon's influence, since they were considered weaker in both mind and body. This idea was also connected to a perceived correlation between the timing of the lunar and menstrual cycles. Medicine in the Middle Ages was rudimentary, and the four humors—blood, phlegm, yellow bile, and black bile—were believed to govern health. With blood a liquid humor, and menstrual and lunar cycles supposedly interconnected, it was assumed that the full Moon exacerbated women's emotional state, making them more prone to hysteria, mood swings, and other emotional disturbances.

Diseases of the mind

Cures for illness were given when astrological readings deemed them to be most beneficial, with the Moon's position in the zodiac thought to affect the outcome of treatments. Despite later advancements in medical training and science, the idea that the Moon influenced mental health persisted. In 1845, the Lunacy Act in the UK still linked some mental disorders to lunar phases. Institutions such as London's Bedlam (the Bethlem Hospital), which had been admitting patients with a mental-health condition since the 15th century, continued to use the term "lunatic" ("moonstruck") to describe such individuals. At full Moon, when patients' violent tendencies were thought to be amplified, some of them would be chained to control their behavior.

The Moon and sleep

Although associations between the Moon and mental health have now largely been debunked, some modern research suggests that the Moon's brightness may influence sleep patterns, with sleep being shorter and starting later around the full Moon. Sleeping less on moonlit nights could be a throwback to humanity's preindustrial past, when bright, moonlit nights made nocturnal activities—such as social gatherings, traveling, hunting, and fishing—more feasible, as well as safer.

Under the influence
This 17th-century engraving depicts dancing women wearing crescents. These are directly linked by rays to the full Moon, which was thought to make women temperamental and changeable.

Protective pendant
In ancient Rome, women wore crescent-Moon gold pendants to ward off the "evil eye." The Moon was believed to avert illness brought about by malevolent glares.

MEDICAL PRACTICE

In medieval Europe, the timing of treatments was often linked to the Moon's phases. The beneficial effect of bloodletting was said to be magnified during the waxing Moon, when the body was in a phase of growth and accumulation, removing bad humors and restoring good ones. Medical guides and calendars advised the optimum location on the body for bloodletting, alongside the best lunar phase to attempt it.

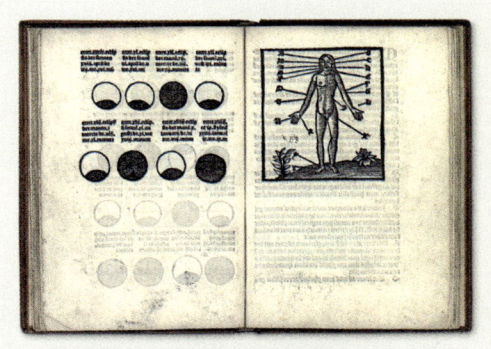

Guide relating lunar phases to bloodletting points

THE DIVINE ORDER

In Europe during the Middle Ages, Christianity and science were closely aligned. Understanding of the Moon was largely shaped by beliefs about its place in the divine order of the universe.

Flight to the Moon
This 15th-century Tuscan illustration of Dante's *Divine Comedy* shows Dante (in blue, shown twice) and his beloved Beatrice (in red) flying to the Moon, aided by Apollo (in orange). The Moon was seen as a place for those who have waxed and waned in their vows to God, and is depicted as heaven's lowest level.

Influenced by classical Greek ideas such as a geocentric model (see pp.52–53) of the universe, medieval scholars believed that the Moon was one of the seven planets orbiting Earth. The Catholic Church also shaped astronomy in Western Europe at the time. Lunar observations were essential for setting the ecclesiastical calendar, which determined the dates of Christian festivals such as Easter, and some in the clergy practiced astronomy. For monastic scholars, such as the 9th-century English monk Bede, studying the Moon was a sacred duty as well as a scientific pursuit. Bede recorded the lunar phases, linked them to the tides, and calculated the Moon's movement through the zodiac's constellations. His investigations were framed by the belief that the Moon's predictable cycles were evidence of God's grand design.

A sophisticated understanding of different lunar cycles is evident from medieval artifacts such as the 15th-century astronomical clock at Exeter Cathedral in the UK—which indicated the day of the lunar month and the phase of the Moon, as well as the hour of the day—and the intricately illustrated Metonic-cycle manuscript from St. Emmeram's Abbey in Germany (see opposite). Such objects embodied reverence for the Moon's role in God's cosmic plan, as well as serving practical purposes.

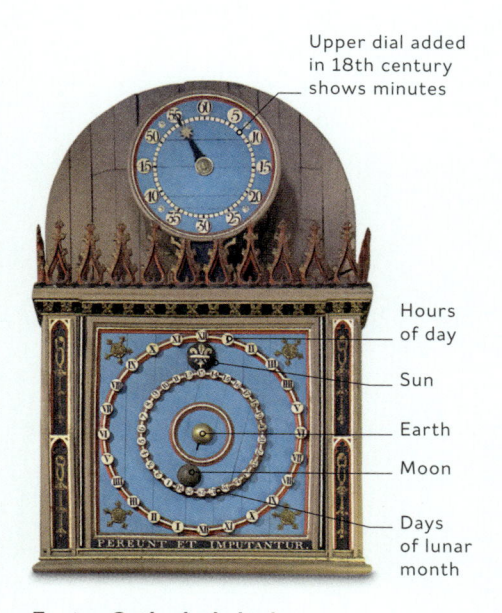

Upper dial added in 18th century shows minutes

Hours of day

Sun

Earth

Moon

Days of lunar month

Exeter Cathedral clock
The Moon, half-black and half-silver, turns on its axis to show the correct phase as it travels around the lunar month dial. The fleur-de-lis on the Sun disk points to the hour as the Sun goes around the outer dial.

WOMEN AND THE MOON

In the medieval era, the Moon was often associated with female fertility. Its phases were thought to align with menstrual cycles and pregnancy, with the crescent Moon being linked to conception and birth. The Moon also symbolized the perceived purity of women. In religion, it was associated with the Virgin Mary.

***Maria Lactans on a Crescent Moon* (c. 1485)**
In the Bible, the Book of Revelation tells of a pregnant woman with "the Moon under her feet" (chapter 12, verse 1). The medieval Catholic Church identified this person as the Virgin Mary, and the crescent Moon came to symbolize her miraculous conception and chastity.

THE METONIC CYCLE

Discovered by Meton of Athens in the 5th century BCE, the Metonic cycle is a period of almost exactly 19 years, after which the Moon's phases repeat on the same days of the solar calendar as at the cycle's start. The 235 synodic months (see pp. 28–29) of the Metonic cycle are just 2 hours and 5 minutes longer than 19 solar years. This discovery became a crucial tool in the Christian calendar. The 9th-century manuscript from St. Emmeram's Abbey in Regensburg, Bavaria, shows the meticulous work involved in calculating the cycle and coordinating lunisolar celebrations such as Easter.

St. Emmeram's manuscript
The large wheel at the top of this manuscript page depicts the Metonic cycle. The calculations in the illustration begin with the Julian calendar date for the Easter new Moon in the 9th century.

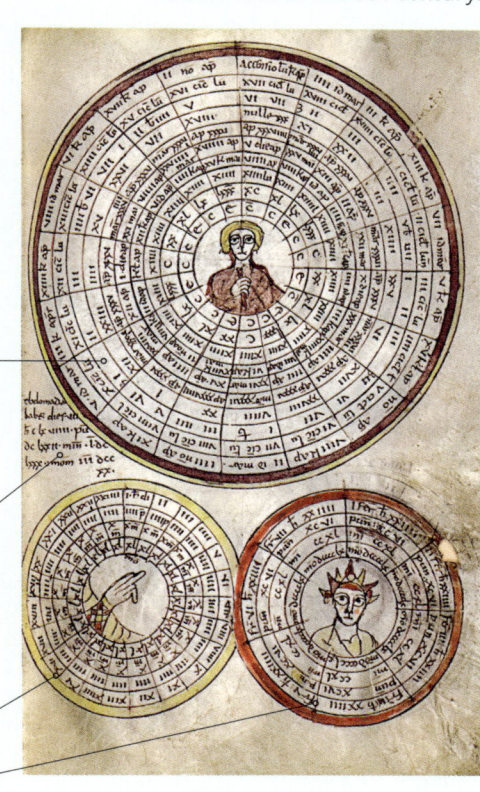

Each segment of Metonic cycle "wheel" represents one year; Christ is shown at wheel's center

Summary of number of medieval time units in week: 7 days, 168 hours, 672 "points" (quarter-hours), 1,680 "minutes" (tenths of hour), and 6,720 "moments" (fortieths of hour)

Time divisions of day

Time divisions of week

CREATURES OF THE NIGHT

Animals, mythical creatures, and magical beings have long been connected to the Moon via folklore and fairy tales. Different cultures offer unique interpretations of the Moon's influence on life. In Japan, for example, bats flying in front of the moon are celebrated as symbols of good fortune, but in European folklore, bats flitting on moonlit nights were often equated with devilry and witchcraft.

The Moon is sometimes associated with melancholy, and its appearance is linked to malevolent creatures like mermaids and the water spirit called a *rusalka*, which mesmerizes its victims by singing haunting songs under moonlit skies. Stories of werewolves, which transform from people into wolves under the full Moon, embody the raw, untamed aspects of human nature, drawing on the belief that people behave strangely during a full Moon.

Yet the Moon also brings a sense of joy and energy, such as in scenes where fairies or elves dance under moonlit skies, adding a touch of magic and enchantment to the night.

2

1

3

1. Good omen
Silhouetted against the Moon, the dark body and wings of a bat sharply contrast with the pearly lunar disk. In Japan, such a sight would be a good luck omen, thought to bring happiness and longevity to those who witness it. This woodblock print, from around 1905, is by Takahashi Biho.

2. Siren songs
Rusalki gather beneath the Moon in this dreamy, ethereal scene from a Slavic folk tale, as realized by Russian artist Ivan Kramskoi in 1871. Emerging from the water, with white dresses glowing softly in the moonlight, the spirits' beauty and singing will lure unsuspecting men to their peril.

3. Elfin dance
Three red-haired elf princesses dance with shawls "woven of mist and moonshine" for the Goblin King in front of the full Moon. They were drawn by Irish artist Harry Clarke to illustrate the story *The Elf Hill* for a 1916 edition of Hans Christian Andersen's *Fairy Tales*.

5

6

4. Mythical night creature

In some tales, the unicorn is a nocturnal beast with a horn derived from the crescent Moon. This Flemish tapestry from around 1500 shows a woman with a staff and flag bearing crescent Moons. Her hand rests on the horn of a unicorn wearing a tabard adorned with the same motifs.

5. Witches' coven

In *The Witches' Sabbath* (1798) by Spanish artist Francisco Goya, a coven of witches make offerings to the Devil, represented as a creature with horns wreathed in oak leaves and with hooves extended toward an infant. Bats fly overhead, their shapes mimicking that of the crescent Moon.

6. Curse of the werewolf

With the full Moon rising on the horizon, a man runs along a desolate country track with a werewolf clinging to his back. Bitten by the monster, he will then be cursed to transform into a wolf himself at every full Moon. This 1857 engraving is based on a painting by French artist Maurice Sand.

A FORCE ON NATURE

The Moon shapes animal behavior in subtle but significant ways through its light, control of tides, and role as a celestial clock. The lunar cycle also determines when some animals mate, migrate, and hunt.

Predators and prey alike adjust their behavior according to the Moon's illumination. Lions, which hunt by ambushing their quarry, are more successful on moonless nights: they use the darkness to cloak their movements, which makes it harder for prey to detect their stealthy approach. Lions' hunts are less effective on bright nights, since their pale underbelly can betray their presence. After the less successful full-Moon hunts, daytime attacks by lions increase.

Prey animals, aware of the danger of being seen, are more likely to limit their movements at full Moon in order to avoid detection. Keenly aware of the threats posed by a bright moonlit night, voles and other small mammals tend to reduce their activity to avoid detection by predators such as barn owls, which thrive under the full Moon.

Guiding light

The Moon's light serves as a beacon for many birds, guiding them during flight, feeding, and on long migrations. The black swift spends

Nocturnal hunter
Bright nights provide ideal hunting conditions for barn owls. Moonlight reflects off the owl's brilliant-white feathers, momentarily dazzling prey and allowing the owl to swoop in for the kill.

Loggerhead turtles prefer to lay their eggs at new and full Moons, when the high tides give them access to nesting grounds higher up the beach.

eight months of the year airborne, and on nights with a full Moon it rapidly ascends to heights of 13,000 ft (4,000 m) to better view its prey. Nightjars also take advantage of the bright moonlight to feast before embarking on lengthy migrations. Some insects, too, respond to moonlight. Dung beetles use it to help them travel in swift, straight lines (to avoid predators). Likewise, bull ants of the species *Myrmecia mida*s rely on it to guide them back to their nests.

Effects on marine life

Acting as a natural clock, the Moon triggers biological processes in some marine creatures. Bristle worms, for example, rise to the ocean's surface to mate a few days after full Moon. They have only one chance to breed before they die, so it is vital that they coordinate their actions. Their bodies produce a light-sensitive protein called a cryptochrome that helps the worms discern moonlight from sunlight and detect variations in illumination from the Moon, so the worms can rise to breed at the correct time. Corals also use cryptochromes to register changes in moonlight (see below).

The Moon's gravitational pull drives the tidal rhythms that shape life in coastal habitats. Mollusks and crabs synchronize their behavior with the tides: at low tide, they bury themselves in the sand to hide from predators; when the sea returns, they emerge to feed and reproduce. Turtles use the tidal flow to choose when to lay their eggs on beaches (see opposite).

Premigration feast
Nightjars fly more than 4,700 miles (7,500 km) when migrating between northern Europe and Africa. They gorge on insects under a full Moon to build up energy reserves before leaving and migrate as a group when the Moon wanes.

One species of **Australian bull ant** uses the Moon like a **nocturnal compass**, following polarized moonlight to find its way **back to its nest**.

SPAWNING TOGETHER

Biological processes in corals enable these reef animals to use the Moon as a clock. Light-sensitive cryptochrome proteins build up within the corals from exposure to moonlight under a waxing Moon. At full Moon, protein levels reach a maximum, signaling the corals to release their spores simultaneously. This synchronized act gives the best chance of reproductive success.

Reef spores
A coral releases its spores into the dark ocean, triggered by moonlight and an internal biological mechanism. This perfectly timed event lasts a few hours and ensures mass reproduction across vast reef systems.

A MOON OF MANY NAMES

The Moon holds sacred significance for many North American Indigenous Peoples, with nations and tribes having their own rich traditions, stories, and many names for the Moon and its phases.

For North American Indigenous communities, the Moon symbolizes life's natural rhythms. Groups commonly divide their calendars using the phases of the Moon, giving each cycle and each full Moon of the year its own name. These names give important information about the communities' seasonal activities and cultural practices. For example, the Ojibwe, who are from an area stretching from present-day Ontario in eastern Canada to Montana in the US, call January's Moon "the Great Spirit Moon" because January is a challenging time in winter set aside for people to reflect on creation. The Lakota people, who are part of the Great Sioux Nation, also observe the Moon's phases to guide their activities. They follow a lunar calendar divided into four seasons. Because the number of full Moons in each season can change, each has its own name. For example, the first Moon in spring is "Moon When Ducks Come Back," and the second Moon of winter is "the Hard Moon." May's Moon has the practical name of "Corn Planting Moon."

Stories of the cosmos

The Moon is central to the cosmology of the Cherokee Nation, who originate from the southeastern United States area but are now

Moon mask
A shaman from the Haida people of the Pacific Northwest would wear this colorful wooden mask of the Moon during traditional rituals.

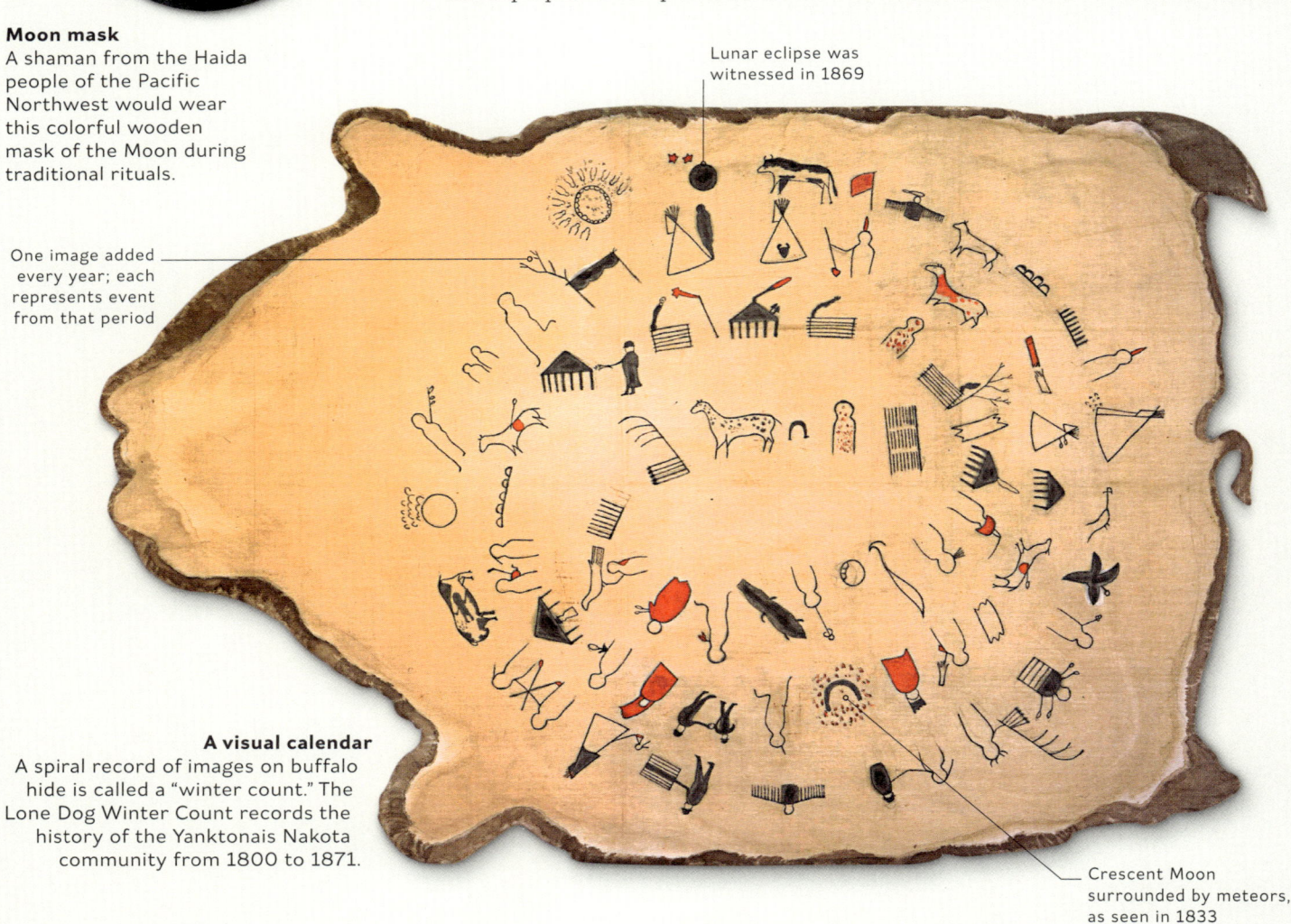

Lunar eclipse was witnessed in 1869

One image added every year; each represents event from that period

A visual calendar
A spiral record of images on buffalo hide is called a "winter count." The Lone Dog Winter Count records the history of the Yanktonais Nakota community from 1800 to 1871.

Crescent Moon surrounded by meteors, as seen in 1833

> # "**Everything** the **power of the world** does is **in a circle** … The Sun **comes forth** and **goes down** again in a circle. The Moon **does the same**, and both are round."
>
> Black Elk, wičháša wakháŋ (medicine man) of the Oglala Lakota people, *Black Elk Speaks* (1932)

based in eastern Oklahoma. They tell a story of the Moon visiting his lover, the Sun, anonymously under the cover of darkness. The Sun discovers the Moon's identity only when she wipes ash on his face, which is then revealed when the Moon next rises.

During an eclipse, the Cherokee believe a giant frog is trying to swallow the Sun and Moon, so they beat drums, pots, and pans to chase it away with loud noises.

The "Moon Grandmother"

The Navajo Nation, who are from what is now the southwestern United States, call the Moon "the One Carried at Night" and regard it as female. Also often referred to as the "Moon Grandmother" or "Mother Goddess," the Moon is seen as a symbol of life, fertility, and nurturing. Navajo spiritual leaders objected to the first interment of human remains on the Moon in 1998, and they have opposed recent attempts by commercial companies to place more there. The Navajo Nation consider this a desecration of a sacred space.

Medicine wheel
The 28 divisions of the Bighorn Medicine Wheel in Wyoming are thought to be based on the days in the lunar cycle. The stone monument was a sacred space used for ceremonies by the Plains tribes.

LUNAR CALENDAR

Many North American Indigenous Peoples, including the Cherokee, divide a year into 13 months of 28 days. Coincidentally, a turtle's shell has 13 large scales and an outer ring of 28 smaller ones, so the shell is used as a calendar. The beginning of each month is celebrated with a named festival, such as Cold Moon, Flower Moon, Ripe Corn Moon, or Harvest Moon.

Timekeeping
The turtle-shell calendar is used with two stones. One is placed to track the day and moves around the 28 outer spaces. The other stone is placed on the 13 larger scales to mark the current month.

Around edge, 28 small scales represent 28 days of lunar month

Each of 13 large scales represent one lunar month of year

Two dancers in Burkina Faso are shown wearing lunar headpieces studded with cowrie shells, a symbol of fertility, and holding poles for vaulting towards the Moon.

AFRICAN LUNAR TRADITIONS

After the Sun's intense heat has ruled the vast African skies by day, the Moon brings light and cool relief at night. More than just a celestial body, the Moon shapes many of the continent's mythologies, rituals, and beliefs.

Many sub-Saharan traditions attribute spiritual energy to the Moon. The Songye people of the Democratic Republic of the Congo (DRC) publicly display a sacred Nkisi statue as a conduit for supernatural forces. When its influence wanes, the figure is anointed with chicken blood and palm oil, then reenergized by moonlight, to signify the Moon's role as cosmic rejuvenator.

The Luba and Tabwa peoples, also of the DRC, believe moonlight enhances the link between earthly and spiritual realms. Shamans wear white pigment on their faces to represent the Moon, while quartz crystals capture the lunar light and harness its powers.

To the northwest, in Bondouku, Burkina Faso, men wear Moon masks and dance in white costumes, illuminating the darkness, in an annual ritual at the start of a new Moon. They vault into the air on long poles to try to capture the Moon's transformative powers.

The Moon and the gods

Some African cosmologies associate the Moon with certain deities. The San of southern Africa tell how the trickster god Kaggen created it by throwing his shoe into the sky. As it was produced by this maverick deity, the Moon is clever and has the ability to talk to people. Farther to the north, Mali's Dogon people believe their creator god made the Moon from an earthenware bowl. They also follow a lunar calendar.

Nature, fertility, and growth

Cultures worldwide link the symbolism and myth of the Moon to natural phenomena. To explain its slow journey across the sky compared to the Sun, West Africa's Hausa-speaking peoples tell stories of its path being filled with thorns. If the Moon catches the Sun, an eclipse occurs; people bang drums and make noise to help the Sun escape. It is also widely identified with fertility and renewal. In Luba tradition, twins are "children of the Moon," blessed due to their abundance. Meanwhile, to the northwest, the Ngas of Nigeria believe the Moon regulates women's fertility and the agricultural cycle, with crop growth dependent on its phases.

Moon-eyed mask
Worn by the Côte d'Ivoire's Baule people in a *gbagba* dance, this mask has semicircular eyes, representing the crescent Moon.

Nkisi figure
This 3-ft (1-m) wooden Nkisi carving was paraded on wooden poles under moonlight by the Singye people of the DRC to tap the Moon's powerful life force.

A SCIENTIFIC REVOLUTION

During the Scientific Revolution (1500–1700) in Europe, astronomers disproved many ideas that had been thought to be true for millennia, including the belief that Earth is at the center of the cosmos. They did so by studying the planets through telescopes—a recent invention. Advances in mathematics swiftly followed, including Johannes Kepler's laws of planetary motion, as well as detailed maps of the lunar surface and the discovery that the Moon "wobbles" in its orbit.

THE HEAVENLY SPHERES

New, more accurate astronomical observations, especially of the Moon, caused 16th-century astronomers to question existing ideas about the solar system's structure and the Moon's place within it.

Weighing the merits of theories
In this engraving of 1651, mythological figures weigh the planetary models of Copernicus (left) and Brahe (right); Ptolemy's model has already been discarded (bottom).

A geocentric (Earth-centered) scheme of the solar-system model that was proposed in the 2nd century CE by mathematician and astronomer Ptolemy of Alexandria became the accepted framework for the cosmos for more than a millennium. In Ptolemy's model, as the celestial bodies (including the Moon) orbited Earth, they also traced out smaller circles called epicycles (see below).

Ptolemy's model helped account for some of the "irregularities" in the observed data, but it failed to explain all aspects of the Moon's behavior. To solve this, Ptolemy adjusted the Moon's orbit so that occasionally the Moon came twice as close to Earth than it did at other times. However, this would have doubled the Moon's apparent size—a phenomenon that clearly did not occur.

Copernicus and Brahe

An alternative model, with the Sun located at the solar system's center, was formulated by Polish astronomer Nicolaus Copernicus (1473–1543). In Copernicus's model, all the planets—including Earth—orbited the Sun; only the Moon orbited Earth. Like Ptolemy's scheme, the Copernican system employed epicycles to address observed anomalies.

Copernicus's ideas were not immediately adopted due in part to the prevailing religious emphasis on Earth as the center of a divinely ordered universe. Danish astronomer Tycho Brahe (1546–1601) offered a compromise. His model—in which the Sun and Moon orbited Earth and the planets orbited the Sun—blended heliocentric (Sun-centered) and geocentric ideas. Neither model worked perfectly with what astronomers actually observed, however.

The breakthrough came with Johannes Kepler's laws of planetary motion (see opposite). Kepler discarded the idea of circular orbits and epicycles, showing that planets follow elliptical paths around the Sun. His work confirmed the heliocentric theory.

EPICYCLES

In Ptolemy's solar system, the Sun and each planet traced out a smaller circle, called an epicycle, as it moved along a larger circular orbit of Earth, called a deferent. The Moon, which was considered a planet, also followed an epicycle. Epicycles were first suggested in around 200 BCE by Apollonius of Perga, a Greek mathematician, to explain variations in the way the Moon traveled across the sky and cyclical variations in its apparent size. Ptolemy refined the epicycle idea.

Simple epicycle
Epicycles accounted for the apparent looping motion of the planets against the background of stars, as well as changes in their size and brightness (since they are closer to Earth when inside the deferent).

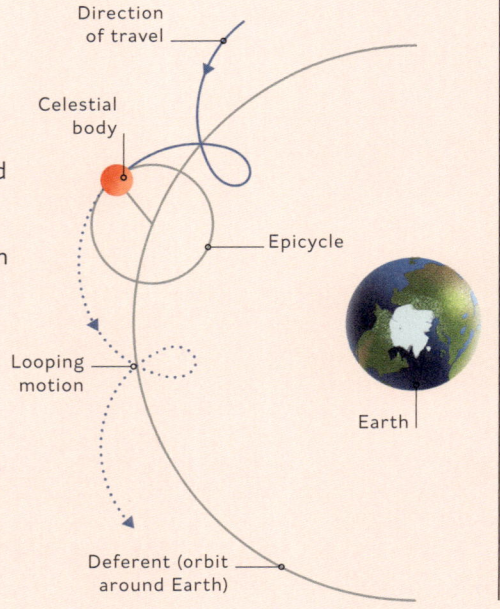

Direction of travel

Celestial body

Epicycle

Looping motion

Earth

Deferent (orbit around Earth)

KEPLER'S LAWS OF PLANETARY MOTION

German mathematician Johannes Kepler (1571–1630) published his ideas on planetary motion between 1609 and 1619. Using Brahe's observations of Mars, Kepler concluded that Mars's orbit was not perfectly circular, but elliptical. He formulated his laws on planetary motion based on this orbit. An ellipse has two focal points (foci), which are equidistant from the ellipse's center. Kepler's laws also applied to the motion of moons around planets.

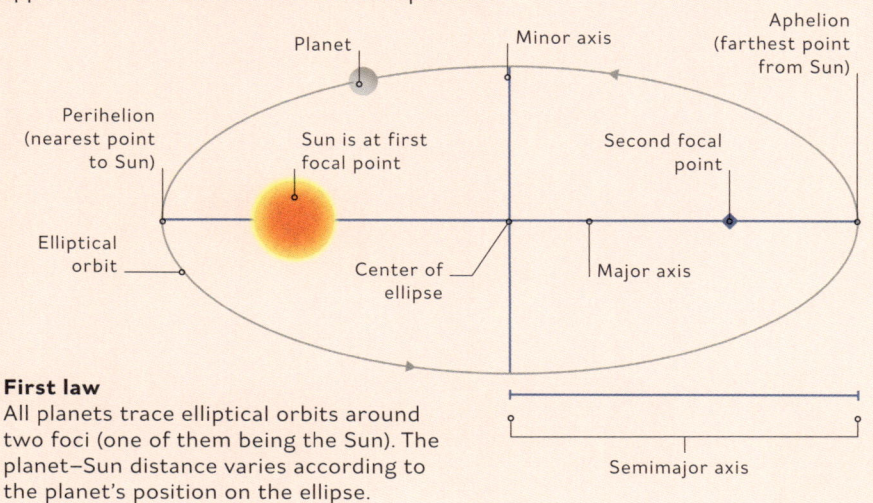

First law
All planets trace elliptical orbits around two foci (one of them being the Sun). The planet–Sun distance varies according to the planet's position on the ellipse.

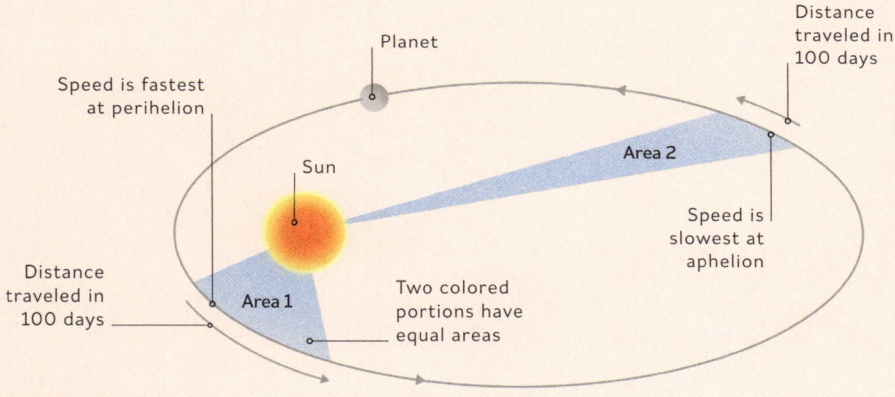

Second law
A planet sweeps out equal areas between itself and the Sun in equal periods of time. Planets move faster when closer to the Sun and slower when farther away.

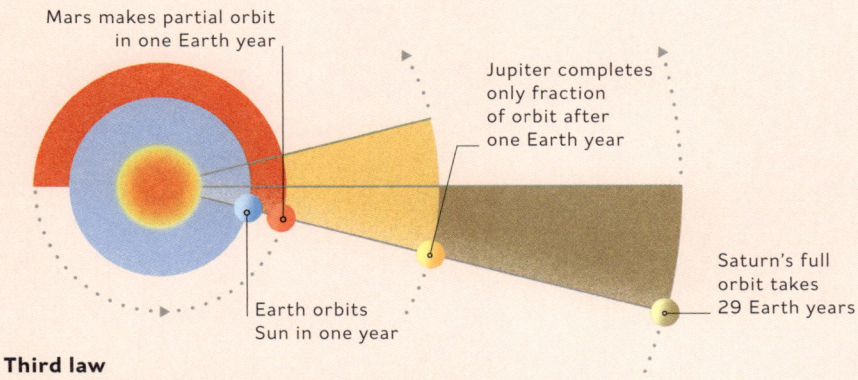

Third law
Planets take longer to orbit the farther they are from the Sun. Kepler devised a mathematical formula linking their orbital periods with the size of their orbits.

COMPETING MODELS

The solar-system models of Copernicus and Brahe vied for acceptance. Brahe's complex geo-heliocentric model, which maintained Earth's central position, was more palatable to religious beliefs of the time, but it lacked the accuracy needed to predict planetary and lunar motion. In contrast, Copernicus's simpler model, despite relying on uniform circular orbits and containing anomalies, marked a pivotal shift in scientific thought: it was a stepping stone to modern astronomy.

Engraving of Brahe's model, 1660
All orbits are circular, including those of the Sun and Moon, which orbit the central Earth. The constellations of the zodiac encircle the entire system.

Copernicus's solar system
This 1693 illustration shows only the Moon orbiting Earth and the planets orbiting the central Sun, all within an outer sphere of stars and clouds.

THE INVENTION OF THE TELESCOPE

The telescope stands as one of the most revolutionary instruments in the history of astronomy. By magnifying distant objects, such as the Moon, it allowed astronomers to explore far-off worlds as never before.

The telescope was invented in the early 17th century, possibly by Dutch spectacle-maker Hans Lippershey (c.1570–1619), who tried to patent a design in 1608. By 1609, the Italian astronomer Galileo Galilei (1564–1642) had heard of it and decided to build one himself. Galileo ground and polished the lenses and made a tube to house them, producing a telescope that magnified objects to just a few times their usual size. After refining it, he created a version that magnified 20 times; he turned this improved telescope to the skies to view stars, planets, and the Moon.

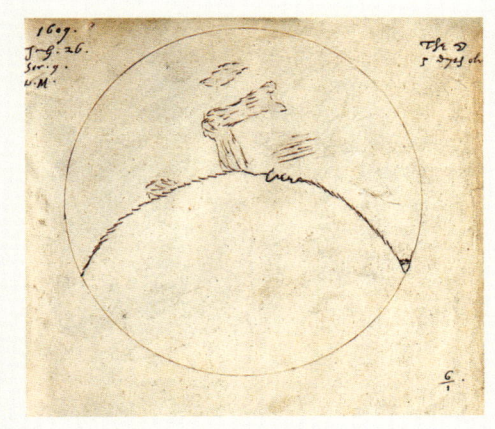

First telescopic observation
On July 26, 1609, using his homemade telescope with 6x magnification, Thomas Harriot became the first person to draw the Moon through a telescope. He sketched the 5-day-old crescent Moon, marking its edge and shading the dark patches on its surface.

Galileo's Moon observation
These six different phases of the Moon were drawn by Galileo using a brown ink wash on watercolor paper. Galileo made these observations starting in late November 1609 and discovered an uneven and mountainous lunar surface.

Imperfect Moon

Through his telescope, Galileo discerned that the Moon's surface was covered with craters and mountains. These features became especially visible along the terminator—the line dividing lunar night from day—where their tops glinted in the sunlight. He noticed that over a number of days, the terminator's position shifted, revealing different features.

Galileo recorded his findings in a series of drawings, which he published in 1610 in *Sidereus Nuncius*. His work disproved the idea that the Moon was a perfectly smooth object; it was marked by numerous "imperfections," making it more akin to Earth than a flawless heavenly body. Galileo's discoveries led to conflicts with the Catholic Church, however, and he remained under house arrest for much of his later life.

First map

Like Galileo, English astronomer Thomas Harriot (c.1560–1621) built his own telescope. In July 1609, he used his new instrument to study the Moon. From his observations, Harriot produced a series of drawings, including a map of its surface (see pp.110–111). His work remained unknown until it was discovered in 1784.

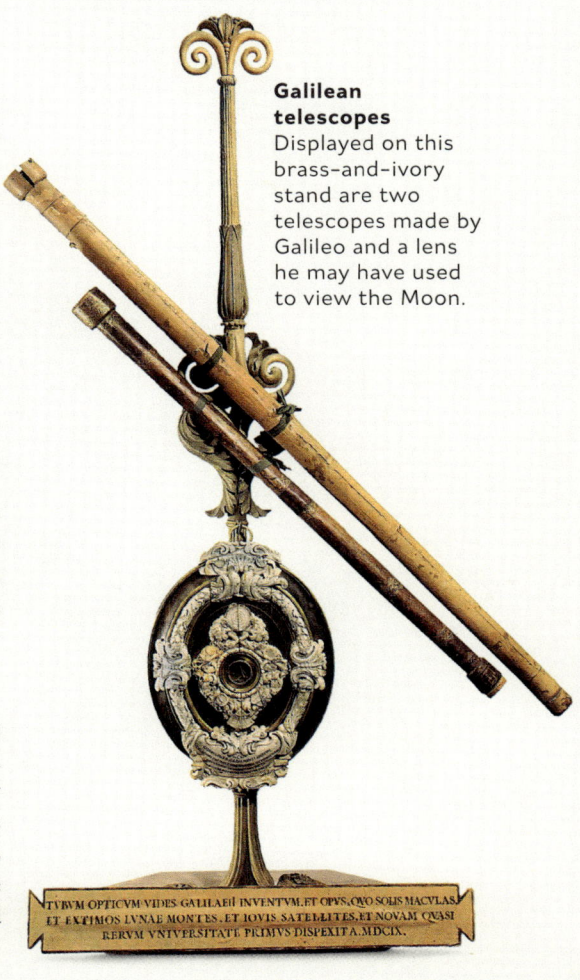

Galilean telescopes
Displayed on this brass-and-ivory stand are two telescopes made by Galileo and a lens he may have used to view the Moon.

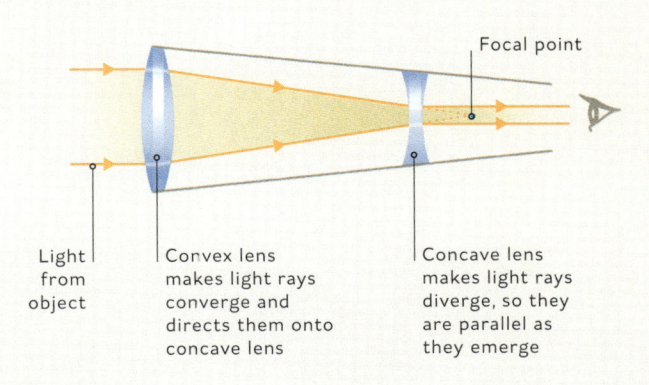

TELESCOPE DESIGN

Galileo's telescopes used two lenses, one positioned at either end of a long tube. A convex (outward-curving) lens gathered light entering the tube and passed it through a concave (inward-curving) eyepiece lens. The design was simple and prone to image distortion, but it enabled groundbreaking discoveries.

Focal point

Light from object

Convex lens makes light rays converge and directs them onto concave lens

Concave lens makes light rays diverge, so they are parallel as they emerge

PLENILVNIUM,
pinxit ad Archetypum M.C.Eimmarta Norimb:

One the first-known female astronomers, German illustrator Maria Clara Eimmart (1676–1707) produced hundreds of influential drawings of the Moon, many of which were published in her book *Micrographia stellarum phases lunae ultra 300.*

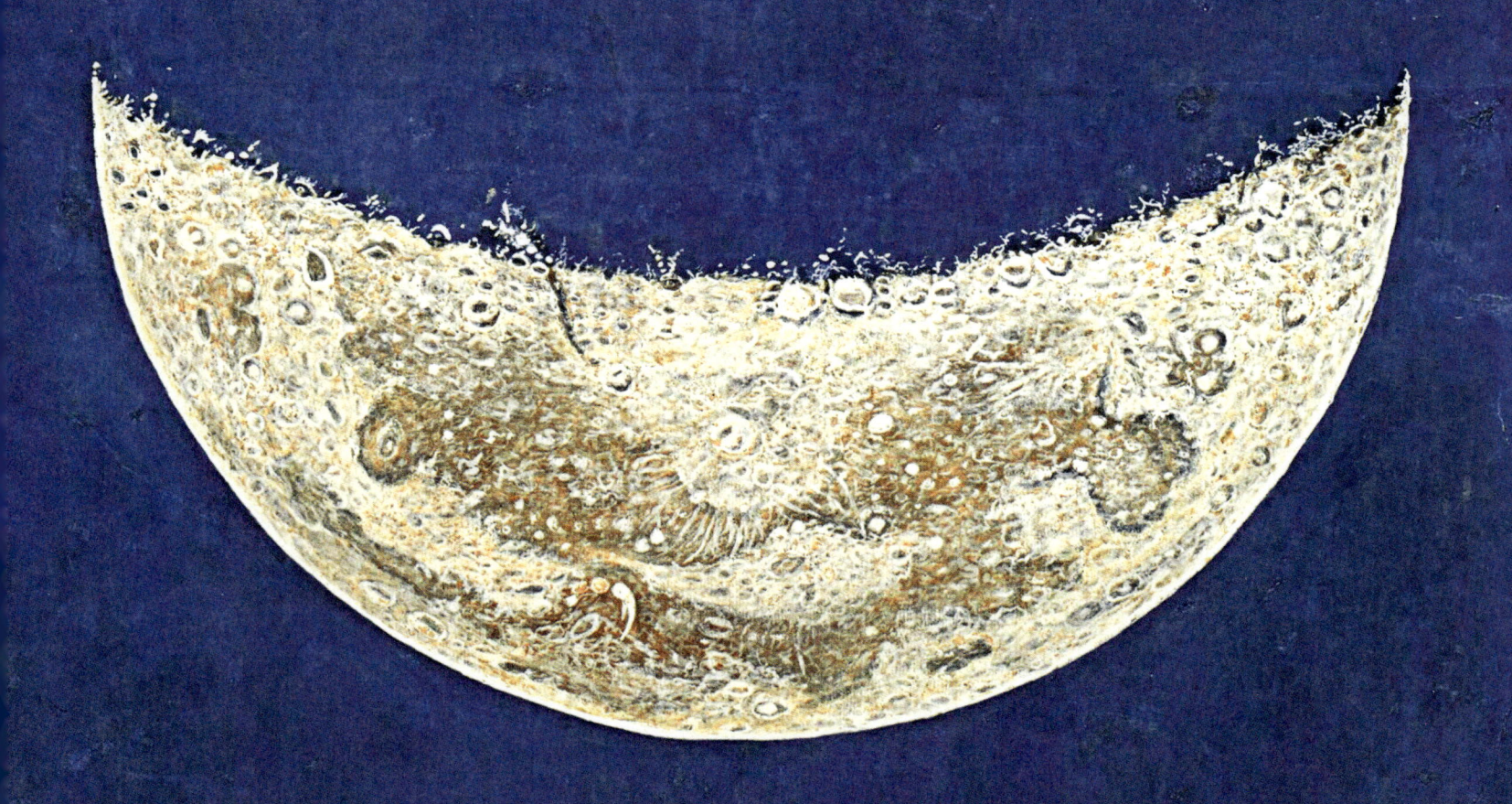

PHASIS LVNÆ POST □ VLT. DECR:

Anno Chr: 1697. Die 29. Aug: st: v.

ad Archetyp: depinxit M.C.Eimarta Norimb:

IMPACT CRATERS

Like Earth, the Moon has been bombarded with rocks ever since its formation. However, without plate tectonics and erosion to erase them, impact craters are still its dominant feature.

In the early 20th century, scientists debated whether the Moon's many circular craters were volcanic in origin, like most of the craters on Earth, or the result of meteors falling onto the lunar surface. Since then, observations and laboratory experiments have shown that most of the Moon's craters are caused by meteor strikes. Only a tiny fraction of lunar craters are relics of the Moon's volcanic past (see pp. 124–125). These are either extinct volcanoes or the remains of volcanic vents that flooded the surface with lava

Explosive impact

The Moon has no atmosphere to slow down a meteoroid, asteroid, or comet. On hitting the surface, the kinetic energy of these very high-speed impactors is almost instantly converted into a huge amount of explosive energy. A 100-ft (30-m) impactor hitting at 33,500 mph (54,000 km/h) could release the equivalent of four megatons of high explosive. The impact releases a shockwave with millions of times Earth's standard atmospheric pressure, enough to compress even dense rock to one third of its usual volume. The structure of the Moon's surface material breaks down under the stress and it flows almost like a liquid. Parts of the impactor and the surface will melt or even vaporize. A typical cratering event might take less than a minute. An impactor of up to 1,600 ft (500 m) in diameter will produce a simple bowl-shaped crater, but above that size a crater's shape becomes more complex. The largest craters of all are known as "basins" (see pp. 112–113). »

Giordano Bruno
This 4-million-year-old crater has a steep, sharply defined rim and pools of impact melt among the uplifted ridges of its floor.

SIMPLE CRATER

Craters up to about 4.3 miles (7 km) in diameter are bowl-shaped and have raised rims and smooth outer walls. Parts of their floors may be level where ejecta (debris thrown by the impact) has accumulated and rocks have rolled down from their walls.

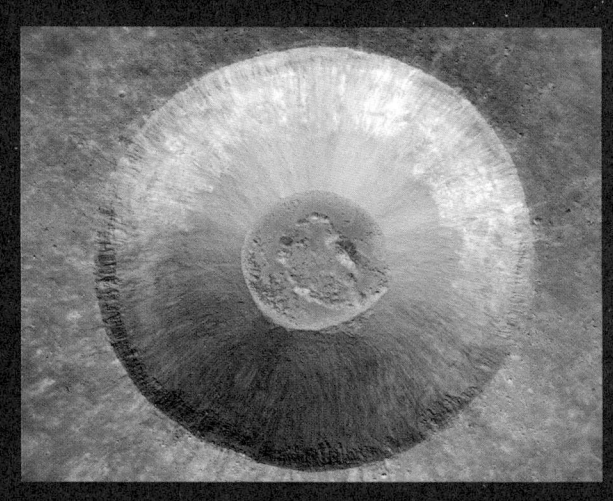

Lichtenberg B
At around 1.68 billion years old, Lichtenberg B is a young crater. It is bowl-shaped with a 3-mile (5-km) diameter and a bright ejecta ring. Impact melt and wall debris lie on its floor.

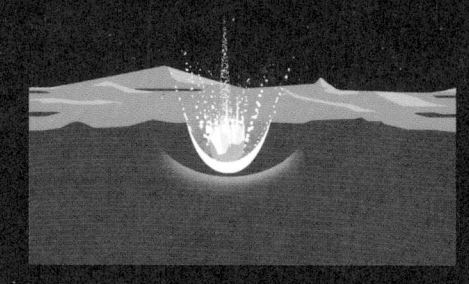

Contact
As the impactor rapidly decelerates, its kinetic energy is converted into shockwaves and heat, melting it and some of the surface rock.

Melt
Shockwave

Excavation
As the shockwave passes through the surface, a decompression wave follows, throwing ejecta up and out of the expanding cavity.

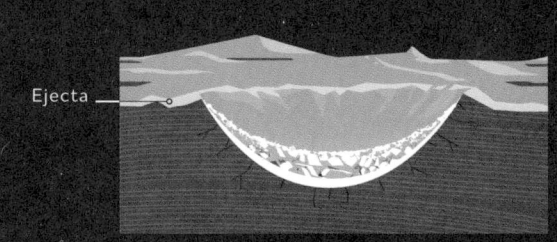

Ejecta

Modification
As the decompression wave dissipates, the ejecta falls back to drape the surface (see overleaf) and loose material slumps into the center of the crater.

COMPLEX CRATER

Craters wider than about 4.3 miles (7 km) tend to be more complex, with terraced walls and central peaks or plateaux, or a ring of peaks if the crater is 90 miles (150 km) or wider. After several stages of slumping, the crater rim is less well defined and may not rise above the surrounding terrain at all.

Tycho
A large, complex crater about 53 miles (85 km) in diameter, Tycho has a central peak, a flat floor, and terraced walls.

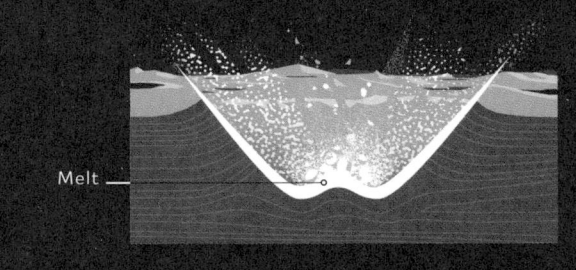

Melt

Excavation
A larger impactor releases more energy (sometimes enough to vaporize it), digs out a larger cavity, and penetrates deeper into the lunar surface.

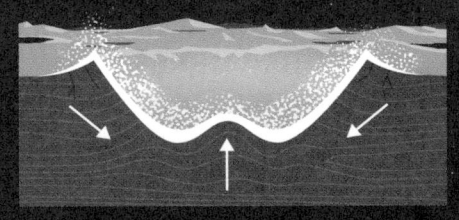

Modification
Deeper rock layers rebound to create a central peak. Steep walls slump inward, expanding the crater's diameter.

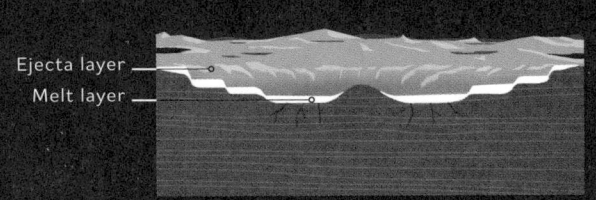

Ejecta layer
Melt layer

Final structure
A large crater may have several terraces instead of a raised rim, a flat floor partly covered by impact melt, and one or more central peaks.

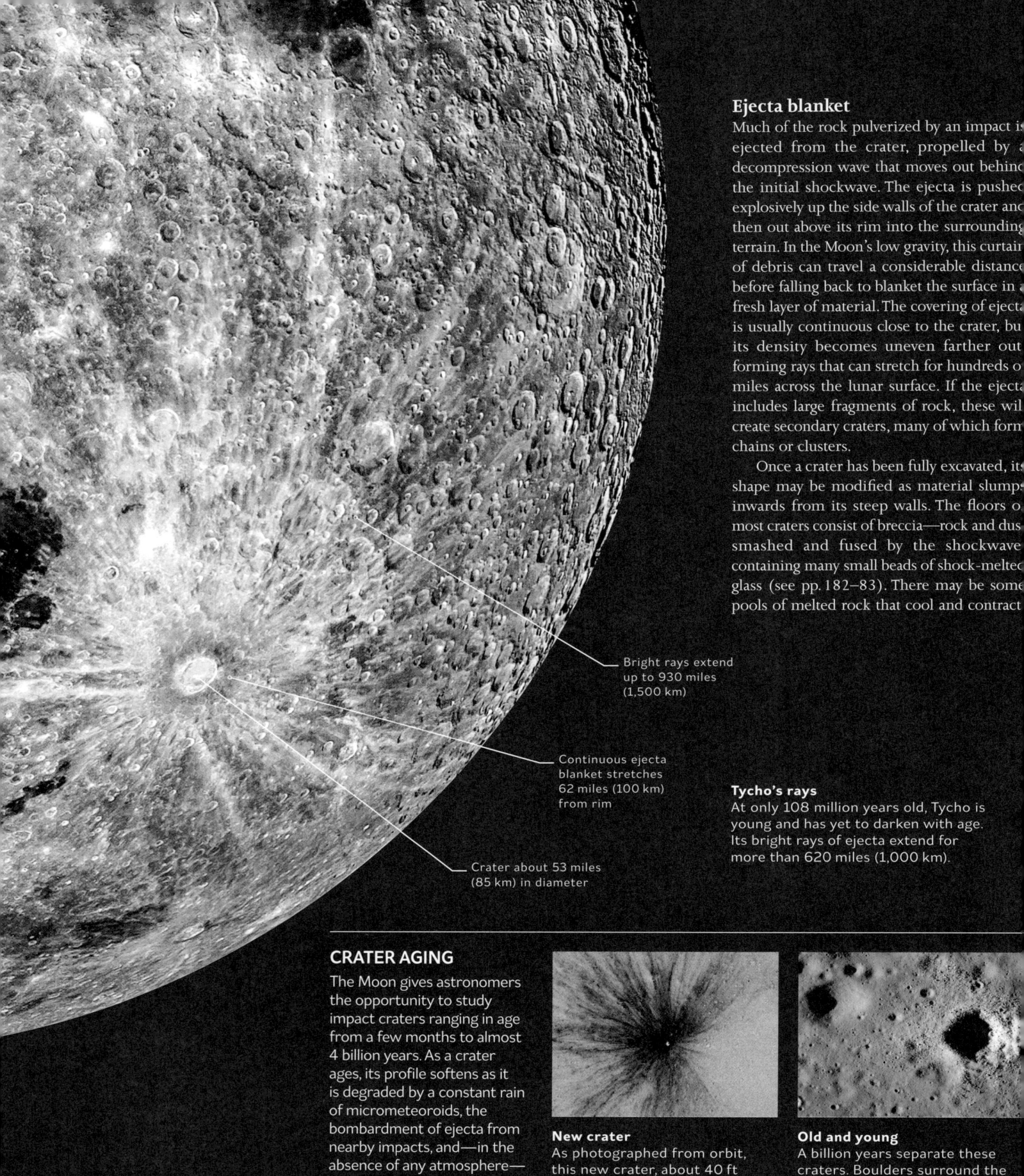

Ejecta blanket

Much of the rock pulverized by an impact is ejected from the crater, propelled by a decompression wave that moves out behind the initial shockwave. The ejecta is pushed explosively up the side walls of the crater and then out above its rim into the surrounding terrain. In the Moon's low gravity, this curtain of debris can travel a considerable distance before falling back to blanket the surface in a fresh layer of material. The covering of ejecta is usually continuous close to the crater, but its density becomes uneven farther out, forming rays that can stretch for hundreds of miles across the lunar surface. If the ejecta includes large fragments of rock, these will create secondary craters, many of which form chains or clusters.

Once a crater has been fully excavated, its shape may be modified as material slumps inwards from its steep walls. The floors of most craters consist of breccia—rock and dust smashed and fused by the shockwave, containing many small beads of shock-melted glass (see pp. 182–83). There may be some pools of melted rock that cool and contract

Bright rays extend up to 930 miles (1,500 km)

Continuous ejecta blanket stretches 62 miles (100 km) from rim

Tycho's rays
At only 108 million years old, Tycho is young and has yet to darken with age. Its bright rays of ejecta extend for more than 620 miles (1,000 km).

Crater about 53 miles (85 km) in diameter

CRATER AGING

The Moon gives astronomers the opportunity to study impact craters ranging in age from a few months to almost 4 billion years. As a crater ages, its profile softens as it is degraded by a constant rain of micrometeoroids, the bombardment of ejecta from nearby impacts, and—in the absence of any atmosphere— by the high-speed particles carried on the solar wind

New crater
As photographed from orbit, this new crater, about 40 ft (12 m) across, formed between October 2012 and April 2013

Old and young
A billion years separate these craters. Boulders surround the young one (right) but have been worn away around the older one

leaving cracks across part of the crater floor. The floors of larger craters may even rebound after the impact, creating a central peak—sometimes made of rock lifted from deep beneath the lunar surface. Some larger craters may be flooded by later eruptions of lava—a small-scale version of the process that created the Moon's maria (see pp. 118–19).

Counting craters

Craters are useful in determining the age of the Moon's surface, since the longer a surface is exposed to bombardment, the more craters it should display. However, this can only establish the relative ages of different parts of the Moon. In order to determine the absolute age of an area, physical samples, such as those gathered by the Apollo missions, have to be brought back to Earth and dated radiometrically (see pp. 182–183). Having done so, scientists have concluded that most of the Moon's craters are from the Late Heavy Bombardment period (see pp. 14–15), more than 3.9 billion years ago (BYA). The cratering rate dropped dramatically between 3.9 and 3.3 BYA and has been declining at a steady rate ever since.

Messier and Messier A
Even impacts from angles as low as 15° produce symmetrical, circular craters. This pair of elongated craters are thought to be the result of the extremely low-angle impact of a body that may have bounced across the lunar surface.

ANATOMY OF A CRATER

Euler is a 17-mile- (28-km-) wide crater in Mare Imbrium (the Sea of Showers) with features typical of a moderately large impact. It has a raised rim surrounded by a smooth area of continuous ejecta, a rougher area of discontinuous ejecta, crater rays, and secondary impact clusters. Within the crater, parts of the wall have slumped to form terraces, and the floor has lifted to form a central peak.

Euler
This image of Euler is part of Apollo photograph AS17-M-2923. It was taken by the Metric Mapping Camera in the Apollo 17 Service Module on December 16, 1972.

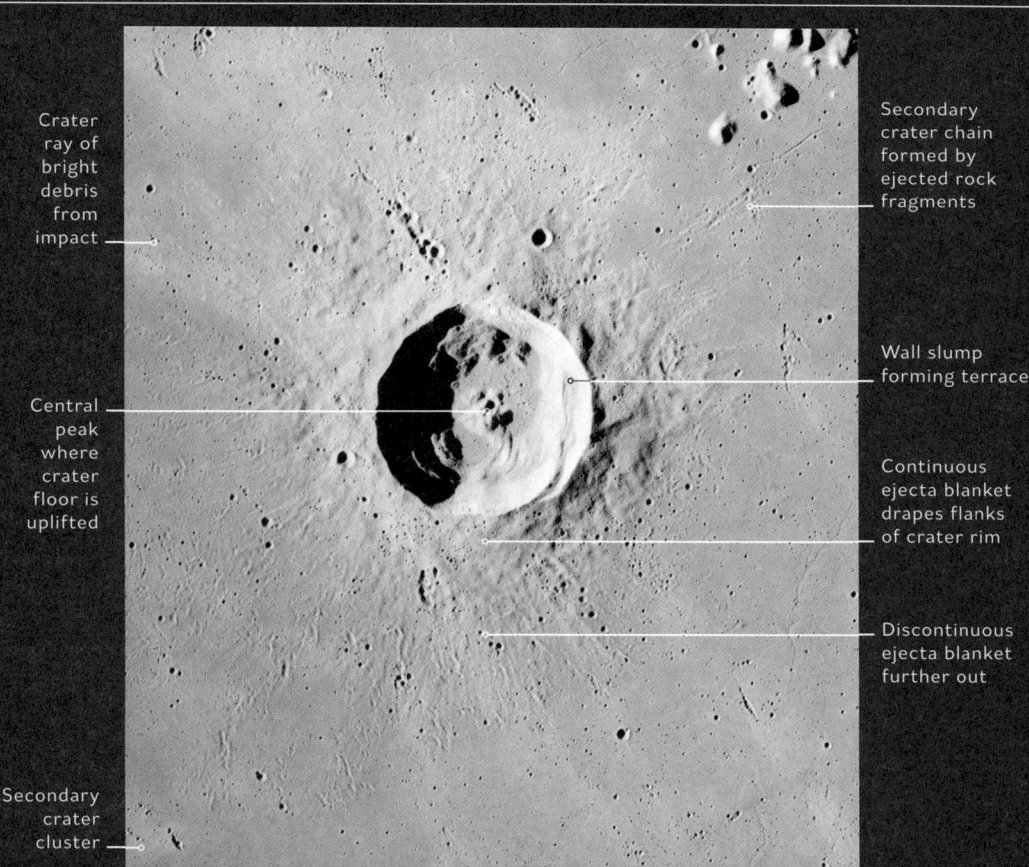

Crater ray of bright debris from impact

Secondary crater chain formed by ejected rock fragments

Central peak where crater floor is uplifted

Wall slump forming terrace

Continuous ejecta blanket drapes flanks of crater rim

Discontinuous ejecta blanket further out

Secondary crater cluster

EARLY LUNAR MAPS

As telescopes increased in power, astronomers began to see distinct features on the Moon: mountain peaks; craters; valleys (also called rilles); and vast smooth expanses that became known as maria, or seas. Thomas Harriot made the first simple Moon map around 1611. With more detailed and accurate lunar maps came the need for a formalized method of naming features and a reliable coordinate system.

Lunar libration (see pp.116–117), caused by the Moon's slightly tilted and elliptical orbit around Earth, allowed astronomers to observe more than 50 percent of the Moon's surface. At times, they could see beyond the north and south poles, effectively increasing the visible area to about 59 percent. Only when space probes circumnavigated the Moon was the far side imaged and mapped for the first time.

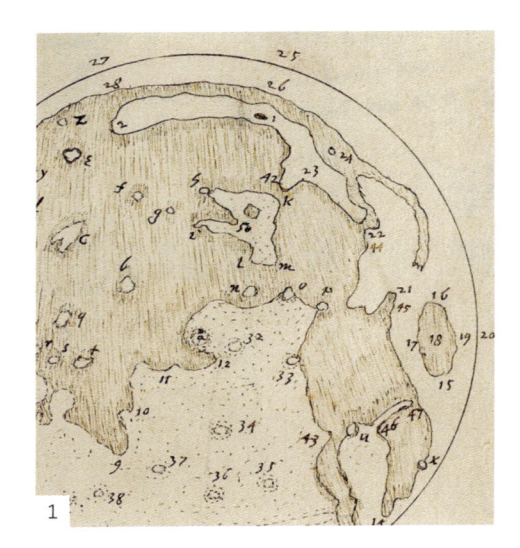

1. Sketch map
Thomas Harriot, an English astronomer, was one of the first people to turn the newly invented telescope toward the sky. Carefully studying the Moon's surface, he began sketching its lunar seas and craters in 1609. He produced the first map of the Moon about two years later.

2. Named features
Flemish astronomer Michael van Langren thought it might be possible to determine longitude on Earth from observations of illuminated features on the Moon. This led him to make the first detailed lunar map with individually named features, published in 1645; many of his names did not endure.

3. First atlas
From his rooftop observatory in Poland, Johannes Hevelius created more than 130 engravings of the Moon. His maps formed the first lunar atlas, *Selenographia*, published around 1647. For the first time, they depicted areas of the Moon visible due to libration and covered all its phases.

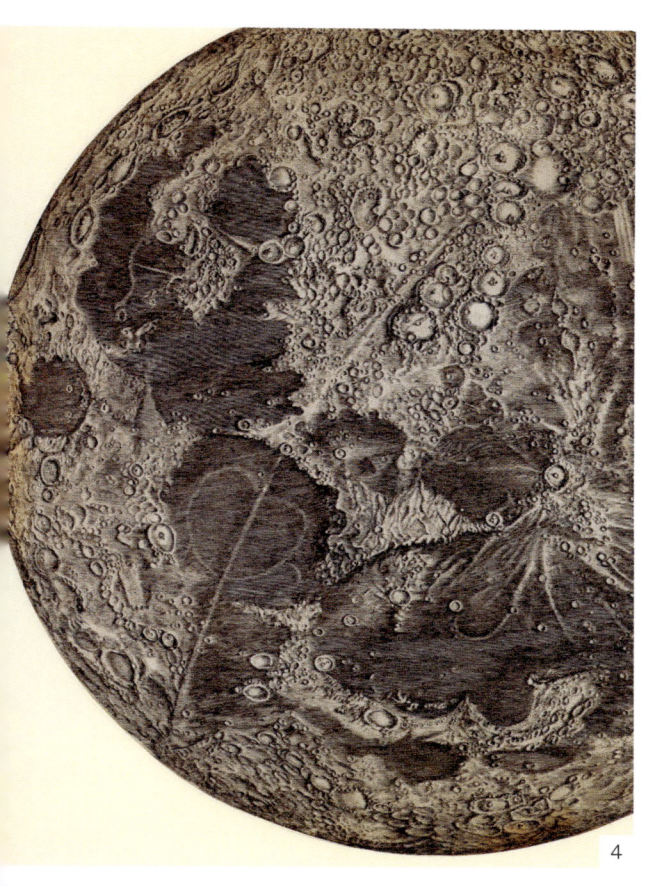

MOON GLOBE

A moon globe plots lunar features on a spherical surface. It represents their sizes and the distances between them more realistically than a flat map or chart. English astronomer John Russell crafted the first globe in 1797. Moon globes, now rare, also demonstrate the effects of libration.

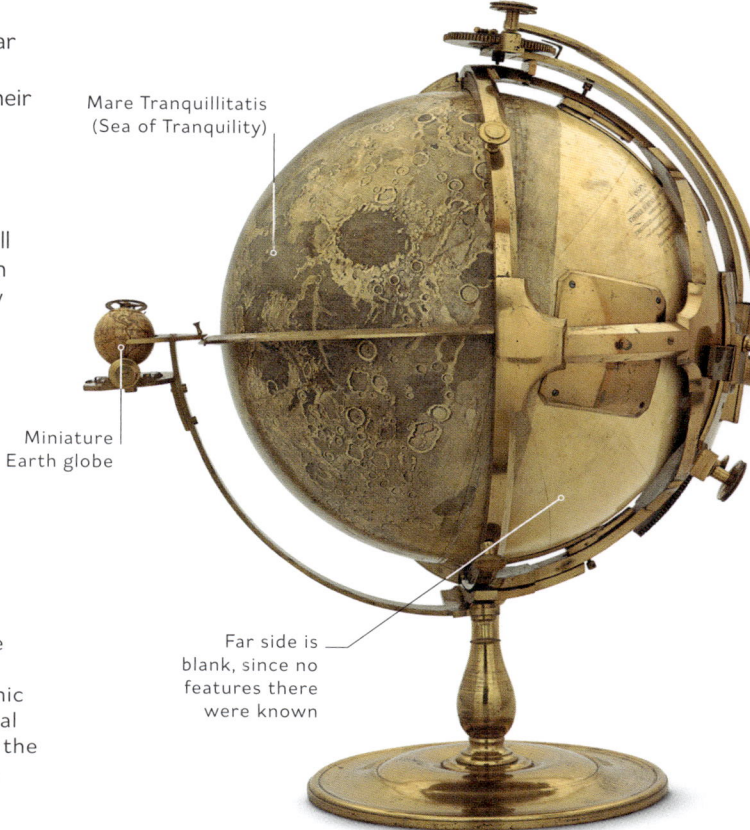

Mare Tranquillitatis (Sea of Tranquility)

Miniature Earth globe

Far side is blank, since no features there were known

Russell's Moon globe
The 12-in- (30-cm-) diameter globe is made of papier-mâché and hand-engraved. To mimic libration, the mechanical brass mounting moves the globe so that different parts of the surface become visible.

4

5

6

4. Scientific detail
Italian astronomer Giovanni Domenico Cassini, founder of the Paris Observatory in France, made around 60 Moon drawings between 1671 and 1679. His drawings were the basis of this engraved map, the first scientific map of the Moon, which gives the lunar features a three-dimensional quality.

5. Maps compared
Around 1742, Johann Gabriel Doppelmayr made this comparison of the maps of Hevelius (left) and Jesuit priest Giovanni Battista Riccioli (right). Riccioli's naming system, which included Mare Tranquillitatis (the Sea of Tranquility, the site of the first Apollo landing), ultimately prevailed.

6. Lunar coordinates
German cartographer Tobias Mayer created this highly accurate lunar map in 1749, though it was not published until 1775. The map's inclusion of lines of longitude and latitude ensured that, for the first time, lunar features were given precise coordinates as well as names.

IMPACT BASINS

When an impact is large enough to fracture the Moon's crust, it creates a basin. These huge impacts have punctuated the Moon's history, delineating its geography and geological timescale.

Basin wall rises 9,850ft (3,000m) above floor

Peak ring 93 miles (150 km) in diameter

Schrödinger basin gravity map
A mascon (in yellow) lies at the center of the basin. Yellow and red show excess mass (high gravity); blue shows mass deficit (low gravity).

Basins are impact craters that are greater than 185 miles (300 km) in diameter (see pp.106–109). They typically display concentric rings of mountains. In fact, all the Moon's named mountain ranges are segments of basin rims. Basins are some of the Moon's oldest features and dominate its surface, owing to their sizes and far-flung ejecta deposits. Ejecta from the Imbrium basin spreads over much of the Moon, lying up to 3,280 ft (1,000 m) deep as far as 370 miles (600 km) from the basin's edge.

The largest of all is the South Pole-Aitken basin on the Moon's far side, at some 1,550 miles (2,500 km) across. Imbrium is the largest near-side basin, at 720 miles (1,160 km) in diameter. It was formed by the impact of a small protoplanetary-sized body of up to 150 miles (240 km) wide. This event marks the start of the Imbrian period 3.85 billion years ago (BYA). Before then, lunar history is separated into the Pre-Nectarian and Nectarian periods, divided by the creation of the Nectaris basin 3.92 BYA.

Mascons

To date, some 45 impact basins have been identified. Their surface structural features can often be obscured by subsequent impacts and mare flooding (see pp.118–119). Impact basins also perturb the Moon's gravity field. Areas with slightly stronger gravity can be detected by orbiting spacecraft and reveal the location of mass concentrations ("mascons"), where dense rock from the mantle has squeezed up to fill a basin's original impact cavity.

HOW BASINS ARE FORMED

An impactor wider than about 30 miles (50 km) may have enough energy to break through the Moon's crust, which is about 30 miles (50 km) thick on the near side. In such cases, the crust fractures over a wide area, extending the basin rim far beyond the initial crater and resulting in multiple concentric rings of mountains, but no central peak. The explanation of basin formation given here is based on extensive orbital observation and geological mapping of the Imbrium basin, as well as rock samples taken from the area.

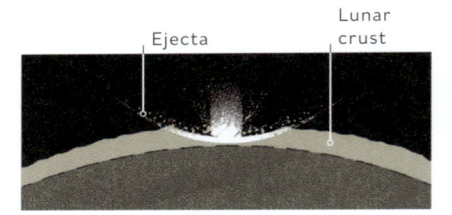

Ejecta

Lunar crust

1. Impact
A large asteroid or protoplanetary body hits the lunar crust at high speed, excavating a large crater and depositing ejected material across a huge area.

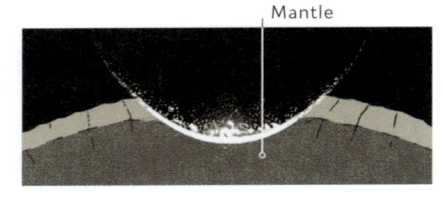

Mantle

2. Maximum cavity growth
The cavity created by the impact penetrates the Moon's crust and into the mantle. The crust weakens and fractures over an area far larger than the cavity itself.

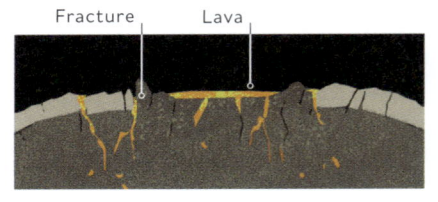

Fracture   Lava

3. Postimpact modification
Mantle material rises up to largely fill the cavity, while surrounding crustal blocks shift, tilt, and collapse to new positions. Lava may erupt through some fractures.

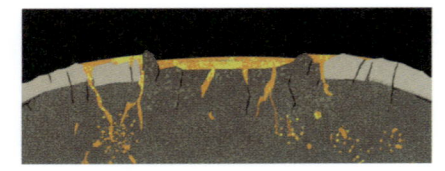

4. Basin development
The broad, shallow basin may undergo further structural adjustment, including from other large impacts. Later eruptions of lava may flood its lowest-lying parts.

MULTI-RING BASIN

Orientale is a 580-mile- (930-km-) wide impact basin containing multiple rings of mountains surrounding a lava-flooded plain (Mare Orientale). The youngest basin on the Moon, it is thought to have formed 3.8 BYA, when a 40-mile- (65-km-) wide asteroid slammed into the lunar surface. Only a small part of Orientale was subsequently flooded by mare lava, so most of the original structure remains exposed to view. Its study has added much to scientists' understanding of the geology of impact basins.

The Moon has at least **45** impact basins: **30** formed in the Pre-Nectarian period, **12** in the Nectarian, and **3** in the early Imbrian.

Orientale basin
This view combines imagery and surface height data from NASA's Lunar Reconnaissance Orbiter. It shows Orientale, which is about the size of Texas, as it would appear from lunar orbit.

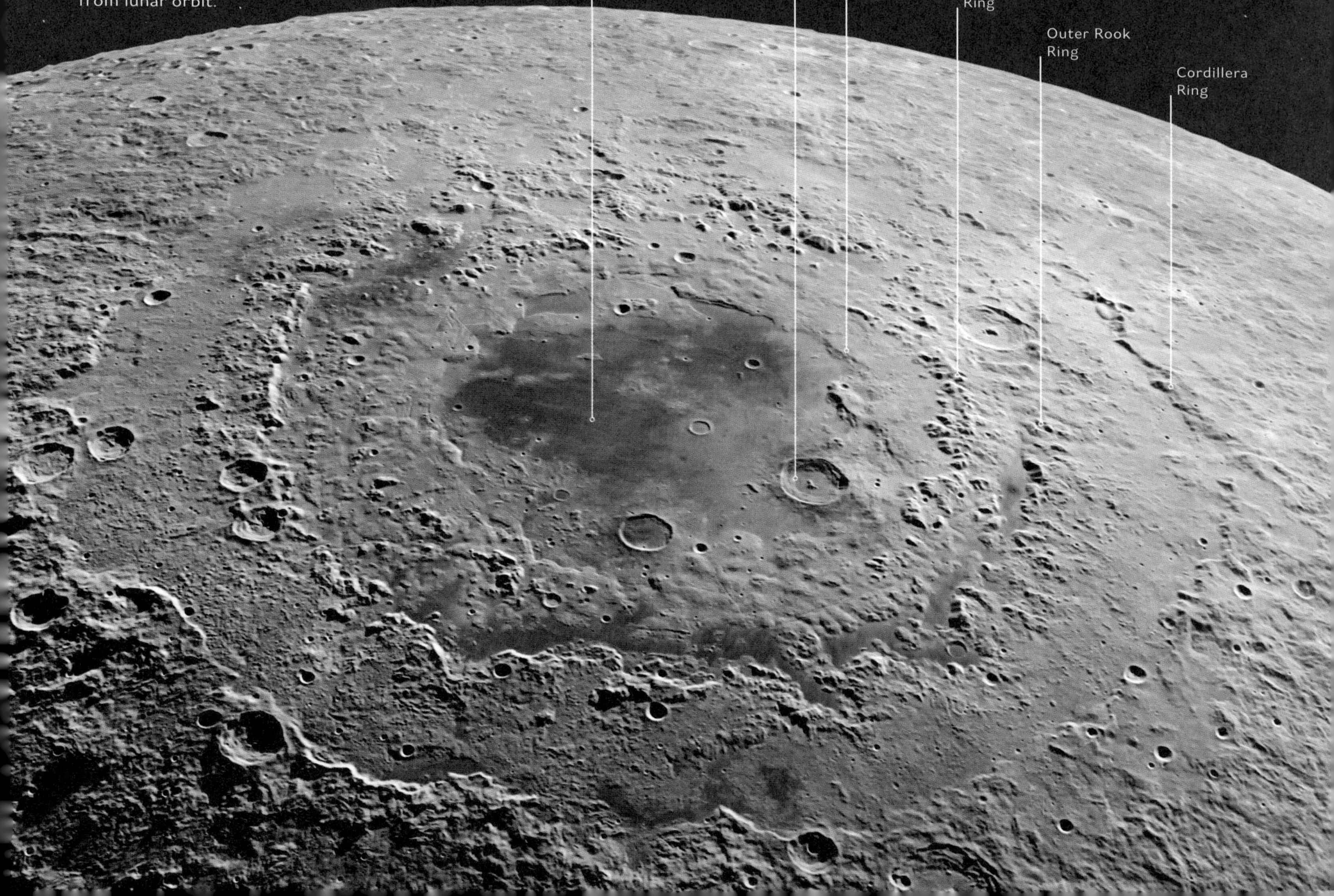

Numerous impact craters postdate formation of basin

Mare Orientale—central part of basin, flooded by basalt lava

Inner scarp

Inner Rook Ring

Outer Rook Ring

Cordillera Ring

THE APPLE AND THE MOON

Early scientists wondered why the Moon continued to orbit Earth instead of falling toward it. Studies to solve the mystery resulted in the discovery of gravity and a fresh understanding of how the universe works.

In 1665, a bubonic plague outbreak compelled English polymath Isaac Newton (1642–1726) to leave Cambridge and return to his family home in the Lincolnshire countryside. According to legend, while sitting in the garden at Woolsthorpe Manor and observing an apple fall, Newton experienced a moment of revelation, which led him to develop his law of universal gravitation. He reasoned that the same force that pulls an apple to the ground—gravity—also acts on the Moon. The Moon must constantly be falling toward Earth due to gravity, but its forward motion prevents it from colliding with Earth, keeping it in a near-circular orbit. Newton further theorized that every object in the universe attracts every other object with a gravitational force proportional to their masses. His final breakthrough came when he deduced his inverse-square law (see below).

Two decades later, Newton published his work on celestial mechanics in the book *Principia* (1687), which set out a mathematical explanation of how gravity acts on the Moon. Additionally, Newton used his gravitation theory to explain tides, demonstrating how they are caused by the gravitational pull of the Moon and the Sun on Earth's oceans.

HOOKE'S *MICROGRAPHIA*

Robert Hooke (1635–1703), one of Newton's contemporaries, played a key role in scientific research during the mid-17th century. His skill in building and improving instruments, such as the first reflecting telescope, aided his lunar studies, and he added crosswires to optics to make precise observations. Hooke published his detailed lunar drawings in *Micrographia* (1665), along with illustrations of insects and plants as seen through microscopes.

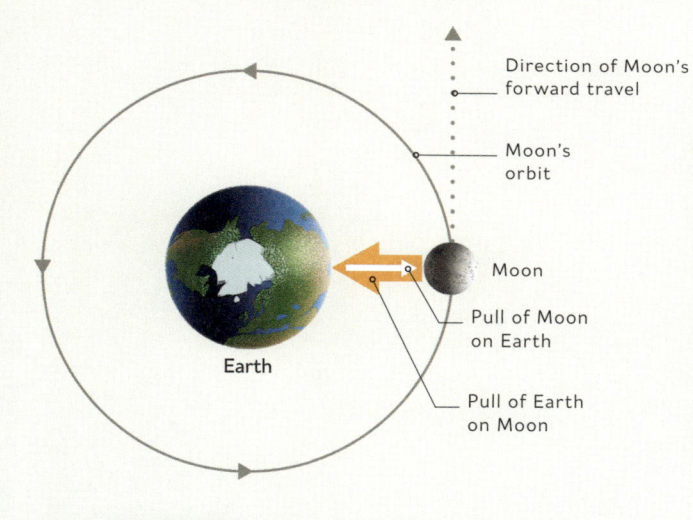

Gravity and the Moon
As the Moon circles Earth, its apparent size remains constant, suggesting that the radius of its orbit doesn't change. Earth's gravity tries to pull the Moon toward Earth, while the Moon's own motion tries to propel it in a straight line. Together, these two forces keep the Moon orbiting Earth.

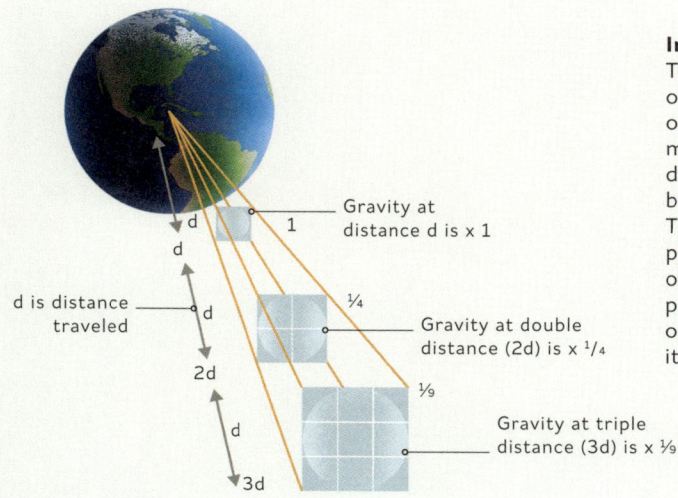

Inverse-square law
The gravitational pull of a body, such as Earth, on another object with mass, such as the Moon, decreases as the distance between them increases. The force is directly proportional to the mass of the body and inversely proportional to the square of the distance separating it from the object.

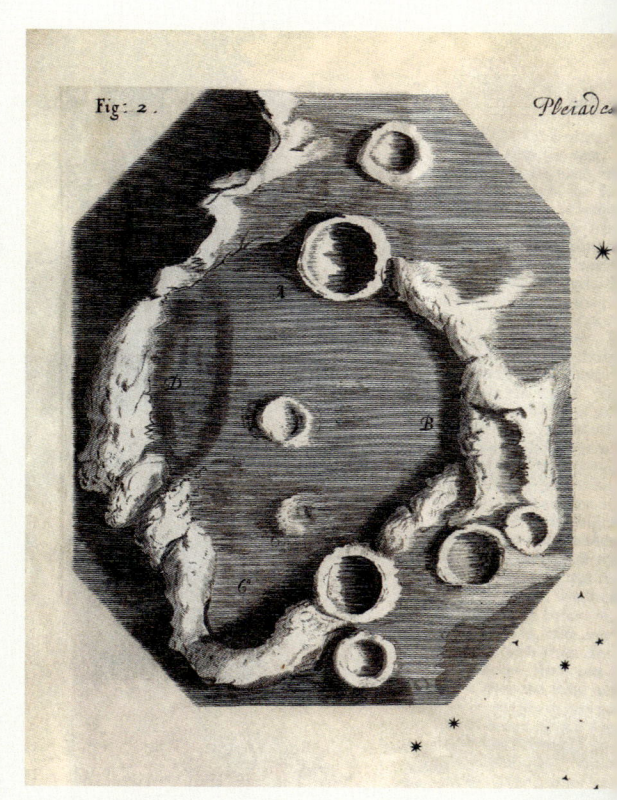

Crater drawing
This *Micrographia* drawing of the crater now called Hipparchus, which Hooke observed through his 30-ft (9.1-m), self-built refractor telescope, captures light and shadows on the Moon's surface. Hooke thought that craters were formed from volcanic activity and subsurface bubbling.

THE MOON'S "WOBBLE"

Although only one side of the Moon is ever visible from Earth, the Moon's apparent "nodding" and "wobbling" motion, known as libration, enables us to see 59 percent of its surface features.

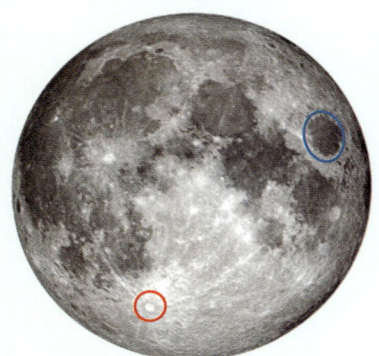

In the early stages of lunar history, Earth's gravitational pull caused the Moon to bulge slightly in Earth's direction, so one side of the Moon remained more strongly attracted to Earth. The geological asymmetry put a brake on the Moon's axial spin, which decelerated until, about 3.9 billion years ago, the Moon remained revolving just once per orbit, with one face permanently pointing toward Earth. This is called tidal locking (see pp. 16–17).

Despite Earth and Moon being tidally locked, the nature of the Moon's orbit means that observers on Earth get a slightly changing view, both longitudinally (east–west) and latitudinally (north–south); the Moon seems to nod up and down and wobble from side to side. Over time, these apparent motions, called librations (from the Latin for weighing scales, which oscillate until they balance), give glimpses of 9 percent of the far side.

Nodding and wobbling
The Moon's libration causes Mare Crisium (the Sea of Crises), in blue, and Tycho crater, in red, to shift both up and down and from side to side.

Libration in longitude is a consequence of the Moon's elliptical orbit, which gives a peek around its normal eastern and western horizons (see right). Libration in latitude is mainly caused by the difference in the tilt of the Moon's orbit relative to Earth's own orbit around the Sun, which temporarily allows observers on Earth to see beyond the north and south lunar poles. During a month, these phenomena make it look as if the Moon itself is gyrating, but the "nods" and "wobbles" are simply the result of the Moon's position changing relative to Earth (see pp. 28–29).

Libration discoveries

Until the Moon's far side was photographed by the Soviet Union's Luna 3 spacecraft in 1959 (see pp. 134–135), astronomers only knew those far-side features they had managed to identify during librations. For example, in his 1906 book *Der Mond* (*The Moon*), the German astronomer Julius Heinrich Franz (1847–1913) described a huge structure on the Moon's eastern rim that he named Mare Orientale (the Eastern Sea); it later proved to be one of the Moon's largest impact basins.

OBSERVING THE FAR SIDE

Although the Moon's librations can be observed with the unaided eye, they only came to the attention of astronomers after the invention of the telescope (see pp. 102–103). Since then, lunar librations have become a key feature of the stargazer's month. Focusing on a single lunar feature to watch it shift position is a good way to see libration in action. For example, due to librations in both latitude and longitude, Endymion crater gradually moves inward and reveals Mare Humboldtianum (the Sea of Alexander von Humboldt)—a huge impact basin that is otherwise hidden or foreshortened on the horizon.

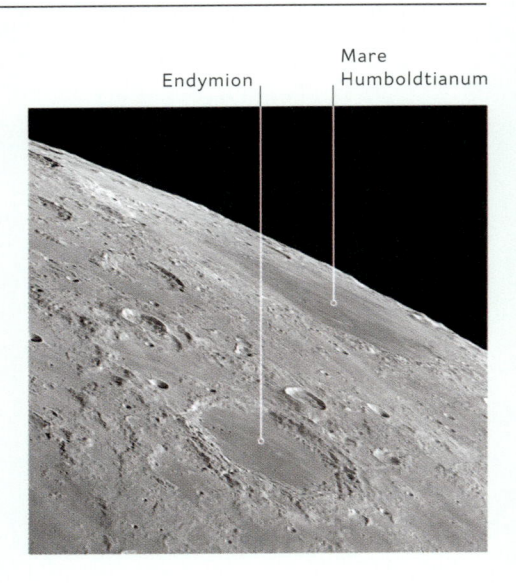

Endymion   Mare Humboldtianum

LIBRATION IN LONGITUDE

Discovered in 1648 by Polish astronomer Johannes Hevelius (1611–1687), the Moon's longitudinal libration—its east-west "wobbling" motion—is due to the Moon's elliptical orbit (see pp. 16–17). Although the Moon revolves once as it orbits Earth, it speeds up as it reaches perigee (its closest point to Earth) and slows down at apogee (its farthest point from Earth). Consequently, the Moon's rotation is sometimes ahead of its orbital position and sometimes behind it, temporarily revealing an extra 6.3° of its western or eastern side.

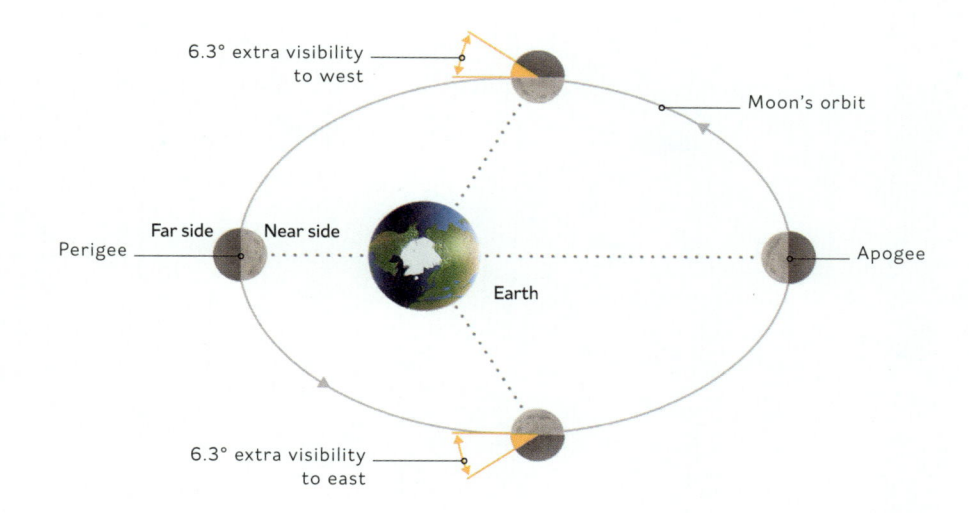

LIBRATION IN LATITUDE

The Moon's north–south "nodding" motion is its latitudinal libration. It is the result of the Moon's orbital path being tilted at 5.15° from the ecliptic (the plane of Earth's orbit around the Sun) and its axis also being tilted at 1.54° to the ecliptic (see pp. 16–17). This enables viewers on Earth to see about 6.7° more of the Moon's north and south polar regions at different times of the month. The discovery of latitudinal libration has variously been credited to the astronomers Galileo Galilei (1564–1642), Thomas Harriot (1560–1621), and William Gilbert (1544–1603).

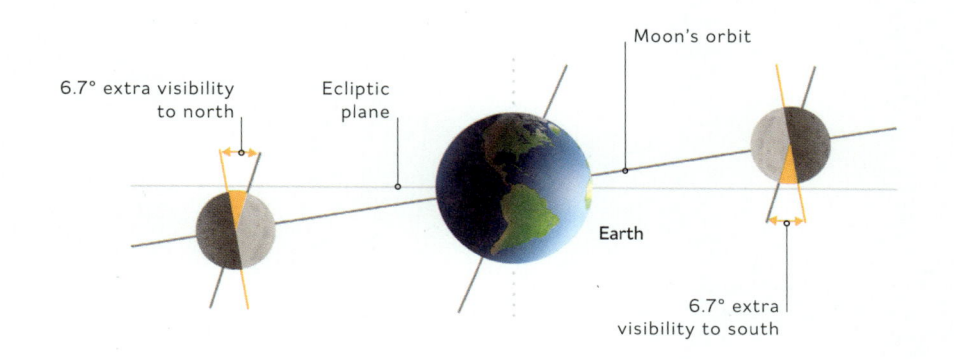

DIURNAL LIBRATION

The Moon's diurnal libration—its apparent daily, sideways, back-and-forth "rocking" motion— is a form of longitudinal libration caused by Earth's rotation. When the Moon rises in the east and sets in the west, observers see more of its eastern and western sides respectively at those times. The same is true of observers viewing the Moon at its high point from different sides of Earth: more of the Moon's western side is visible to someone in the east than to someone in the west, and vice versa. Diurnal libration reveals roughly 1° of the Moon's far side.

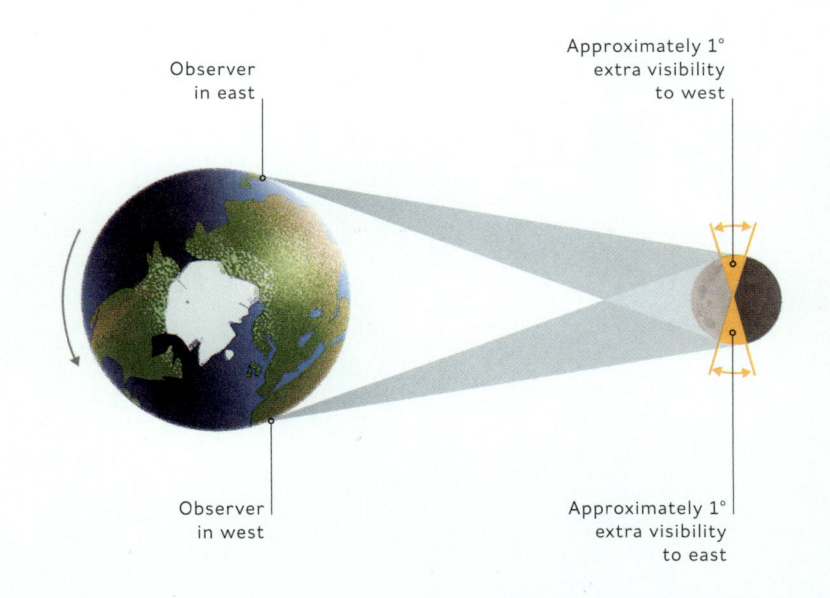

HIGHLANDS AND PLAINS

The Moon's surface can be separated into two distinct terrain types that can be clearly differentiated from Earth: the bright, rugged highlands and the dark, smooth plains of the maria.

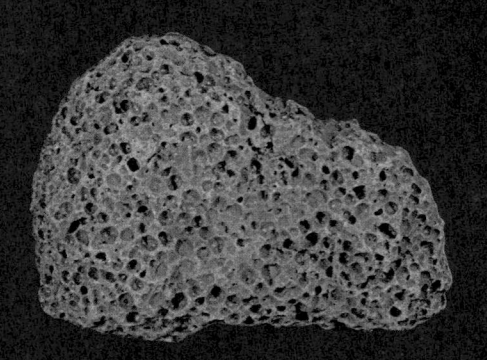

Basalt pumice
Apollo 15 sample 15016 is a mare basalt pumice (a rock with gas bubbles). It is radiometrically dated at 3.38 billion years (BY) old—far younger than the 4–3.8 BY age of highland breccia samples.

Effects of an impact
The dark Mare Imbrium (the Sea of Showers) in the foreground has rays of bright ejecta from Copernicus crater on the far side of Montes Carpatus (Carpathian Mountains).

The highlands—also known as the *terrae* (meaning "lands" in Latin)—are the oldest parts of the lunar surface and represent the original lunar crust. They are heavily cratered, with many overlapping craters of varying sizes and states of degradation. The most recent impacts have obliterated all signs of the oldest. Here, the Moon's crust has been repeatedly pulverized and churned by meteorite impacts, producing a lunar "soil" called regolith—a mixture of mineral grains, rock fragments, and impact-melted glass beads (see pp.182–183).

Bedrock and regolith

The regolith has a chemical composition very similar to the underlying bedrock: In the highlands, it is rich in feldspar minerals. It has been modified by long exposure to gases that flow in with the solar wind—the constant stream of charged particles that emanates from the Sun. The thickness of the regolith depends on the age of the underlying bedrock. In the highlands, the lunar crust is now concealed beneath a layer of regolith up to 45 ft (15 m) deep. In the maria, by contrast, the depth of the regolith is only 6–24 ft (2–8 m).

Impacts larger than about 3 ft (1 m) across can compress and fuse the regolith into a type of rock called breccia. This rock is more resilient to the constant impact-churning, known as "gardening," of the fine regolith and it can be thought of as a "fossilized soil," in that the breccia records lunar characteristics from the time it was formed.

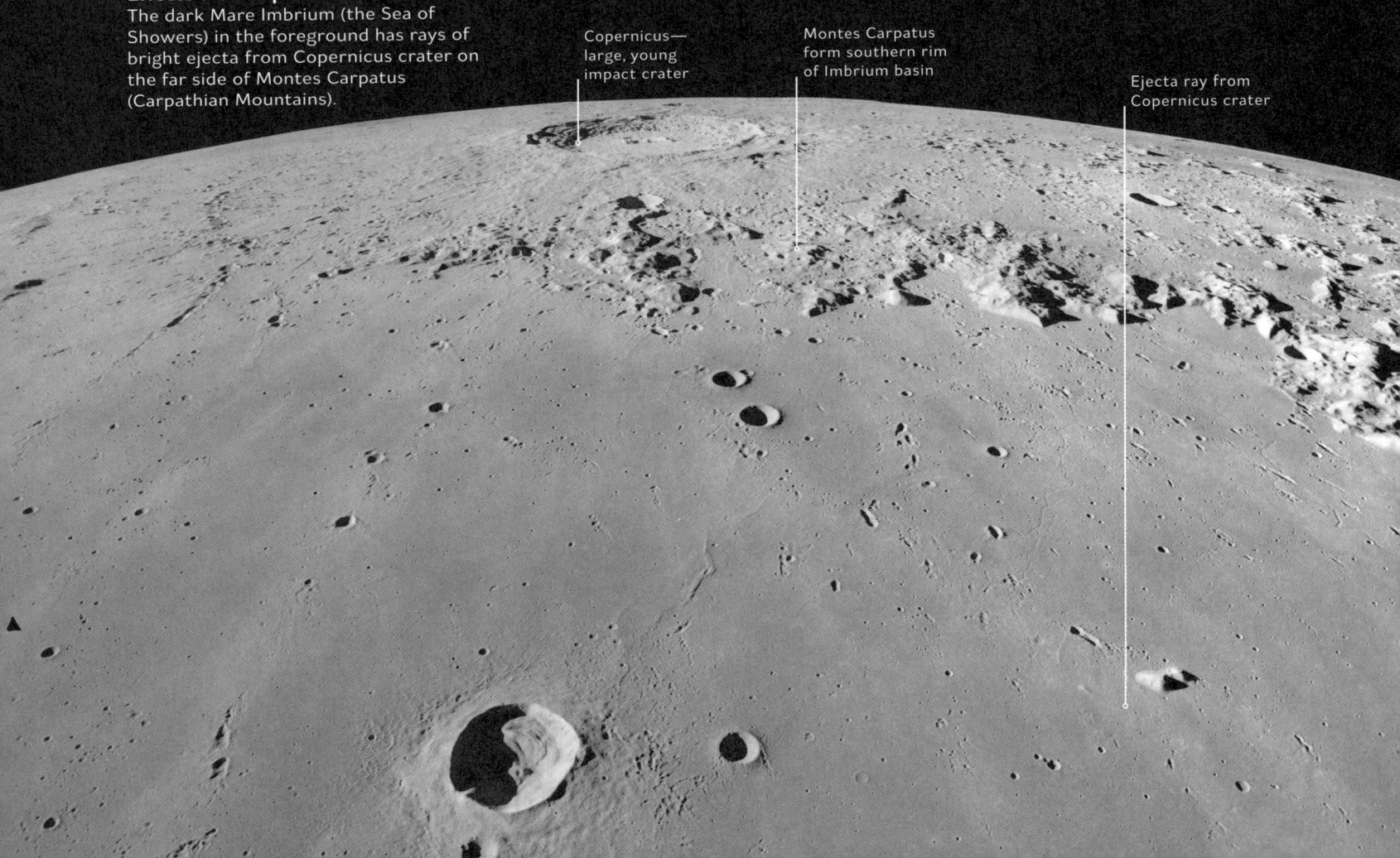

Copernicus—large, young impact crater

Montes Carpatus form southern rim of Imbrium basin

Ejecta ray from Copernicus crater

"We had gotten there, and it looked pretty desolate. But it was magnificent desolation."

Edwin "Buzz" Aldrin, US Apollo 11 astronaut, interviewed in 2019

Seas of lava

The Moon's dark maria ("seas" in Latin; singular: mare) are smooth plains that cover 16 percent of its total surface and 31 percent of the lunar near side. They contain fewer impact craters than the highlands, so they must be younger; the radiometric dating (see pp.182–183) of rock samples confirms this. The maria are basalt, a type of lava originating in the mantle. They formed when volcanic eruptions sent sheets of lava flooding into low-lying basins (see pp.14–15). Like basalt on Earth, lunar basalt is rich in iron and magnesium, but it has fewer volatile elements and no water. Most basalts lie hundreds of feet thick on the original basin floor, but in places they are 2.5 miles (4 km) deep.

Highland mountains
The lunar highlands include the mountain ranges along the rims of large impact basins. The peaks shown below, which rise to a height of 23,000 ft (7,000 m), are on the edge of the South Pole–Aitken basin.

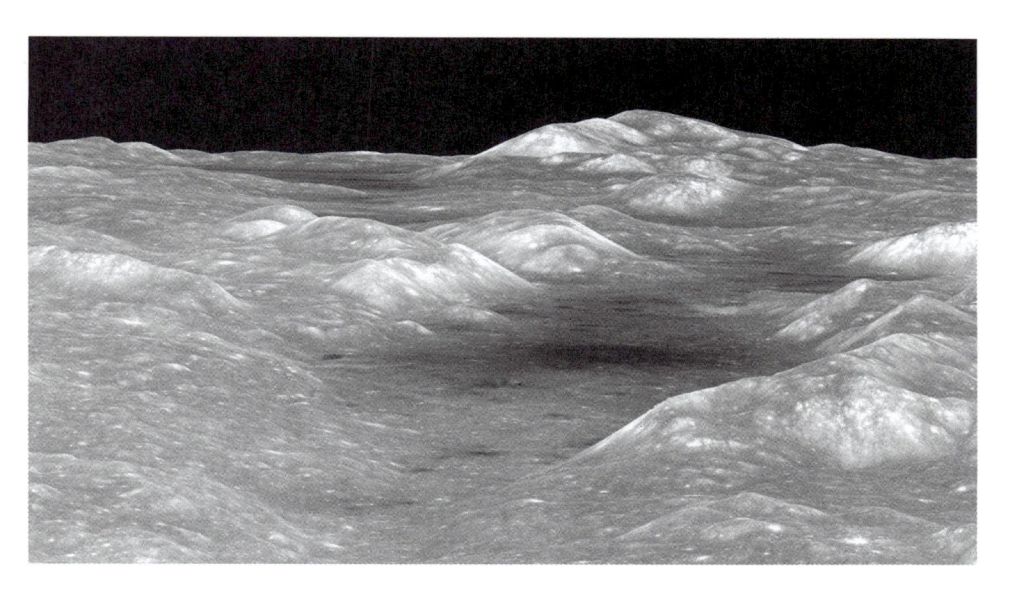

FLOOR-FRACTURED CRATERS

Mare basalts are found across the floors of the Moon's large basins, where the impact of an asteroid was powerful enough to fracture the lunar crust. Crustal fracturing also occurred on a smaller scale in a number of large craters, leading to some mare deposits in highland regions. These floor-fractured craters are themselves cut by fissures shaped like polygons or concentric circles, showing where later volcanic activity uplifted blocks of the crater floor and flooded some of them with mare lava flows (see pp.124–125).

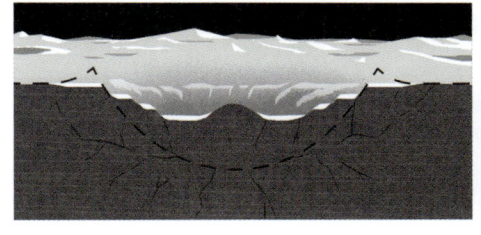

1. Crater excavation
The initial impact excavates a deep, bowl-shaped cavity that penetrates a significant way through the crust and fractures the rock below. As the impact decompresses, the floor rebounds.

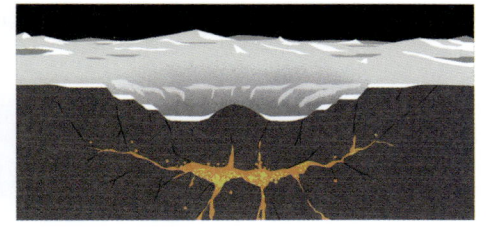

2. Deep fractures
Some of the deep crustal fractures breach a magma chamber, and molten rock begins to intrude into fissures beneath the floor of the crater, raising the level of the floor.

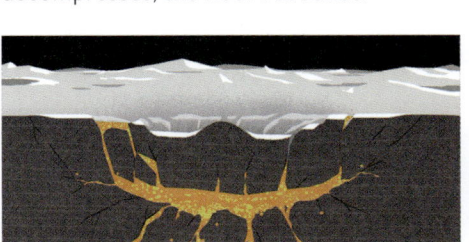

3. Lava flows
Pressure from the magma intrusion lifts the floor further, opening up faults between rising blocks that provide routes for magma to reach the surface, where it erupts as lava.

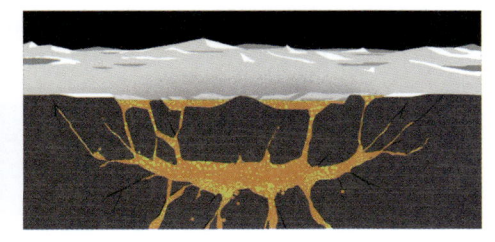

4. Lava sets
Multiple eruptions at fractures across the floor and around the crater edge flood parts of the crater floor with lava. The layers of lava cool and solidify, forming a plain of basalt rock.

Lunar Animals and Other Objects is an illustration from a series of six articles published in *The Sun* (a New York newspaper) in 1835 concerning the latest theories about life on the Moon. Although a hoax, it was taken seriously at the time.

PHOTOGRAPHING THE MOON

In 1839, the French artist Louis Daguerre (1787–1851) introduced photography to the world, and within a year, the Moon had become one of its most famous subjects. Although the pioneering 1840 photograph of the Moon by English scientist John William Draper (1811–1882) was blurred, it marked the beginning of a revolution in lunar cartography, one that swiftly gathered pace as Daguerre's original "daguerreotype" process was superseded by new photographic techniques. Gone were the days when astronomers had to be draftsmen to share their work with the world (see pp.110–111).

By the 20th century, astronomers were taking detailed photographs of the lunar surface, and most of the near side had been published in photographic atlases. The next revolution in lunar photography came toward the end of the millennium, with the advent of high-resolution digital imaging, which revealed the lunar landscape in stunning definition.

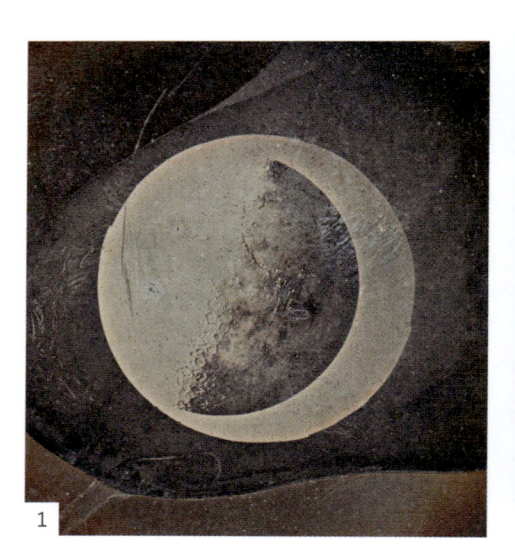

1. First known photograph
In 1840, using the daguerreotype process and a 5-inch (12.7-cm) reflector telescope, John Draper took the first lunar photograph from the rooftop observatory at New York University. With a 20-minute exposure, he captured the image on a silver-coated copper plate treated with mercury.

2. Lunar detail captured
In 1853, English geologist John Phillips took a more detailed view of the Moon using the new collodion wet-plate process. This involved treating a glass plate with photosensitive chemicals, exposing the plate, and then developing the image in a darkroom; the process took 15 minutes.

3. A clearer view
Aided by scientists at Harvard University Observatory, photographers John Whipple and James Black captured the Moon with the collodion wet-plate process, making negatives from which they produced prints on salted paper. This 1857 example shows the Moon in unprecedented clarity.

4. Contrasting terrain

American doctor and amateur astronomer Henry Draper, son of John (see opposite), built an observatory at the family property in New York. This albumen silver print mounted on glass was made in 1863. It shows the great contrast between the dark lunar plains and lighter highlands.

5. Moon atlas

In 1960, the Dutch-American astronomer Gerard Kuiper published the last great lunar atlas composed of Earth-based photographs. While the images it featured were the most detailed at the time, many had been taken more than 40 years earlier, including this one from 1919.

6. Composite photograph

Astrophotographer Andrew McCarthy's 1.3-gigapixel "GigaMoon," made in 2023 from 280,000 individual images, is one of the most detailed Moon pictures ever taken. The images were captured using a 12-inch (30.5-cm) telescope and a full-frame digital camera, then aligned using Photoshop.

VOLCANIC FEATURES

The origins of the eruptions that created the lunar maria are mostly hidden under the vast lava flows themselves. But other features—such as domes, pits, and rilles—testify to the Moon's volcanic past.

Most lunar craters were formed by high-energy impacts, but a few appear to be of volcanic origin. In a few areas, there are groups of broad domes with small craters at their summit. These craters are the mouths of shield volcanoes, created by erupting lava flowing out in all directions and building up a shieldlike mound. Other small craters resemble the cinder cones found on Earth: They were created when explosive eruptions produced fountains of lava droplets that rapidly crystallized and fell to the surface, piling up into a steep cone.

Volcanic glass beads have been identified at most of the Apollo landing sites, suggesting that lava-fountain eruptions were widespread on the Moon. (These droplets are distinct from the glass beads that have been generated by impactors striking the Moon and that are found throughout lunar soil.) Some have a similar composition to mare basalt rock.

While the Apollo glass samples are 3.7 billion years old, the volcanic beads returned by China's Chang'e 5 lunar probe in 2020 were created only 125 million years ago.

Flow features

The Moon has several features created by lava flowing across its surface—notably the vast, smooth maria plains (see pp.118–119). A rille is a valley on a planetary body. Linear rilles are the result of parallel faults in the crust creating a rift valley, but sinuous rilles were formed by surface lava flows or by the collapse of lava tubes (underground channels made by lava flows). Pits in the mare surface show where the roof of a drained lava tube or void has fallen in. In future, astronauts could use caves in these pits to shelter from solar radiation, micrometeoroids, and the great temperature fluctuations from lunar day to lunar night.

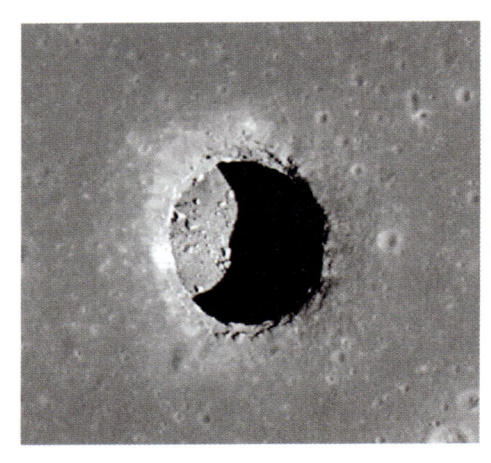

Mare pit
This 328-ft- (100-m-) wide collapse pit on Mare Tranquillitatis (the Sea of Tranquility) has a boulder-strewn floor and may house a large cave sheltered by overhanging rock.

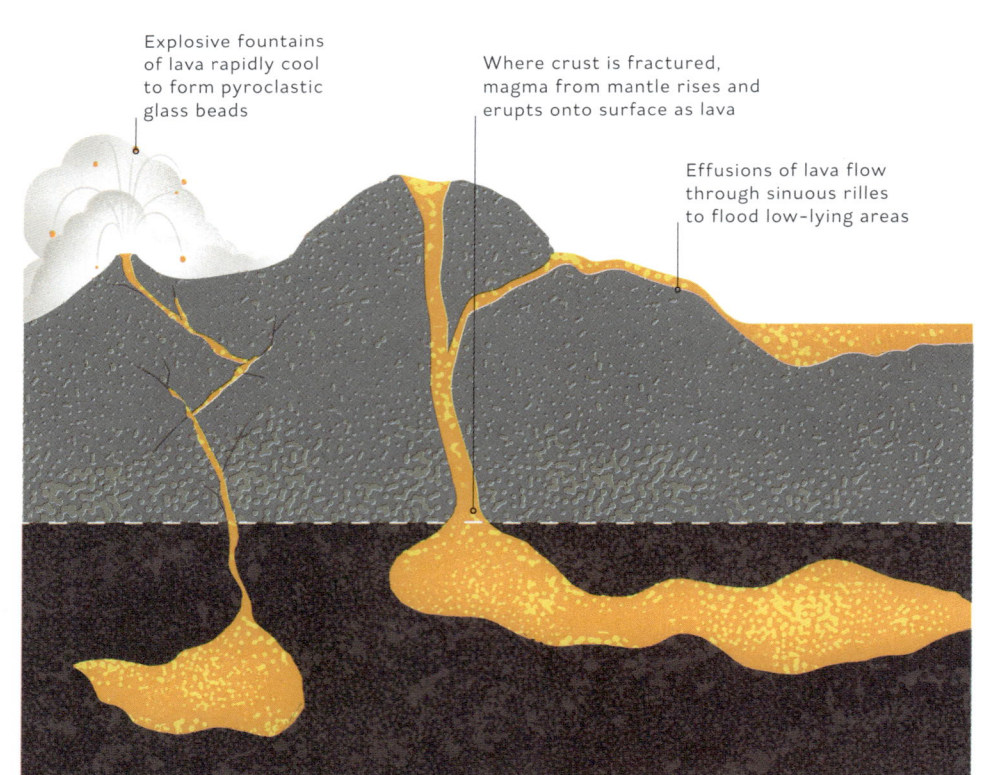

Explosive fountains of lava rapidly cool to form pyroclastic glass beads

Where crust is fractured, magma from mantle rises and erupts onto surface as lava

Effusions of lava flow through sinuous rilles to flood low-lying areas

Broad domes
Six shield volcanoes rise to the north of the Hortensius crater in Mare Insularum (the Sea of Islands). They are 5 miles (8 km) across and 1,300 ft (400 m) high.

Lunar volcanoes
The Moon has witnessed countless explosive and effusive volcanic eruptions. Explosive eruptions throw fountains of lava high above the surface; effusive eruptions send lava flowing across great distances.

SINUOUS RILLES

Formed by flowing lava, the meandering course of these channels is reminiscent of a riverbed. They may stretch for hundreds of miles across the lunar surface—much farther than similar features on Earth. Some cut across earlier lava flows on the maria; others twist across highland terrain on their way to lower ground. Many, such as Schröter's Valley, have a volcanic pit at their origin.

Schröter's Valley

Threading its way 100 miles (160 km) across the Aristarchus Plateau, from a pit called the Cobra's Head down to Oceanus Procellarum, Schröter's Valley is a broad, winding highland rille. A second, narrower rille snakes along the valley floor.

Aristarchus crater

Aristarchus Plateau

Cobra's Head

Herodotus crater

Schröter's Valley

Oceanus Procellarum (Ocean of Storms)

Inner rille

DESTINATION MOON

From 1947 until 1991, the United States and the Soviet Union were engaged in a period of intense rivalry: the Cold War. As an arms race developed between them, each side produced increasingly more sophisticated rockets designed to propel increasingly more destructive warheads at the other. In the 1950s, these rockets were powerful enough to enter Earth's orbit; by the 1960s, they were used to send the first crewed missions to the Moon.

REACHING FOR SPACE

Rocketry came of age in the early 20th century, when engineers demonstrated that space travel—and even lunar exploration—were no longer fantasies, but genuine technological possibilities.

Made in China in the 13th century, the earliest-known rockets were simple gunpowder-filled bamboo tubes, which the Chinese used both as fireworks (for entertainment) and to sow chaos on the battlefield. Over the following centuries, these rockets spread across Asia and Europe and evolved into increasingly more powerful and precise weapons of war.

Theory and practice

In later centuries, writers occasionally imagined rockets being used as space-faring vehicles—the most famous example being French novelist Cyrano de Bergerac in *The Other World: Comical History of the States and Empires of the Moon* (1657). However, the true potential of rockets only became clear in the 19th century, when early balloon experiments showed that Earth's atmosphere grows thinner with altitude and that traveling to other worlds involves crossing an empty void. Several scientists independently realized that rockets were uniquely suited to the task.

One of the pioneers of modern rocketry was Russian schoolteacher Konstantin Tsiolkovsky (1857–1935), who, like many, was inspired by French science fiction writer Jules Verne (1828–1905). From the 1880s, Tsiolkovsky investigated what he called "reaction devices" and published his findings

Soviet icon
As illustrated by this 1957 poster, Konstantin Tsiolkovsky was a popular figure in the Soviet Union due to his theoretical work on rockets, jet propulsion, and space exploration.

in *Exploration of Outer Space by Means of Reaction Devices* (1903). The book contained many revolutionary ideas, including that liquid hydrogen and liquid oxygen would make powerful propellants and that rockets could have multiple stages. He even speculated on the practicalities of spacecraft design.

Rockets take flight

Meanwhile, scientists and engineers elsewhere were thinking along similar lines. In Germany, physicist Hermann Oberth (1894–1989) published *The Rocket into Interplanetary Space* (1923). Three years later, US engineer Robert Goddard (1882–1945) accomplished the first flight of a liquid-fueled rocket in Auburn, Massachusetts.

In 1930s Germany, however, enthusiastic rocket engineers were recruited for military projects by the new Nazi government. Within a decade, German rocketry became the most advanced in the world, culminating late in World War II with deployment of the V-2 missile. This terrible weapon could deliver explosive warheads across hundreds of kilometres, but could also reach the edge of space.

PRINCIPLES OF ROCKET TECHNOLOGY

Rockets harness the fact that every action produces an equal and opposite reaction. The expulsion of exhaust by burning propellant is the action, and the movement of the rocket in the opposite direction is the reaction—a principle that holds both in Earth's atmosphere and in the vacuum of space. Releasing air from a balloon shows the same principle (see right).

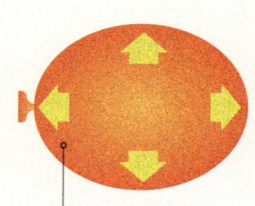

In sealed balloon, gas pushes out in all directions against skin

When nozzle is opened, contracting skin pushes air out

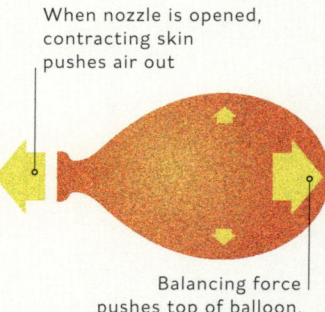

Balancing force pushes top of balloon, driving it forwards

Rocket revolution

The V-2 missile featured powerful engines that burned an ethanol/water mix with liquid oxygen, alongside a gyroscopic guidance system with steering vanes to deflect the rocket exhaust. On June 20, 1944, test launch MW 18014 became the first human object to reach space.

First flight

Robert Goddard with his first successful liquid-fueled rocket, which reached a height of 41 ft (12.5 m) during a 2.5-second flight on March 16, 1926.

THE SILVER SCREEN

The whimsical portrayal by Georges Méliès (1861–1938) of the Man in the Moon in 1902 may be cinema's first truly iconic image. But by the 1920s, modernist directors had developed a more technologically minded vision of lunar exploration, pairing thoughtful science fiction with melodramatic plots.

Spaceflight had become a more plausible prospect by the 1950s. Some filmmakers strove to depict the realities of such missions but struggled to compete with the era's fantastical pulp sci-fi "B movies." The rival approaches were synthesized in *2001: A Space Odyssey* (1968), which depicted a realistic lunar colony months before the first crewed Apollo flights.

With *Star Wars* (1977), sci-fi became a major blockbuster genre, immediately spawning a host of "science fantasy" films that veered away from realistic depictions of spaceflight. A more realistic subgenre also developed, featuring films based either on historical events, such as *Apollo 13* (1995), or on imaginary but plausible scenarios, such as *Gravity* (2013).

1. *Le Voyage dans la Lune* (*A Trip to the Moon*; 1902)
French cinematic pioneer Georges Méliès borrowed from both Jules Verne and H. G. Wells for his tale of astronomers who travel by a bulletlike spacecraft to the Moon. Méliès himself played the Man in the Moon, assailed by unwelcome visitors.

2. *Frau Im Mond* (*Woman in the Moon*; 1929)
Fritz Lang's silent classic portrays a lunar expedition beset by greed and envy. Lang consulted space enthusiasts to create a credible vision of spaceflight. The prelaunch countdown, one of his dramatic innovations, was later adopted for real space missions.

3. *Cosmic Voyage* (1936)
To promote public interest in spaceflight, Russian director Vasili Zhuravlov was authorized to create a scientifically grounded tale of rivalry during a Moon mission. Soviet authorities deemed the completed movie too ideologically unsound for widespread circulation, however.

4. *Destination Moon* (1950)

Irving Pichel's tale of a privately financed first lunar expedition combines Technicolor visuals (with matte paintings by renowned space artist Chesley Bonestell) and realistic thrills. Hazards the crew must overcome include being set adrift on a spacewalk and becoming stranded on the Moon.

5. *2001: A Space Odyssey* (1968)

In Stanley Kubrick and Arthur C. Clarke's epic meditation on the past and future of humanity, lunar colonists exploring Tycho crater unearth an ancient and mysterious alien monolith. It has remained unknown until human intelligence advances enough to master space travel and discover it.

6. *Apollo 13* (1995)

Ron Howard's gripping drama recounts this troubled Apollo mission with documentary realism. Hardware and settings were carefully recreated, actors underwent rigorous technical instruction, and some scenes were filmed in weightless conditions on NASA training aircraft.

A RELATIVE PROOF

In 1919, the Moon played a key role in demonstrating that one of the century's strangest ideas—Albert Einstein's general theory of relativity—had a firm basis in fact.

The crucial moment
An image of the 1919 solar eclipse shows several stars of the Hyades cluster in Taurus. The apparent positions of the stars are displaced by the Sun's gravity.

HR1403

HR1375

67 Tauri 65 Tauri 69 Tauri

Published in 1905, Albert Einstein's (1879–1955) special theory of relativity offered a radical solution to a long-standing problem: the fact that light always travels at the same speed, regardless of how its source or observer is moving. To explain this, Einstein argued that the speed of light is a constant—a kind of cosmic speed limit—and that, because of this, when objects accelerate close to the speed of light, measurements of space and time become distorted. Einstein went on to develop these ideas, and in 1915 published his general theory of relativity. According to this thesis, matter, particularly large objects such as the Moon, bends space and time around it—or rather, bends the fabric of a fourth-dimensional "spacetime" (see opposite)—which, among other things, creates gravity.

Bending light

From the outset, general relativity made predictions that differed from traditional physics, allowing scientists to test its claims. Given the technology of the day, the most obvious test was to see if the light of a distant object, such as a star, was deflected by a closer, massive object, such as the Sun. However, in spite of its size, the Sun only deflects light that passes very close to it, and those rays get lost in its glare. The only option was to study a

> "**One thing is certain** and the rest debate. **Light rays**, when near the Sun, **do not go straight.**"
> Arthur Stanley Eddington, Royal Astronomy Society dinner speech, 1920

total solar eclipse (see pp. 38–39) and measure the positions of the background stars revealed in that moment.

An ambitious expedition

British astrophysicist Arthur Stanley Eddington (1882–1944) collaborated with Astronomer Royal Frank Dyson (1868–1939) to plan a multinational expedition for the eclipse of May 29, 1919. One team, led by Andrew Crommelin (1865–1939), traveled to Sobral, in Brazil, while another, including Eddington himself, based itself at the other end of the eclipse's path, on the West African island of Príncipe. Each team used "astrographs" (telescopes with attached cameras), together with slowly turning mirrors known as "coelostats," to compensate for Earth's rotation and produce steady images of the Sun. The resulting photographs contained enough star images to determine their positions accurately—and to confirm Einstein's theory.

The result was not only a scientific triumph, it was a media sensation that catapulted Einstein to international renown. Today, this key effect of general relativity is known as "gravitational lensing."

The father of modern physics
Einstein delivers a lecture on his theories of relativity in 1921. The phenomena that he predicted, and their implications for our understanding of the universe, are still being explored today.

HOW GRAVITY BENDS LIGHT

According to general relativity, spacetime can be imagined as a grid of four dimensions (the three spatial dimensions, plus time). Large masses warp the spacetime around them, which affects the paths of objects that pass nearby. The distortion becomes a "gravitational well," which deflects light rays.

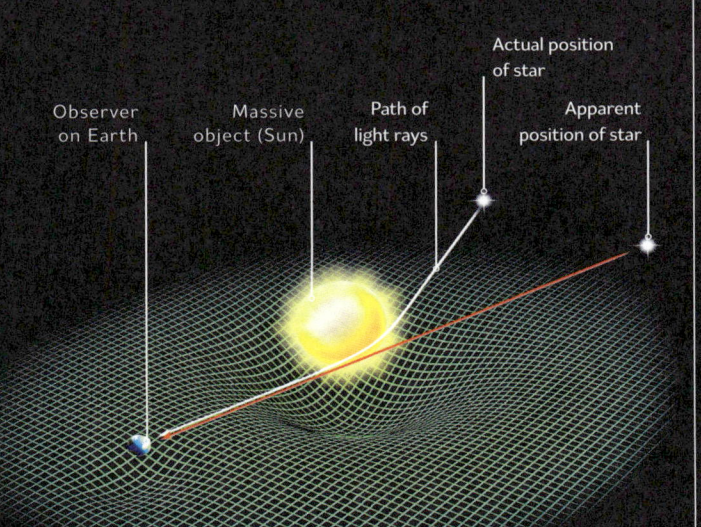

Observer on Earth · Massive object (Sun) · Path of light rays · Actual position of star · Apparent position of star

GRAVITATIONAL LENSES

Concentrations of mass far bigger than the Sun, such as entire galaxies and even galaxy clusters, create huge distortions of space. These act as gravitational lenses, bending and focusing the light of more distant objects behind them. One such lensing effect is an "Einstein ring," which occurs when the viewer, the lens, and the more distant object are perfectly aligned.

The relatively nearby galaxy LRG 3-757 (red sphere) acts as a gravitational lens to focus the light from a more distant galaxy (blue halo), which appears as an Einstein ring

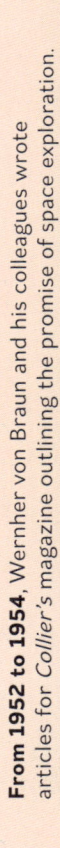

From **1952 to 1954,** Wernher von Braun and his colleagues wrote articles for *Collier's* magazine outlining the promise of space exploration.

Collier's

OCTOBER 18, 1952 • FIFTEEN CENTS

MAN ON THE MOON

Scientists Tell How
We Can Land There
In Our Lifetime

A NEW FRONTIER

The launch of the first artificial satellites in 1957 took the rivalry between the superpowers into orbit and beyond. From the very beginning, the Moon played a key role in this "space race."

Although its deployment came too late to affect the outcome of World War II, the devastating potential of the V-2 missile was clear (see pp.128–129), and in 1945, victorious wartime allies became rivals in a race for German rocket hardware and personnel. The US captured leading German scientists, including aerospace engineer Wernher von Braun (1912–1977), while the Soviet Union took control of key V-2 development centers.

While the Cold War drove postwar rocket development, many people working on the missile programs were spaceflight advocates. Von Braun's team adhered to the priorities of their US Army employers, but also promoted their vision of space travel. Likewise, in the Soviet Union, engineers Sergei Korolev (1907–1966) and Valentin Glushko (1908–1989) lobbied politicians to back space exploration.

Race to orbit

In 1955, the US announced plans to launch an artificial satellite in the forthcoming International Geophysical Year (1957–1958).

Sputnik 1
The world's first artificial satellite was a 23-in (58-cm) aluminum sphere that broadcast signals from orbit for three weeks until its batteries ran down.

Soviet leaders, too, were by now persuaded that space launches not only offered scientific benefits, but could demonstrate a country's technological superiority.

The launch of the Sputnik 1 satellite on October 4, 1957, shocked the world. A second Sputnik, carrying a dog named Laika, followed a month later, and American alarm turned to embarrassment when the Vanguard TV3 satellite blew up on the launch pad in early December. Von Braun—whose own team had been side-lined in order to make way for the navy's Vanguard program—seized the moment to resurrect his plan for a launch using a modified army missile, and a lightweight US satellite, Explorer 1, reached orbit on January 31, 1958.

Onward to the Moon

The US and Soviet Union were locked in a space race, and the Moon now became their mutual target. Reaching it was a challenge even for powerful Soviet rockets, however.

A new upper-stage booster failed on each of its first three launches, but January 2, 1959 saw the Sputnik-like Luna 1 finally hurtling toward the Moon. The spacecraft had been planned to crash onto the lunar surface, but it missed, flying past at a distance of about 3,730 miles (6,000 km).

Luna 2, which launched on September 12, achieved its intended crash landing, however, and the following month, Luna 3 would send back the first photographs of the far side of the Moon.

Explorer 1
Designer William H. Pickering, mission scientist James Van Allen, and Wernher von Braun display a model of Explorer 1. The two-month mission discovered Earth's "Van Allen" radiation belts.

FIRST SIGHT OF THE FAR SIDE

Returning the first images of the far side of the Moon was a huge technical challenge. Luna 3 carried a twin-lens camera that could be pointed by turning the entire spacecraft. Exposed film was then developed into stable negatives using chemicals in an on-board lab. A scanner then shone a light through the negatives, generating electrical signals that were transmitted to Earth.

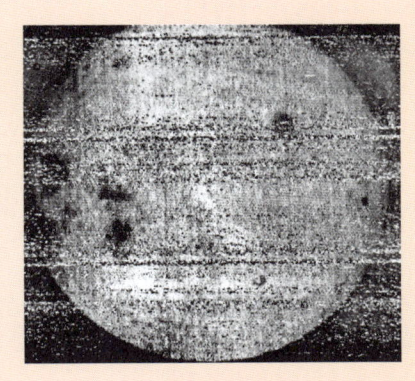

A raw Luna 3 image with distinctive scan lines

THE HIDDEN FACE

Early images of the far side of the Moon revealed a landscape very different from the face that we see. Missions in the decades since have shown that this difference is more than skin deep.

Since shortly after its formation, gravitational forces have ensured that one half of the Moon is turned toward Earth, while the other faces away. Today, the most striking distinction between these two hemispheres is the far side's lack of large maria (seas). Relatively smooth, dark maria cover approximately 30 percent of the Moon's near side, but only 1 percent of the far side.

The far-side terrain is dominated by bright, heavily cratered rock—similar to that of the near-side highlands, but not limited to higher elevations. The hemisphere's major impact basins, including the vast South Pole-Aitken (SPA) basin, mostly avoided the ancient volcanic flooding that formed the near-side maria. This difference was once explained as a result of the far side's thicker crust or the slight offset of the Moon's hot core toward Earth (see pp.198–99). Recent theories attribute it to contrasting chemical composition: the far side has much smaller quantities of the heat-producing KREEP minerals thought to be linked to lunar volcanism.

1 The view from orbit
Captured by Apollo 16 in 1972, this image reveals a lunar surface saturated with craters of all sizes, including ancient Kohlschütter (top left). The far side is generally more heavily cratered than the near side, but calculations show that the oldest areas of the near side have a similar crater density. Volcanic activity in several phases of early lunar history wiped out many near-side craters.

A WORLD OF TWO HALVES

Combining gravity data from NASA's GRAIL mission and topographic detail from its Lunar Reconnaissance Orbiter (LRO), these maps reveal major differences in crustal thickness. On average, the crust on the far side is 12.4 miles (20 km) thicker than the near-side crust.

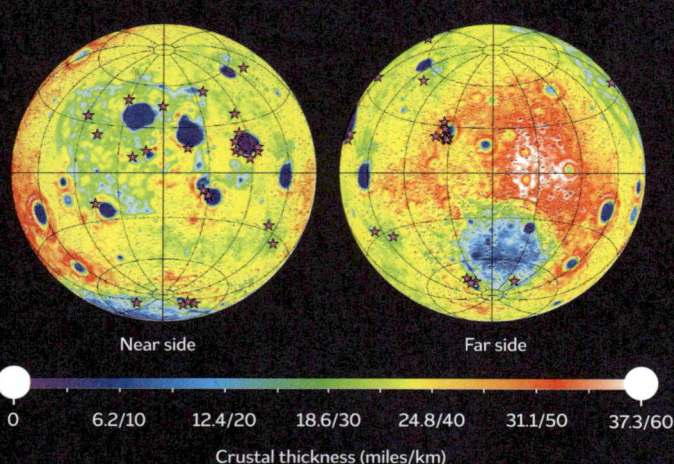

Near side   Far side

0	6.2/10	12.4/20	18.6/30	24.8/40	31.1/50	37.3/60

Crustal thickness (miles/km)

Near side
Crustal thickness on the near side is relatively uniform except for impact basins (blue).

Far side
The far side has regions with both the thickest and the thinnest lunar crust.

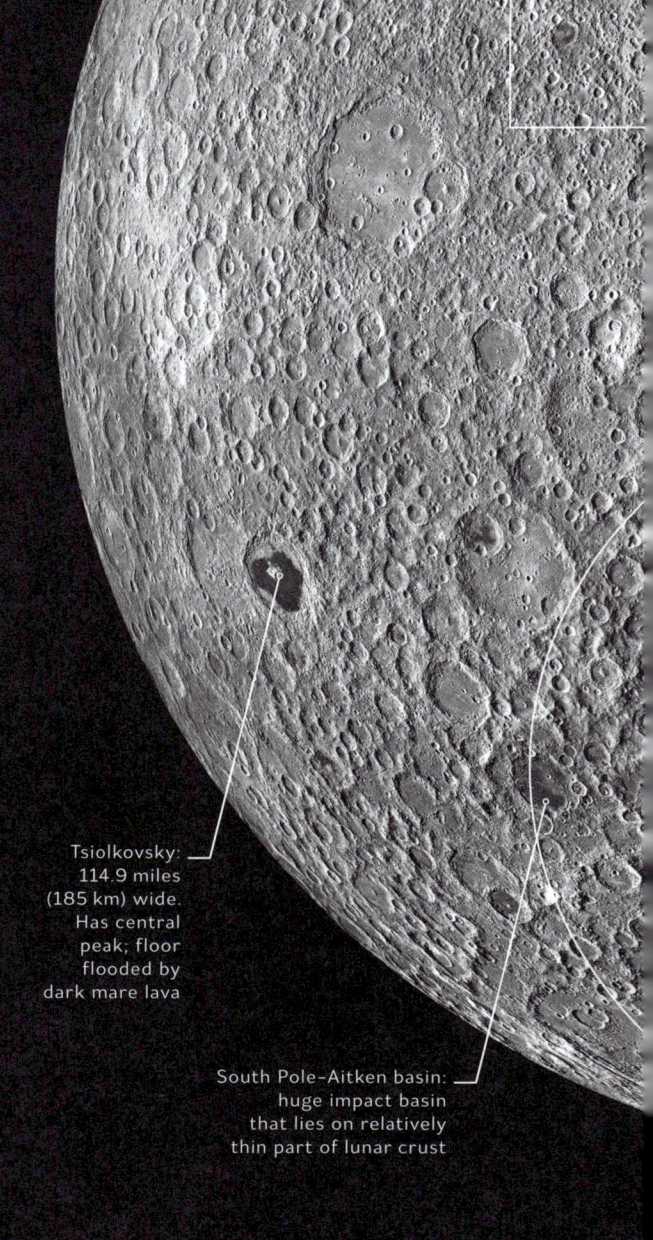
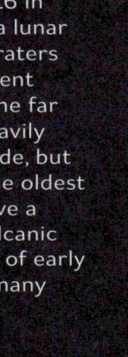

Mare Moscoviense (Sea of Moscow): 171 miles (276 km) wide: volcanic plain partially filling impact basin

1

Tsiolkovsky: 114.9 miles (185 km) wide. Has central peak; floor flooded by dark mare lava

South Pole–Aitken basin: huge impact basin that lies on relatively thin part of lunar crust

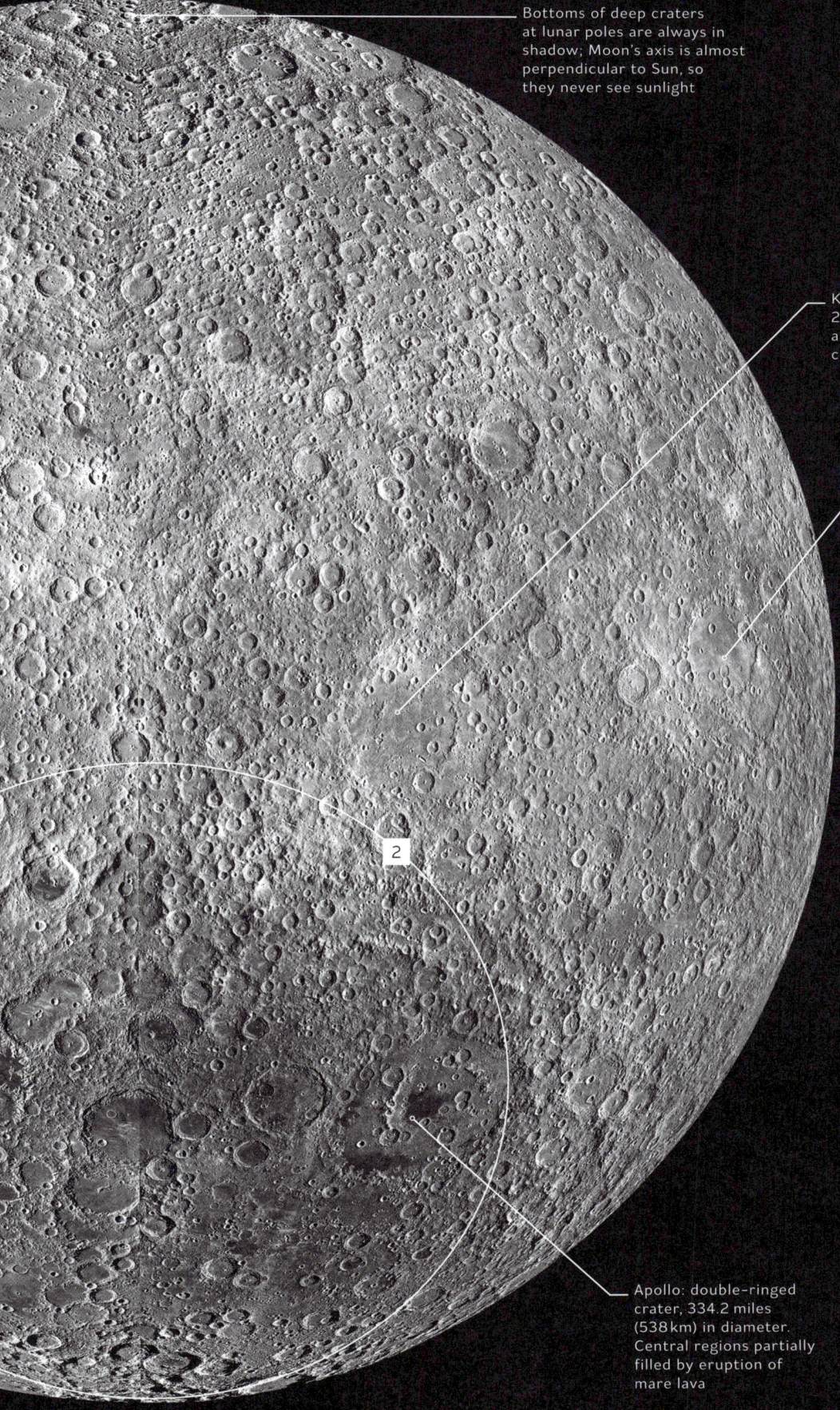

Bottoms of deep craters at lunar poles are always in shadow; Moon's axis is almost perpendicular to Sun, so they never see sunlight

Far side of the Moon
This map of the Moon's unseen hemisphere, compiled using data from NASA's Lunar Reconnaissance Orbiter, shows far less variation than is seen on the near side. Dark maria are almost entirely absent; the major brightness differences on the far side are due to ejecta from impact crater formation.

Korolev: impact basin around 272 miles (437 km) in diameter and pockmarked with many craters; outer rim is heavily eroded

Hertzsprung: 355-mile- (570-km-) wide impact basin of similar age and comparable size to basins of several near-side maria

Aitken crater marks SPA basin's northern rim

Apollo crater

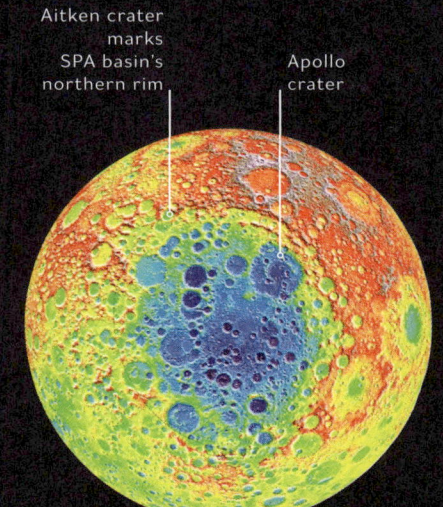

2

Apollo: double-ringed crater, 334.2 miles (538 km) in diameter. Central regions partially filled by eruption of mare lava

2   **South Pole–Aitken basin**
About 1,550 miles (2,500 km) across and more than 5 miles (8 km) deep, the SPA basin—shown in blue and violet—is the solar system's second largest confirmed impact structure. Its formation around 4.2–4.3 billion years ago may have influenced the history of the entire Moon; heat from the impact may have driven KREEP-rich minerals toward the opposite face, later encouraging volcanic activity.

> "**We choose** to go to **the Moon** in this decade and do the other things, **not** because **they are easy**, but because **they are hard**."
>
> John F. Kennedy, Rice University, September 12, 1962

President John F. Kennedy inspects NASA's Saturn I launch vehicle (a forerunner of the giant Saturn V) at Cape Canaveral, Florida, on 16 November 1963.

THE RACE FOR THE MOON

As the Soviet Union cemented its early lead in the space race by launching the first cosmonauts into orbit, the US responded by announcing an even more challenging goal: a landing on the Moon.

The first phase of the space age was dominated by the launch of automated satellites and space probes. Both the US—through its newly founded space agency, National Aeronautics and Space Administration (NASA)—and the Soviet Union were also busy developing crewed spacecraft, however.

First into orbit

NASA's Mercury program progressed in the glare of publicity, but the spacecraft's size and weight were restricted by the relatively low power of US rockets. In contrast, Soviet launch vehicles, derived from the enormous R-7 Semyorka missile, possessed an abundance of thrust to launch heavy payloads into orbit. It was Russian cosmonaut Yuri Gagarin who became the first human in space on April 12, 1961, completing a single orbit of Earth aboard the 5.2-ton (4.7-tonne) Vostok 1. NASA's Mercury-Redstone 3 launched less than four weeks later, but its 2-ton (1.8-tonne) Freedom 7 spacecraft could only carry astronaut Alan Shepard on a brief "suborbital" flight into space.

Moonshot

US politicians and engineers now devised a daring plan: setting a goal far beyond the existing capabilities of either side that might, with sustained political, scientific, and economic effort, secure the US a lasting victory. On May 25, 1961, President John F. Kennedy called for a programme that would place humans on the Moon's surface by the end of the decade.

Almost all other US space efforts were diverted to support Project Apollo. These included fast-tracking an intermediate two-seat spacecraft (Project Gemini) to gain experience of more complex and longer spaceflights, and building the enormous rocket needed to reach the Moon. Automated spacecraft would also be used, to learn about lunar conditions and pave the way for the Apollo landing.

First man in space
Vostok 1 carried Yuri Gagarin on a single 108-minute orbit. After the spacecraft reentered Earth's atmosphere, an automatic ejector seat fired and Gagarin parachuted the final 4.3 miles (7 km) to Earth.

WHICH WAY TO THE MOON?

A key debate early in Apollo planning was about the mission profile: how best to reach the Moon. Each option had different benefits and drawbacks. "Direct ascent" took a heavy spacecraft to the Moon and back on a relatively simple mission, but required a truly enormous rocket. Earth–Orbit Rendezvous (EOR) would use multiple launches by smaller rockets, but demanded numerous complex maneuvers in space. The Lunar–Orbit Rendezvous (LOR) approach that was ultimately chosen offered a compromise, balancing rocket power and payload size against mission complexity.

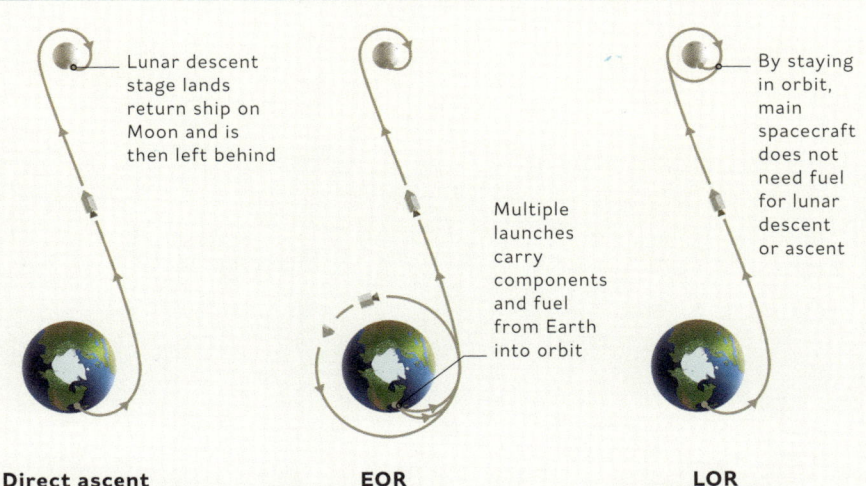

Lunar descent stage lands return ship on Moon and is then left behind

Multiple launches carry components and fuel from Earth into orbit

By staying in orbit, main spacecraft does not need fuel for lunar descent or ascent

Direct ascent
The heavy spacecraft flies directly to and from the lunar surface without entering lunar orbit.

EOR
The heavy spacecraft is assembled in Earth orbit before the "direct ascent" flight to the Moon.

LOR
Launched atop a rocket, the main spacecraft orbits the Moon. The module detaches for lunar landing.

SPACE-RACE PROPAGANDA

In the wake of World War II, the communist Soviet Union and the democratic capitalist United States vied with each other to extend their global influence and promote their own ideology. Both sides wielded nuclear weapons, so direct confrontation had to be avoided. Instead, the Cold War became the new status quo: an uneasy rivalry that sometimes involved conflict via proxies, but more often played out in political maneuvering and propaganda.

The space race became a vital Cold War propaganda tool for both sides, since victories on this new front demonstrated technical superiority over their adversary. Space-race triumphs played a key role at home, too, inspiring national pride and dreams of a bright technological future. Vast expenditure on space programs was justified on the grounds that selling optimistic visions of space travel was vital to achieving political goals and sustaining public support. As a result, space-race propaganda infiltrated many areas of daily life.

1. Apollo toy
From astronaut figures to model spacecraft, toys inspired a fascination with space in children, as well as encouraging a lasting interest in space exploration among the broader population.

2. Poster
"In the name of peace and progress!" proclaims this Soviet poster. It salutes Luna 3, seen here soaring past the Moon, as a worker raises the hammer-and-sickle emblem in a laurel wreath.

3. Pennant
More often associated with American sports teams, pennants were also used to support or celebrate notable causes or events. This one commemorates the Apollo 11 Moon landing in 1969.

4. Heroic sculpture
Moscow's *Monument to the Conquerors of Space* (1964) is a rocket-topped titanium obelisk. A frieze along its base memorializes Soviet space milestones; Yuri Gagarin, the first man in space, is at left.

5. Badge
Space-themed pin badges, such as this one celebrating Luna 9's soft landing on the Moon, were collectible items in the Soviet Union. They offered a simple way of showing support for the Soviet space effort.

6. Art
In 1962, NASA began enlisting celebrated American artists to record key US events in space exploration. *Grissom and Young* (1965) by Norman Rockwell shows the Gemini 3 crew preparing for launch.

7. Postcard
Space found its way onto all kinds of printed ephemera. This postcard commemorating Luna 3's 1959 flight past the far side of the Moon was an affordable souvenir of a historic national event.

8. Stamp
Postage stamps announced success in space to the wider world. This 1969 stamp bearing Paul Calle's artwork of the Apollo 11 landing was printed with a steel die that the astronauts had carried to the Moon and back.

Photographing the lunar surface
As it hurtled toward its impact with the northern edge of Mare Nubium, Ranger 7 returned 4,308 images of the lunar surface.

Guericke crater

Mare Cognitum (the Known Sea)

Bonpland G crater

Altitude 1,311 miles (2,110 km)

Alphonsus crater

Altitude 588 miles (946 km)

Altitude 289 miles (465 km)

Altitude 140 miles (225 km)

EARLY INVESTIGATIONS

In the 1960s, the US and the Soviet Union sent a series of satellites and robot landers to the Moon. Their purpose was to pave the way for crewed expeditions to explore the lunar surface.

Luna 9
The ball-shaped Luna 9 used retrorockets and airbags to make the first soft landing on the Moon. It arrived at Oceanus Procellarum (the Ocean of Storms) on February 3, 1966.

When the Apollo program was announced, only a handful of probes had flown close to or hit the Moon, and these had produced only limited images and scientific results. Throughout the 1960s, therefore, NASA launched three new series of robot spacecraft to learn more about the lunar surface. Meanwhile, the Soviet Union continued its exploratory Luna program (see pp.134–35).

From hard to soft landing

The American probes swiftly became more sophisticated. The "Ranger" program was designed to send probes to hard land on the Moon and beam back television pictures up to the point of impact. After a series of failures, in July 1964, Ranger 7 finally delivered the first close-up images of the Moon.

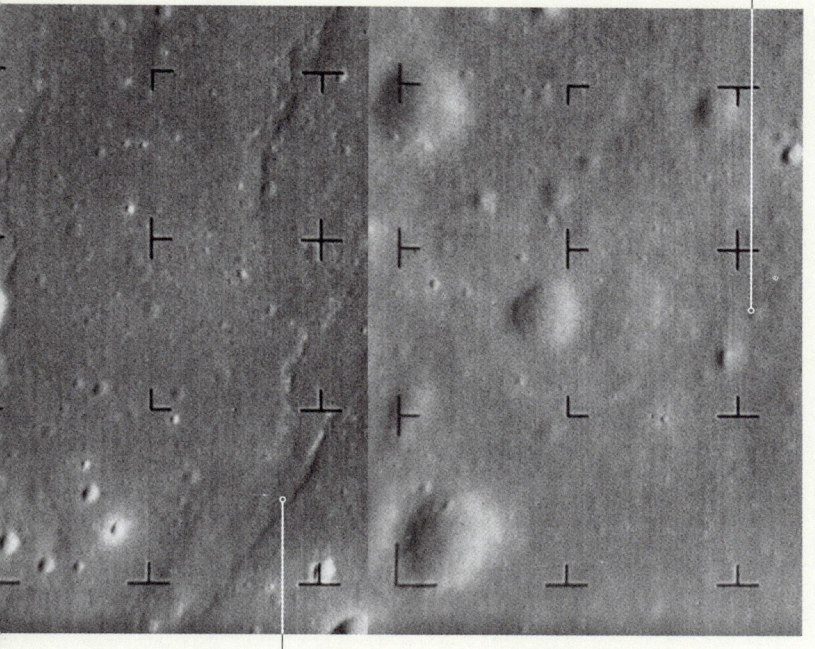

Smallest craters c. 33 ft (c. 10 m) diameter

Altitude 59.4 miles (95.5 km)

Wrinkle ridge in lava plain (see pp.170–171)

Altitude 7.4 miles (11.9 km)

The extraordinary cratering revealed by the images seemed at first to threaten the safety of crewed landings, but the subsequent program, designated "Surveyor," proved that such landings were possible. They showed that rocket thrusters could be used to adjust a lander's trajectory and to slow its descent. Improving on Ranger's record, five out of seven Surveyors were successful, beginning with the first in June 1966.

Scanning the surface

NASA's final series of robotic missions were the Lunar Orbiters. Over roughly a year, from August 1966, five of these spacecraft circled the Moon. They photographed 98 percent of the lunar surface, including 20 potential Apollo landing sites. However, neither Surveyor 1 nor Lunar Orbiter 1 were space-race "firsts": Lunas 9 and 10 made the first soft landing and the first entry to lunar orbit respectively a few months ahead of their American counterparts. However, as it turned out, these were to be some of the last Soviet victories in the race to the Moon.

LUNAR SURVEYOR

NASA's Lunar Surveyor probes traveled directly from Earth to the Moon without entering lunar orbit. They then used retrorockets to slow their descent before dropping the final few feet to the surface. Their landings confirmed that the Moon's soil was stable enough to support the landing pads of future visiting spacecraft.

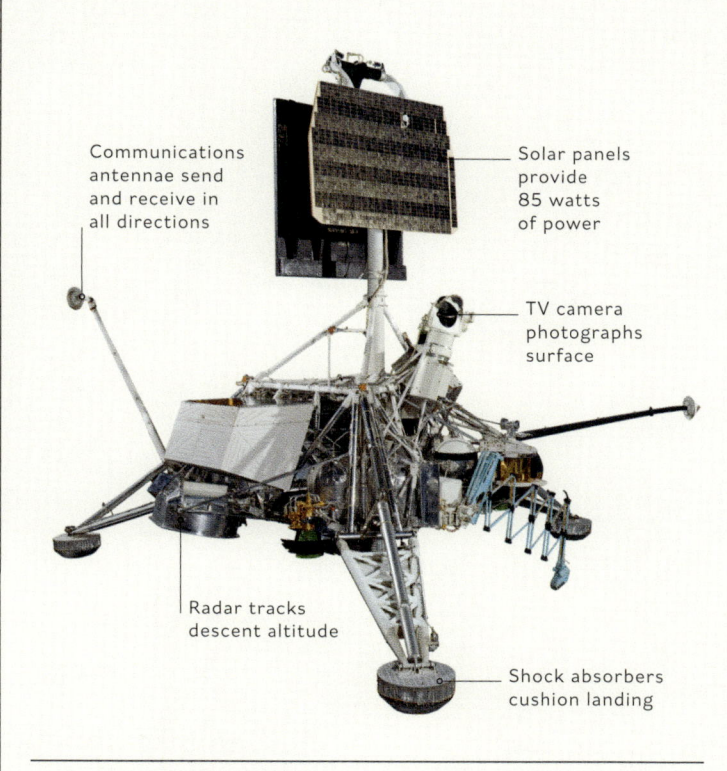

Communications antennae send and receive in all directions

Solar panels provide 85 watts of power

TV camera photographs surface

Radar tracks descent altitude

Shock absorbers cushion landing

MAPPING THE MOON

In the early 1960s, astronomers and geologists made great efforts to map the Moon to improve their understanding of lunar geology. The maps they made played a key role in identifying scientifically interesting landing sites for both robot probes and crewed missions.

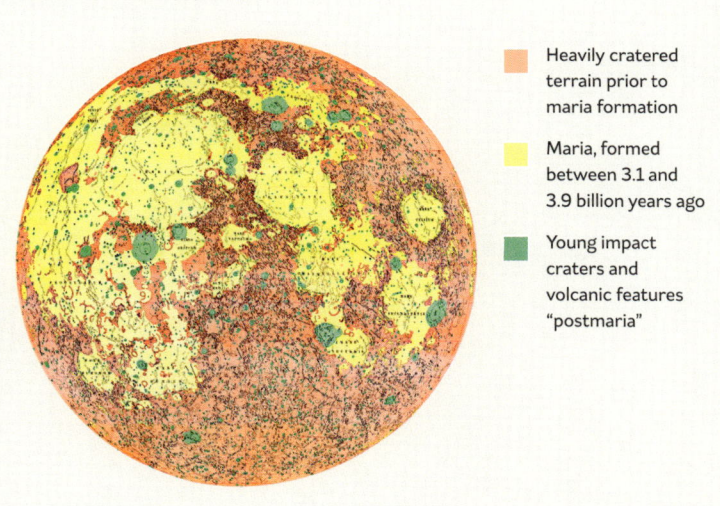

Heavily cratered terrain prior to maria formation

Maria, formed between 3.1 and 3.9 billion years ago

Young impact craters and volcanic features "postmaria"

US Geological Survey lunar map of 1961

APOLLO TECHNOLOGY

The Apollo moonshot required an enormous rocket, more powerful than any before, and new spacecraft to deliver three astronauts to the Moon and carry them safely home.

The height of a 36-story building, the colossal Saturn V launch vehicle was a three-stage rocket. The first two stages' engines burned until they were out of fuel and were jettisoned. The third stage engines ignited to put the spacecraft in Earth orbit, and reignited to send it Moonwards. Atop the third stage was the Apollo spacecraft, comprising the Command and Service Module (CSM) and Lunar Module (LM). The CSM consisted of the Command Module—the crew's quarters—and Service Module, which housed propulsion systems and supplied power, oxygen for breathing, and drinking water. The LM took astronauts from lunar orbit to the surface and back. Once Apollo was on a lunar trajectory, the CSM separated, turned 180°, and docked with the LM. After discarding the third Saturn V stage, Apollo continued on to the Moon.

SATURN V

The first stage was the most powerful, since it had to lift the fully fueled Saturn V off the ground. The second stage took the rocket almost into orbit. The third stage put Apollo into Low Earth Orbit (LEO) and then propelled it toward the Moon.

ROCKET ENGINES

Saturn V used two different Rocketdyne engines. The first stage's F-1 engines burned RP-1 (a refined kerosene) with liquid oxygen. The upper stage's J-2 engines burned liquid hydrogen with liquid oxygen, generating more energy and thrust per pound.

Rocketdyne engines
Both engines worked on the same principle: burning propellants (fuel and oxidizer) in a combustion chamber and ejecting the exhaust at supersonic speeds to push the rocket in the opposite direction.

Escape rocket could carry away astronauts to safety in a launch emergency

Apollo CSM

Fairing (streamlined cover) around LM for protection during launch

Third stage: 1 x Rocketdyne J-2 engine

Interstage adapter links second and third stages

SATURN V DATA

Height
363 ft (111 m)

Diameter
33 ft (10 m)

Mass
6.2 million lb (2.8 million kg)

Payload to Earth orbit (CSM, LM, and fuel)
311,000 lb (141,000 kg)

Rocket burns
First stage: 168s
Second stage: 360s
Third stage to LEO:165s
Third stage translunar: 335s

Second stage: 5 x Rocketdyne J-2 engines

Interstage adapter links first and second stages

First stage: 5 x Rocketdyne F-1 engines

F-1 engine nozzles

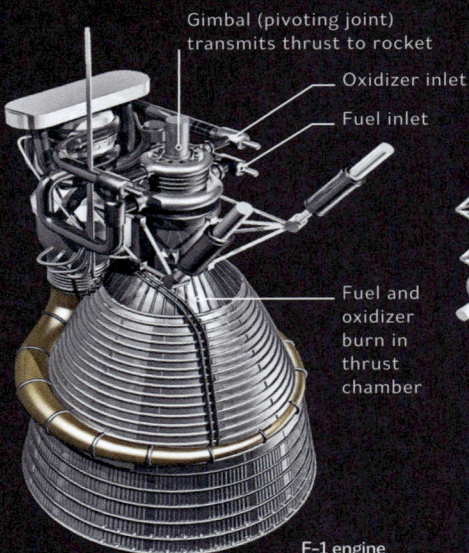

Gimbal (pivoting joint) transmits thrust to rocket

Oxidizer inlet

Fuel inlet

Fuel and oxidizer burn in thrust chamber

F-1 engine

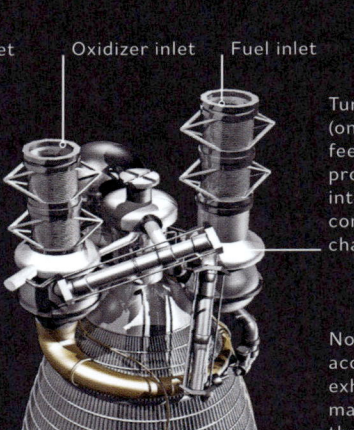

Oxidizer inlet

Fuel inlet

Turbopump (one of two) feeds propellant into main combustion chamber

Nozzle accelerates exhaust to maximize thrust

J-2 engine

COMMAND AND SERVICE MODULE

The Command Module was a conical crew capsule with a side entry/exit hatch, a forward docking tunnel, and a heat shield to withstand reentry to Earth's atmosphere. Until just before reentry, the Command Module remained attached to the cylindrical Service Module, which supplied it with power and carried fuel tanks and a rocket engine for entering and leaving lunar orbit.

In-flight control
In addition to its main engine, the CSM had four sets of Reaction Control System (RCS) thrusters, which helped it maneuver in space and make minor midcourse velocity corrections.

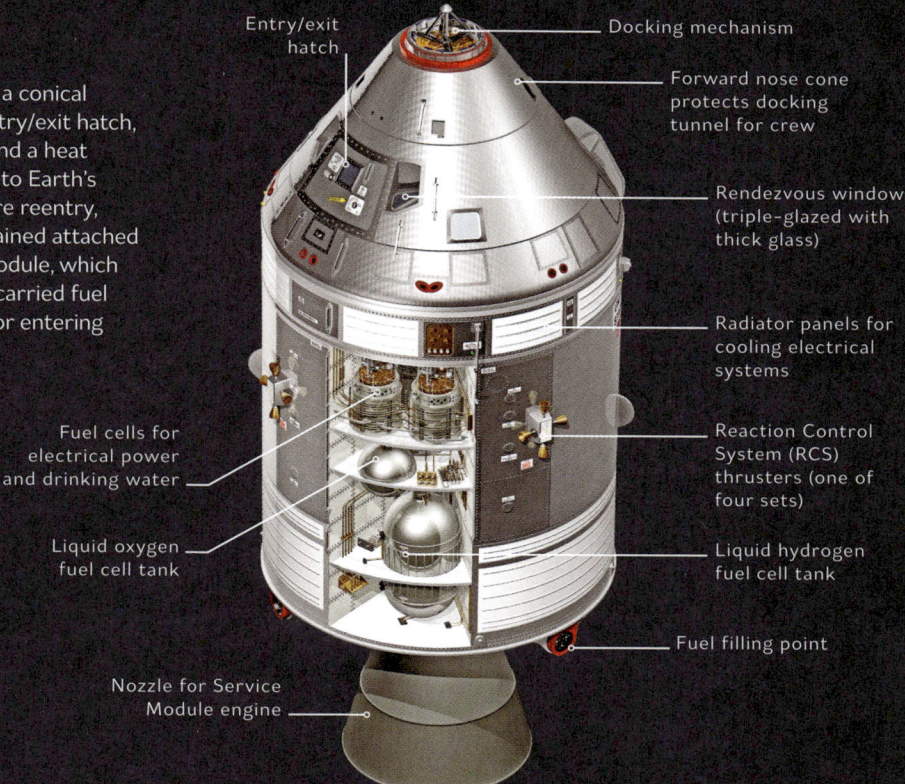

Entry/exit hatch

Docking mechanism

Forward nose cone protects docking tunnel for crew

Rendezvous window (triple-glazed with thick glass)

Radiator panels for cooling electrical systems

Reaction Control System (RCS) thrusters (one of four sets)

Fuel cells for electrical power and drinking water

Liquid oxygen fuel cell tank

Liquid hydrogen fuel cell tank

Fuel filling point

Nozzle for Service Module engine

CSM DATA

Height
36.2 ft (11m)

Diameter
12.8 ft (3.9 m)

Launch mass
63,500 lb (28,800 kg)

Propellants
Aerozine 50
Dinitrogen tetroxide

Rocket burns
Up to 750 seconds
Up to 50 restarts

LUNAR MODULE

This was a two-stage vehicle. The octagonal descent stage carried rocket engines, fuel tanks, four unfolding landing legs, and radar for gauging the distance to the lunar surface. Atop this sat the ascent stage, a multifaceted crew cabin carrying multiple thrusters for maneuvering in space, with its own rocket engine and fuel tanks beneath for blasting free of the Moon's gravity.

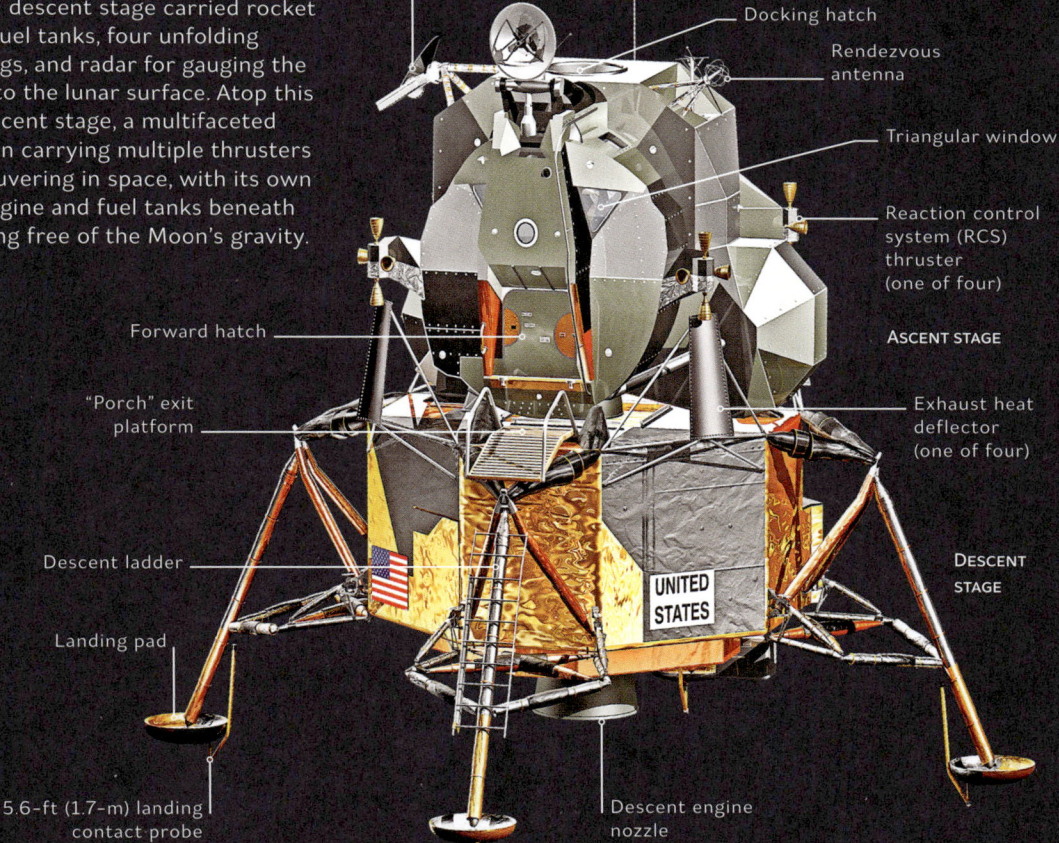

Steerable Earth antenna

Antenna for communication with astronauts on lunar surface

Docking hatch

Rendezvous antenna

Triangular window

Reaction control system (RCS) thruster (one of four)

ASCENT STAGE

Forward hatch

"Porch" exit platform

Exhaust heat deflector (one of four)

Descent ladder

DESCENT STAGE

Landing pad

5.6-ft (1.7-m) landing contact probe

Descent engine nozzle

UNITED STATES

LM DATA

Height
23.1 ft (7.04 m)

Max width
31 ft (9.4 m)

Apollos 11–14 mass
33,300 lb (15,100 kg)

Apollos 15–17 mass
36,220 lb (16,430 kg)

Habitable volume
160 ft³ (4.53 m³)

Vacuum flyer
Since the entire mission of the Lunar Module was to be conducted in a vacuum, it could be designed without concern for aerodynamics, hence its complete lack of streamlining.

BEHIND THE SCENES

America's moonshot was a huge mobilization of people, technology, and expertise. Only 12 astronauts ever reached the surface, but the entire Apollo program involved more than 400,000 personnel.

In May 1961, when President Kennedy announced America's intention to put people on the Moon, NASA (see pp.138–139) was still a young organization. Founded in 1958 as successor to the National Advisory Committee for Aeronautics (NACA), it had inherited the NACA's facilities and absorbed the US Navy's Vanguard satellite program, the California Institute of Technology's Jet Propulsion Laboratory. It also took in the rocket team at the Army Ballistic Missile Agency in Alabama, which brought German rocketry pioneer Wernher von Braun firmly into the NASA fold, where he would play a leading role in the Apollo program.

A national effort

Fueled by substantial funding from Congress, NASA progressed the Apollo project at breakneck speed. It established a new center for astronaut operations at what is now the Johnson Space Center in Texas and a dedicated Florida launch facility named the Kennedy Space Center in 1963.

Many aspects of the Apollo effort, including the construction of rockets and spacecraft, were subcontracted by NASA to around 20,000 commercial firms; in stark contrast, design and manufacture in the Soviet Union were undertaken directly by state agencies. After early schedule and budget overruns, Air

Force General Samuel C. Phillips was named Director of the Apollo program. Bringing a wealth of experience from his previous work in missile development, he instituted rigorous systems to keep progress on track.

Setback and reset

Many aspects of NASA's organization were found wanting after the Apollo 1 fire of 1967 (see pp.148–149). This tragic event led to a review of procedures and the introduction of risk protocols that ultimately allowed the program to proceed successfully with no further casualties, even in the face of unexpected malfunctions and near-disasters.

Engineers test the aerodynamic performance of a model Mercury capsule in the full-scale wind tunnel at NASA's Langley Research Center in Hampton, Virginia.

Mock-up mission

Astronauts Joseph Kerwin, Vance Brand, and Joseph Engle inspect their Apollo Command Module in June 1968, prior to an eight-day earthbound "mission" to test safety and rehearse flight procedures at the Space Environment Simulation Laboratory in Houston, Texas.

Sergei Korolev

The Soviet space effort's "chief designer" headed OKB-1, one of several design bureaux that were often at odds with each other, ultimately costing the Soviet Union its lead in the race for the Moon.

Wernher von Braun

As head of Alabama's Marshall Space Flight Center, von Braun was a powerful voice in Apollo's design. His cautious approach drew criticism for slowing progress but delivered reliable results.

HIDDEN FIGURES

The heyday of the space race coincided with major social changes in the US following World War II, including struggles for women's equality and the Civil Rights movement. Although NASA was far from perfect, its need for large numbers of talented engineers, scientists, and mathematicians created opportunities in fields where both women and people of color had traditionally experienced discrimination. Aspects of this story—in particular the experience with three black female "computers" working at the NACA/NASA Langley Research Center—are told in Margot Lee Shetterly's 2016 book *Hidden Figures* and the film of the same name.

Engineering pioneer

Pictured with other staff at Langley's Supersonic Pressure Tunnel in 1956, Mary Jackson (bottom right) went from work in the racially segregated West Area Computing Section in the early 1950s to become NASA's first female black engineer in 1958.

Apollo 4 was the first test of a "fully stacked" Saturn V launch vehicle. The rocket carried an uncrewed Command and Service Module into Earth orbit.

APOLLO TAKES FLIGHT

After a tragic accident almost derailed it at the outset, the Apollo program rapidly progressed through uncrewed tests to daring missions that brought humans within touching distance of the Moon.

Once Apollo's mission profile had been decided upon (see pp.138–139), NASA engineers and contractors embarked on designing, building, and testing the ambitious hardware required. At the same time, the Gemini program endeavored to improve their understanding of the challenges involved in spaceflights lasting several days.

Loss and recovery

With work on the Saturn V rocket and Apollo spacecraft proceeding apace, NASA targeted February 1967 for a crewed test of Apollo in Earth orbit. Then tragedy struck: on January 27, as the crew carried out tests of their spacecraft's performance on internal power, faulty wiring sparked a fire that spread rapidly due to the pure-oxygen atmosphere. The three Apollo astronauts—Gus Grissom, Edward White, and Roger B. Chaffee—died within moments.

Crewed flights were grounded while the accident was investigated and safety improved. The uncrewed Apollo 4 launched in November 1967 and two further tests followed before the first crewed flight—to evaluate the Command and Service Modules—took place in October 1968.

Rehearsal missions

With its deadline looming and amid uncertainty over the secretive Soviet lunar effort, NASA accelerated its schedule. Apollo 8 carried three astronauts to orbit the Moon in December 1968, and Apollo 9 tested the full spacecraft (with Lunar Module) in Earth orbit in March 1969. After Apollo 10's "dress rehearsal" flew a Lunar Module within 8.9 miles (14.4 km) of the surface in May, the way was clear to put people on the Moon.

In Earth orbit
During the Apollo 9 mission, astronaut Rusty Schweickart carried out a space walk to inspect the outside of the Lunar Module, snapping this view of his crewmate David Scott waiting at the Command Module hatch.

Command Module hatch was redesigned to open more easily following Apollo 1 disaster

EARLY CREWED MISSIONS

Mission:	Apollo 1
Crew:	Gus Grissom, Ed White, Roger B. Chaffee
Date:	February 21, 1967 (planned)
Mission:	Apollo 7
Crew:	Wally Schirra, Walt Cunningham, Donn Eisele
Date:	October 11–22, 1968
Mission:	Apollo 8
Crew:	Frank Borman, James Lovell, William Anders
Date:	December 21–27, 1968
Mission:	Apollo 9
Crew:	James McDivitt, David Scott, Russell Schweickart
Date:	March 3–13, 1969
Mission:	Apollo 10
Crew:	Thomas Stafford, John Young, Eugene Cernan
Date:	May 18–26, 1969

Apollo 1 crew
Gus Grissom (left) was a Mercury and Gemini veteran, while Ed White (center) had made the first US spacewalk from Gemini 4. Roger B. Chaffee (right) was in training for his first spaceflight.

Earthrise is one of a series of photographs taken by astronaut William William Anders as Apollo 8 emerged from the Moon's far side on December 24, 1968. Capturing our planet's isolation in space, it became an icon for the growing environmental movement.

A **mock-up N1 rocket** stands on the test pad at the Baikonur Cosmodrome during a trial run of the launch process in 1967. The N1's conical base was considerably broader than the Saturn V's.

A SECRET RACE

For decades, the Soviet Union denied that it had ever seriously engaged in a "space race" with the US, but in reality, it had its own equally ambitious version of the Apollo program.

As early as May 1961, Soviet chief designer Sergei Korolev was planning a mission that would take cosmonauts into Earth orbit, where they would rendezvous with a spacecraft that would take them to the Moon—a method known as Earth-Orbit Rendezvous. The mission called for a powerful new rocket, designated N1, and an advanced crewed spacecraft, known as Soyuz, which would be attached to a separate lunar lander.

Lone explorer

Work progressed slowly, partly because engineers had to focus on other areas of the space program. Delays were also caused by the Korolev team's conflicts with rival designers who cast doubts on various aspects of the N1 or advocated a Direct Ascent lunar mission with an even larger rocket. Changes in the overall Soviet leadership added further delays, and as a result, the design of the Lunar-Orbit Rendezvous mission, designated N1-L3, was not agreed on until 1965. It consisted of a two-person lunar orbiter (a simplified Soyuz)

and a lightweight lander capable of carrying a single cosmonaut to the lunar surface and back. Various other pieces of hardware were to be sent ahead of the main mission, adding to its complexity. These included a backup lander and a pair of Lunokhod rovers that were capable of both automated and human-operated excursions.

Further events soon robbed the Soviet lunar program of any chance of catching up with the Americans. Korolev's unexpected death in 1966 led to a lengthy hiatus before his successor was appointed. Meanwhile, the complexity of both rocket and spacecraft created a rift between rival engineering teams. A number of Zond missions sent prototype orbiters around the Moon, but the N1 was not ready for its first (failed) test flight until February 1969. By this time, anticipating a successful Apollo landing, Soviet leaders had already decided to cancel and disavow their crewed lunar program.

Lunokhod 1
Two Lunokhod vehicles designed for use by Soviet cosmonauts were repurposed as remote-controlled rovers, landing on the Moon in 1970 and 1973.

Celebrating success
A Soviet stamp celebrating 1968's Zond 5—the first spacecraft to circle the Moon and return to Earth in September of that year.

Soviet lunar lander
The 17-ft- (5.2-m-) tall Soviet Lunniy Korabl spacecraft was designed to carry one cosmonaut to the lunar surface and back.

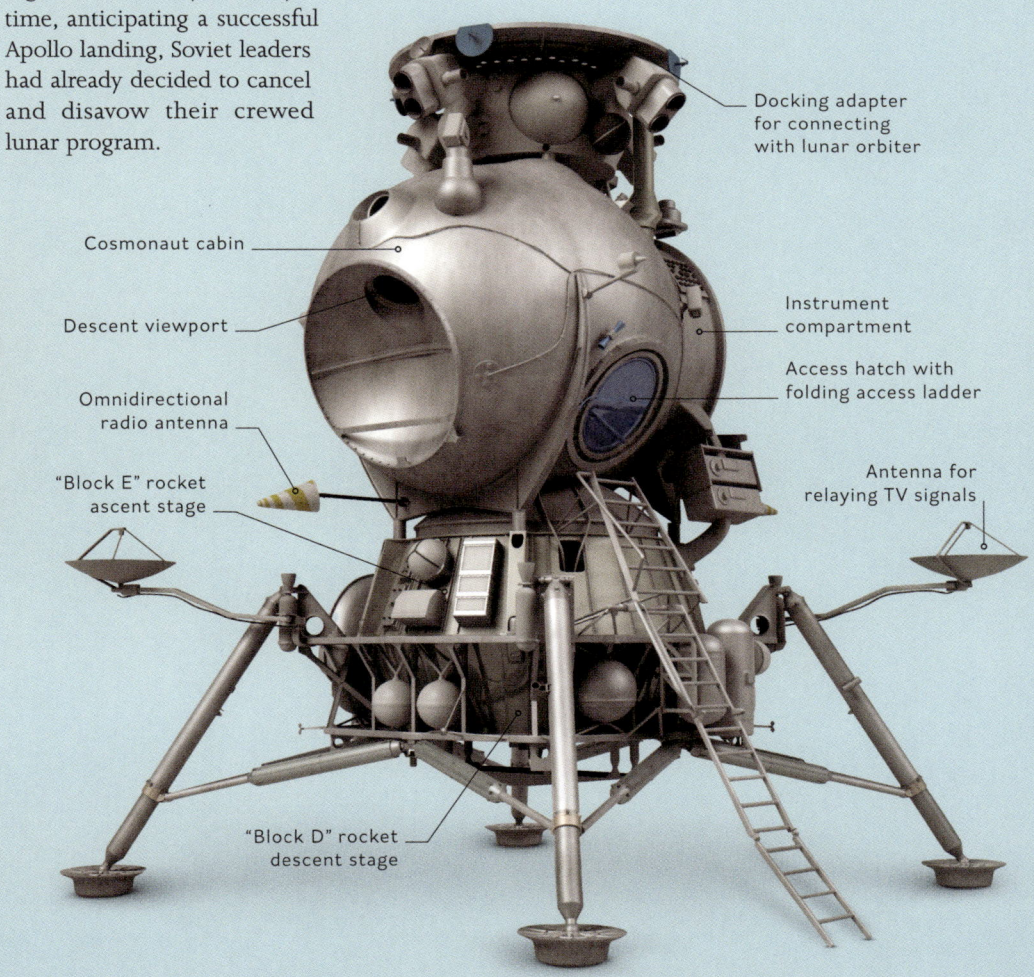

Docking adapter for connecting with lunar orbiter

Cosmonaut cabin

Descent viewport

Omnidirectional radio antenna

"Block E" rocket ascent stage

Instrument compartment

Access hatch with folding access ladder

Antenna for relaying TV signals

"Block D" rocket descent stage

ONE GIANT LEAP

In July 1969, the race to the Moon culminated in the successful landing of Apollo 11 on Mare Tranquillitatis (the Sea of Tranquility) and the first human exploration of the lunar surface.

Apollo 11 begins its journey perched atop the towering Saturn V launch vehicle. This photograph was taken from Kennedy Space Center's LC-39A launch pad.

At 09:32 EDT (Eastern Daylight Time) on July 16, 1969, the mighty engines of a Saturn V rocket burst into life, pushing clear of the launch pad at NASA's Kennedy Space Center in Florida. Strapped into the craft at the top of the 363-ft (111-m) rocket sat Neil Armstrong, Edwin "Buzz" Aldrin, and Michael Collins. All three astronauts had experience from Gemini flights, which had been used to test technology and procedures for the Apollo missions. However, their Apollo 11 mission had only been selected as the first crewed lunar landing in late 1968; NASA's hasty decision to send Apollo 8 around the Moon (see pp. 148–149) meant fewer remaining test flights were needed before a landing could be attempted.

Journey to the Moon

After 12 minutes, Apollo 11 reached Earth orbit and spent just over two hours circling Earth one-and-a-half times before Saturn V's third-stage rocket fired to propel the craft on a trajectory to the Moon. Collins took the controls for the crucial "transposition, docking, and extraction" maneuver, to reconfigure the craft (see pp. 144–145). The combined spacecraft then pulled free of its upper-stage rocket,

which was then steered to a separate flightpath at a safe distance. After a three-day weightless cruise, the Service Module (SM) engines fired to put Apollo 11 into lunar orbit. The next day, Armstrong and Aldrin separated the *Eagle* Lunar Module (LM) from the Command and Service Module (CSM). Collins remained in the Command Module (CM), called *Columbia*.

Final approach

The LM's engines fired to put the craft in a descent path. When this unexpectedly overshot the landing target, Armstrong took the controls to find a safe landing place while managing a dwindling fuel supply. At 16:17 EDT on July 20, 1969, *Eagle* came to rest on the Moon. »

Final moments on Earth

Apollo 11 Commander Neil Armstrong leads Michael Collins and Buzz Aldrin toward the transport van that would take them to their waiting spacecraft.

APOLLO 11 MISSION DATA

Objective: First human lunar landing
Crew: Neil Armstrong
Michael Collins
Edwin "Buzz" Aldrin
Date: July 16–24, 1969

APOLLO 11'S FLIGHT PATH

The three elements of the Apollo 11 spacecraft formed various configurations during the mission. The combined Command and Service Module (CSM) stayed in lunar orbit while the Lunar Module (LM) descended to the surface and its ascent stage returned. Once crew and samples were transferred, the CSM returned to Earth, separating on final approach so the Command Module could safely re-enter Earth's atmosphere.

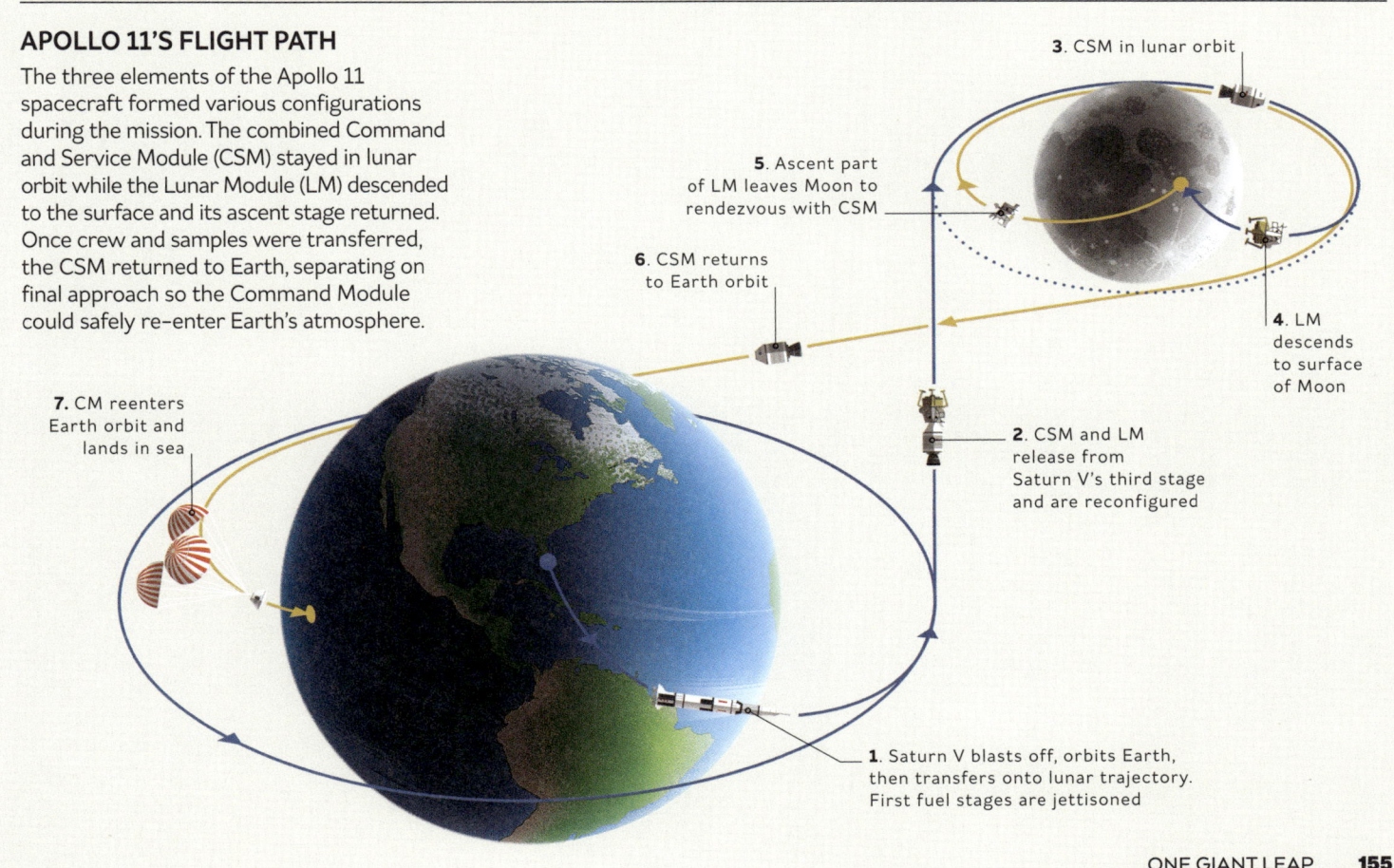

3. CSM in lunar orbit

5. Ascent part of LM leaves Moon to rendezvous with CSM

6. CSM returns to Earth orbit

4. LM descends to surface of Moon

7. CM reenters Earth orbit and lands in sea

2. CSM and LM release from Saturn V's third stage and are reconfigured

1. Saturn V blasts off, orbits Earth, then transfers onto lunar trajectory. First fuel stages are jettisoned

Walking on the Moon

Initial plans had involved a sleep period for the astronauts in between *Eagle*'s touchdown and their extra-vehicular activity (EVA) to explore the lunar surface. However, Armstrong and Aldrin assured mission control that they were not tired and were adjusting well to the weak gravity. As a result, EVA preparations began after the rigorous postlanding checks.

Final descent

Neil Armstrong's historic step was caught by the camera of the LM's Modularized Equipment Stowage Assembly.

Less than seven hours after landing, Armstrong opened the hatch of a now-depressurized LM and stepped onto the "porch"—a small platform at the top of the LM's "number 1" leg. Armstrong pulled on a lanyard to open the external Modularized Equipment Stowage Assembly (MESA), which stored equipment for use on the Moon's surface and also an automated television camera. He then descended the ladder to the surface.

Armstrong set foot on the Moon at 22:56 EDT on July 20. His first task was to collect a contingency sample from the surface in case the EVA had to be abandoned in a hurry. After Armstrong had taken photos of the surrounding landscape, Aldrin joined him about 20 minutes later.

Lunar explorers

While on the surface, the astronauts carried out a range of activities, including testing movement in lunar gravity, deploying various scientific experiments (see pp.172–173), and collecting further rock and soil samples. They erected a US flag (see pp.186–187) and took a short telephone call from US President Richard Nixon. They also deposited a number of commemorative items to remain on the Moon, including an aluminum capsule with a silicon disk containing goodwill messages from 73 countries.

First steps
Neil Armstrong described the lunar surface as "fine and powdery," and it stuck to his boots. With no weathering on the Moon, his footprints may survive for millions of years.

The entire EVA lasted just 2 hours and 31 minutes; Armstrong and Aldrin returned to the LM cabin with some 47 lb 6 oz (21.5 kg) of lunar samples. After their rescheduled rest period, they began preparations for departure. Leaving behind the descent section of the LM, the ascent part of *Eagle* blasted off and began its return to the orbiting CSM some 21 hours and 36 minutes after it had arrived on the Moon.

Reunited in orbit
From his position in lunar orbit, Michael Collins captured this image of the ascent section of the LM as it approached to redock with the CSM. The spider-legged lower descent stage of the LM was abandoned on the Moon's surface.

"That's **one small step** for [a] man; **one giant leap** for mankind."

Neil Armstrong, July 21, 1969

EXPLORING THE SURFACE

In 1969, Neil Armstrong became the first human to walk on the Moon, inaugurating a brief period—ending in 1972—during which crewed missions explored the lunar surface. Since then, many robotic probes have returned to the Moon, either to land or operate from orbit. These have deployed a wide range of scanning equipment, enabling scientists both to map the Moon in detail and to determine its structure.

APOLLO LANDINGS

The six successful lunar landings of the Apollo program were shaped by a mixture of scientific objectives, political and financial pressures, and technical factors.

Even before NASA settled on Apollo 11 for the first lunar landing attempt in late 1968, it envisaged an itinerary of exploration to follow. In 1966, it ordered 15 Saturn V rockets, potentially allowing for 10 landings.

NASA designated Apollo mission profiles by alphabetical letters. Apollo 11 was the only "G-class" mission. By early 1969, there were plans for four "H-class" flights, with precision landings and more detailed surveys, and five longer "J-class" missions, with longer stays, an upgraded Lunar Module (LM), and the Lunar Roving Vehicle (LRV) (see pp.178–179).

Site selection

Many landing targets were shortlisted, with final decisions made by the Apollo Site Selection Board one or two flights in advance. Prospective landing sites had to be relatively close to the Moon's equator (since reaching higher latitudes from a more inclined lunar orbit would take too much fuel). Safety was another factor: Highland

targets were only attempted once improved LM engines enabled extended hover times to inspect the rougher landscape. Scientific aims included surveying different types of terrain and gathering geological samples linked to key events in the Moon's history.

Once the US had triumphed in the space race by landing the first humans on the Moon, however, its imperatives shifted. Budgets were

cut just as NASA faced demands for other projects, such as an Earth-orbiting space station. The last three planned missions were canceled, with H-class flights curtailed at Apollo 14, followed by three J-class missions. Touchdown sites were reassessed as well, which ensured that by December 1972, six Apollo landings had conducted the fullest possible lunar survey.

Landing sites
Initial Apollo landing targets were limited by factors including engine power, navigational techniques, and the illumination angle of the Sun. On later missions, growing experience and technological advances allowed these requirements to be somewhat relaxed.

Apollo touchdown locations

In 2011, NASA's Lunar Reconnaissance Orbiter (see pp. 192–193) took pictures of the Apollo landing sites that show the traces left by the astronauts.

Landings were timed to take place with the Sun at an elevation of less than 45°, ensuring strong shadows to reveal surface details.

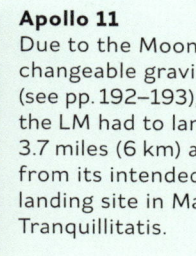

Apollo 11
Due to the Moon's changeable gravity (see pp. 192–193), the LM had to land 3.7 miles (6 km) away from its intended landing site in Mare Tranquillitatis.

Apollo 12
The second lunar landing set down in Oceanus Procellarum, midway between the Head and Surveyor craters and close to the Surveyor 3 robot lander.

Apollo 14
Despite a precise landing in the Fra Mauro region, rough terrain prevented the astronauts from reaching one of their key targets, the nearby Cone crater.

Apollo 15
Landing on the edge of the lunar highlands, Apollo 15 was the first mission to use an LRV, visible to the right here in its final parking place.

Apollo 16
The second J-class mission targeted the bright Descartes region of the central highlands, partly because it was far from most previous landing sites.

Apollo 17
The crew landed north of the small Poppie crater in the Taurus-Littrow Valley to access landslipped rock from different eras of lunar history.

CANCELED PLANS

In addition to the nine missions initially intended to follow Apollo 11 by the early 1970s, NASA developed plans for more advanced exploration and colonization. A second wave of missions in the mid-1970s would have used dual Saturn V launches to put a payload module and a shelter module on the Moon, supporting stays of up to 14 days. Once a new landing vehicle had been developed, bigger components would be landed to build a large lunar base that would eventually become permanent.

Colonization concept
A 1969 artist's impression depicts the future Moon colony envisaged by NASA's Advanced Manned Mission Program Office. It combines modified LM shelters with domelike expansion modules in a semipermanent six-person base.

DRESSED FOR SPACE

The Apollo spacesuit was like a personal spacecraft, capable not only of protecting astronauts on the surface of the Moon, but also—in emergencies—of keeping them alive for up to 14 days in space.

A spacesuit must provide its wearer with safe and comfortable conditions for operating in a range of environments with extreme changes in surrounding temperature and pressure. The key elements of the Apollo suit were the Pressure Suit Assembly (PSA) and the Portable Life Support System (PLSS) backpack. The PSA maintained a layer of air at Earth-like pressure around the astronaut's body, allowing them to breathe and avoid dangerous decompression sickness. Oxygen pumped into the garment inflated the suit, giving it a bulky appearance.

For extra-vehicular activities (EVAs)—trips outside the spacecraft to explore the lunar surface—astronauts also wore a Liquid Cooling and Ventilation Garment. This was an undersuit with embedded flexible pipes that circulated cooled water to regulate body temperature. For less demanding activities, a simpler cotton undersuit called the Constant Wear Garment was worn. The suit of the Apollo Command Module Pilot, who remained in orbit, was similar, but it lacked some of the features required for surface EVAs.

Redesigned spacesuit
Apollos 15 to 17 used a more flexible PSA: the A7LB. It had an entry point at the waist (the original suit had a zipper in the back), which increased comfort on longer EVAs and when driving the lunar rover. It also had a neck joint, enabling the wearer to see more of their surroundings. Drinking water and even snacks were available inside the helmet.

A7L Pressure Suit Assembly
The core of the A7L PSA, which was used on missions up to Apollo 14, was the five-layer Torso Limb Suit Assembly—a rubberized, airtight "bladder" layer with surrounding layers designed for comfort, restraint, and protection. This was covered by the 13-layer Integrated Thermal Micrometeoroid Garment, made of varied materials to protect the inner suit and its occupant from impacts, abrasions, and solar radiation.

Polycarbonate helmet attached to suit by locking ring; under helmet, astronaut wore cap with built-in microphones and headphones

Mission badge: this suit was used on the Apollo 11 mission by astronaut Neil Armstrong

Water connector allowed water from PLSS backpack to circulate through pipes of Liquid Cooling and Ventilation Garment

Extra-Vehicular Visor Assembly (EVVA) with protective polycarbonate visor, gold-coated radiation visor, and movable eye shades

Buckle for attaching PLSS backpack

Electrical connector for attachment of power feed from PLSS

Multiple pockets allowed checklists and tools to be stored around suit

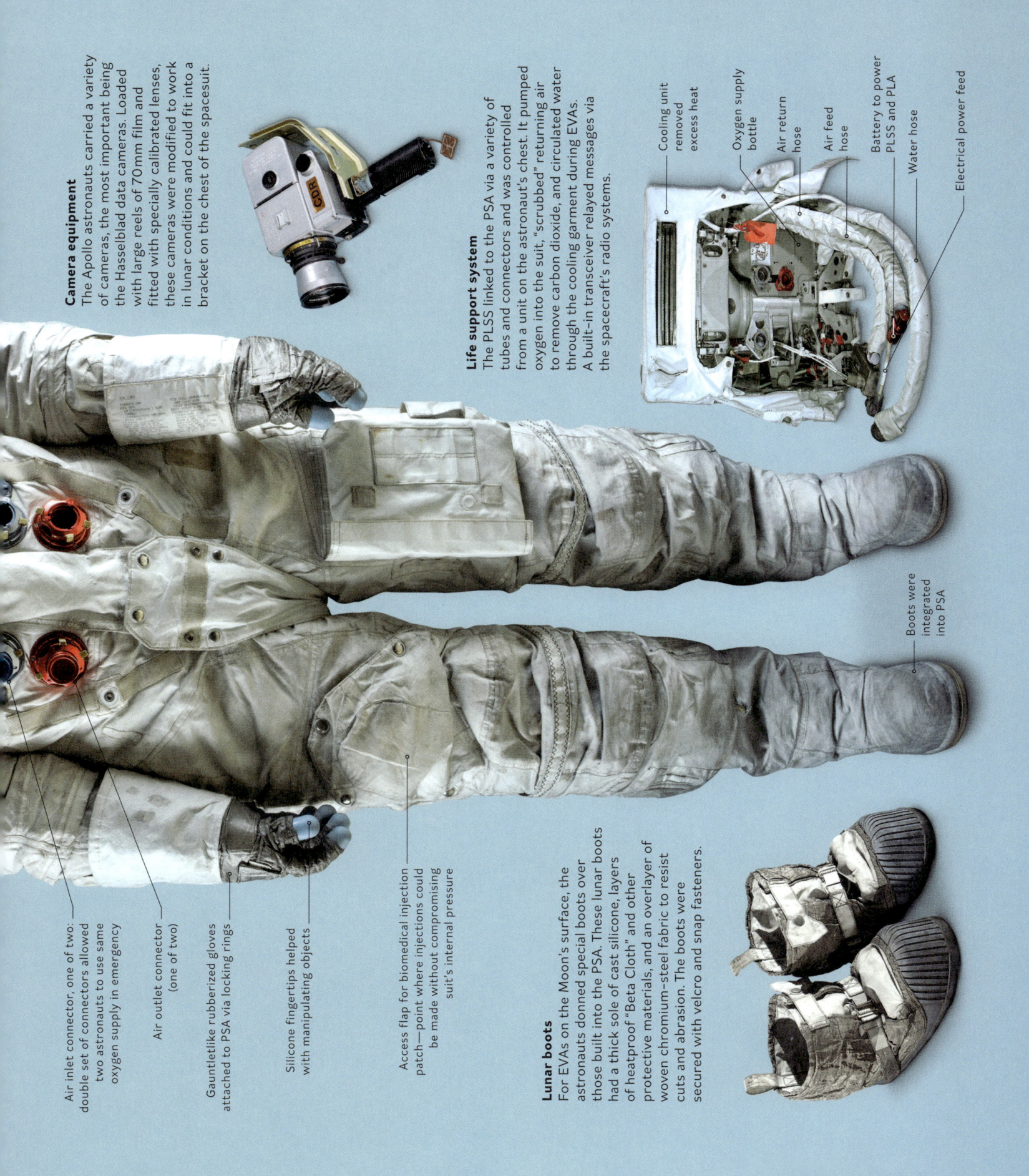

Camera equipment
The Apollo astronauts carried a variety of cameras, the most important being the Hasselblad data cameras. Loaded with large reels of 70mm film and fitted with specially calibrated lenses, these cameras were modified to work in lunar conditions and could fit into a bracket on the chest of the spacesuit.

Life support system
The PLSS linked to the PSA via a variety of tubes and connectors and was controlled from a unit on the astronaut's chest. It pumped oxygen into the suit, "scrubbed" returning air to remove carbon dioxide, and circulated water through the cooling garment during EVAs. A built-in transceiver relayed messages via the spacecraft's radio systems.

Cooling unit removed excess heat

Oxygen supply bottle

Air return hose

Air feed hose

Battery to power PLSS and PLA

Water hose

Electrical power feed

Air inlet connector, one of two: double set of connectors allowed two astronauts to use same oxygen supply in emergency

Air outlet connector (one of two)

Gauntletlike rubberized gloves attached to PSA via locking rings

Silicone fingertips helped with manipulating objects

Access flap for biomedical injection patch—point where injections could be made without compromising suit's internal pressure

Boots were integrated into PSA

Lunar boots
For EVAs on the Moon's surface, the astronauts donned special boots over those built into the PSA. These lunar boots had a thick sole of cast silicone, layers of heatproof "Beta Cloth" and other protective materials, and an overlayer of woven chromium-steel fabric to resist cuts and abrasion. The boots were secured with velcro and snap fasteners.

Apollo 11's **Command Module** made its final descent beneath three huge "ringsail" parachutes, each 25.4 m (83.3 ft) across. Two parachutes were sufficient to ensure a safe splashdown; the third was a safety precaution.

SPLASHDOWN

After a three-day return journey from lunar orbit, each Apollo mission ended with a spectacular landing at sea in the Command Module. A safe splashdown, however, was just the beginning of a complex process.

Apollo's flight path called for a hair-raising direct return to Earth, reentering the atmosphere at speeds of around 25,000 mph (39,600 km/h). Some 15 minutes before the start of reentry, the Service Module was jettisoned with explosive charges, allowing the Command Module (CM) to orient itself with its heat-shielded underside at an entry angle of about 6.5° below the horizon. It then followed a "skip reentry" flight path, skimming the thin upper atmosphere to slow down before falling along a gravity-dominated trajectory that plunged to the landing zone.

Fiery return

As the CM entered the denser part of the atmosphere, some 76 miles (122 km) above the Earth, a layer of hot, electrically charged gas at temperatures of around 5,000°F (2,700°C) blocked off radio communications for several minutes. At the same time, friction with the thickening air rapidly slowed the incoming spacecraft, generating forces six times stronger than Earth's gravity for the astronauts inside.

As the CM's speed fell to about 335 mph (540 km/h), air-pressure sensors triggered the release of its protective nose cone and the

deployment of small drogue parachutes from beneath, followed by three much larger main parachutes that lowered the capsule to a gentle splashdown in the Pacific Ocean. Grumman Tracer aircraft monitored the capsule during the last few minutes of descent. A recovery fleet, including an aircraft carrier, was waiting in the landing zone. After splashdown, the astronauts were winched onto a helicopter and flown to the recovery ship. The CM and its cargo were brought aboard by crane.

In quarantine
The Apollo 11 crew greet President Richard Nixon from inside the Mobile Quarantine Facility—a converted trailer used on early lunar missions as a precaution against possible harmful lunar contamination. The astronauts spent a total of 14 days in quarantine.

First on the scene
After splashdown, swimmers deployed by helicopter attached a flotation ring to keep the capsule upright before helping the crew onto a waiting inflatable.

POSTMISSION PROCESSING

All returning Apollo astronauts, along with their spacecraft and its payload, had to pass through NASA's Lunar Receiving Laboratory (LRL) in Houston, Texas. On early missions, the crew were flown there while still inside the Mobile Quarantine Facility. The LRL stored the Moon rock samples in sterile conditions and managed their distribution for scientific study.

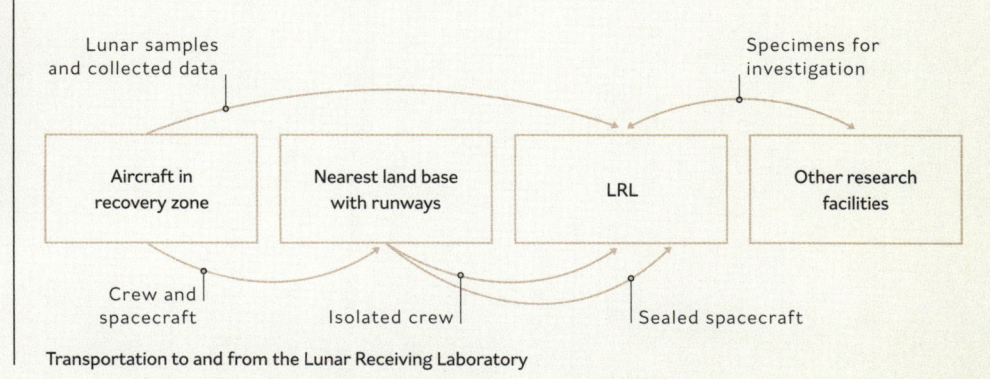

Lunar samples and collected data

Specimens for investigation

| Aircraft in recovery zone | Nearest land base with runways | LRL | Other research facilities |

Crew and spacecraft Isolated crew Sealed spacecraft

Transportation to and from the Lunar Receiving Laboratory

Apollo 11 astronauts Buzz Aldrin, Michael Collins, and Neil Armstrong are given an ecstatic welcome by the people of New York City during a three-hour ticker-tape parade on August 13, 1969.

THE RETURN TO THE MOON

With the first lunar landing safely achieved, NASA eased the timeline of its forthcoming Apollo launches and focused its attention on preparing missions to further explore and study the Moon.

Surveyor 3
Recovering parts from Surveyor 3 enabled engineers back on Earth to study how materials and components were affected by prolonged exposure to lunar surface conditions. Surveyor 3 proved to have been remarkably durable.

APOLLO 12 MISSION DATA

Mission:	Second crewed lunar landing
Crew:	Charles Conrad Jr.
	Richard F. Gordon Jr.
	Alan L. Bean
Date:	November 14–24, 1969

In late July 1969, after the success of Apollo 11 (see pp.154–157), NASA issued a tentative schedule for nine further Apollo missions. Apollo 12's planned September launch was delayed by two months, enabling Commander Charles "Pete" Conrad and his Lunar Module (LM) pilot Al Bean to carry out more geological training. They were joined by Dick Gordon as Command Module Pilot.

Blasting off on November 14, 1969, the mission had an inauspicious start: lightning struck the ascending Saturn V rocket, knocking several instruments temporarily offline and leaving nagging worries about other spacecraft systems. Nevertheless, the hardware performed almost faultlessly, and the LM *Intrepid* made a precision landing in Oceanus Procellarum (the Ocean of Storms) on November 19, within walking distance of the Surveyor 3 spacecraft (see pp.142–143) that had landed on the Moon two-and-a-half years earlier. During two moonwalks, Conrad and Bean gathered rock and soil samples, collected pieces of the Surveyor, and deployed scientific experiments.

> **"Whoopee!** Man, that may have been a **small one for Neil**, but that's a **long one for me**."
>
> "Pete" Conrad, stepping onto the lunar surface, November 19, 1969

Active Seismic Experiment (ASE)
Edgar Mitchell lays out a geophone line during an ASE. Geophones are acoustic detectors that picked up vibrations from the lunar interior.

APOLLO 14 MISSION DATA

Mission: Third crewed lunar landing
Crew: Alan B. Shepard Jr.
Stuart A. Roosa
Edgar D. Mitchell
Date: January 31–February 9, 1971

Testing times

If Apollo 12 eventually ran smoothly, the same could not be said of its immediate successor, and the accident that led to Apollo 13's epic trip around the Moon (see pp.174–175) triggered a lengthy safety review. As a result, it was not until January 31, 1971, that the next Apollo mission was ready for launch. Crewed by Mercury veteran Alan Shepard, Stuart Roosa, and Edgar Mitchell, Apollo 14 was seen as a make-or-break moment for the Apollo program. Updates to the mission's profile heaped pressure on the crew, as did making the Fra Mauro Highlands (Apollo 13's original destination) its new landing destination.

Despite some setbacks during transit and landing, the LM *Antares* successfully touched down near its target of Cone Crater. After an initial walk to deploy an Apollo Lunar Surface Experiments Package and collect samples (see pp.172–173), Shepard and Mitchell set out on a second walk in search of the crater rim, but became disoriented by the terrain. Photographs of their tracks from orbit later showed that they had missed their goal by just 100 ft (30 m). Fortunately, plans were already underway for additions to later missions that would greatly improve NASA's exploration capabilities.

MOONQUAKES

Data from the various seismometers deployed during the Apollo missions enabled lunar scientists to distinguish four different types of tremors on the Moon. Their causes can be identified by a mixture of their seismic characteristics and when they most commonly occur.

Deep quakes
Quakes deep inside the lunar mantle are triggered by tidal stresses as Earth's gravity pulls on different parts of the Moon.

Shallow quakes
As the Moon's core cools and shrinks, its contraction puts stresses on the upper mantle around 180 miles (300 km) deep, creating tremors at the surface.

Thermal quakes
The extreme temperature changes produced by the Moon's long day-and-night cycle cause its surface to expand and contract.

Meteoritic quakes
Space rocks hitting the Moon send shockwaves through its inner layers. Studying these provides information about the Moon's interior structure.

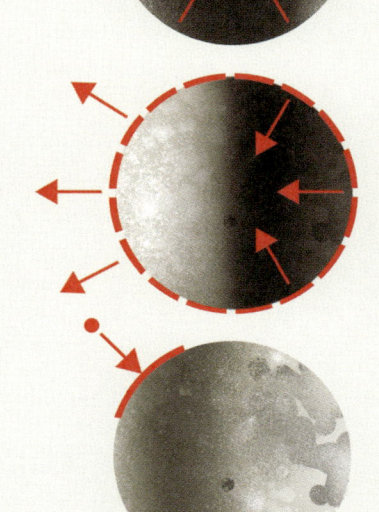

TECTONIC FEATURES

Without an atmosphere or weather to disturb it, the lunar surface remains as it has been for billions of years, revealing traces of its volcanic history, as well as signs of tectonic activity.

Rupes Recta
Stretching for 70 miles (110 km), and 790–985 ft (240–300 m) high, Rupes Recta (the Straight Cliff) is a linear fault caused by one portion of the lunar surface (on the right) dropping relative to another.

Studies have revealed that the Moon is too cold to have tectonic plates—large, grinding landmasses that ride the convection currents rising through planetary interiors. However, its crust shows signs of having shifted in the past, and it may still be moving today. Astronomers believe that these slow movements are caused by two main factors: the Moon shrinking as a result of gradual cooling, and its surface sinking under the weight of volcanic deposits.

The Moon's cooling began soon after it was born (see pp. 12–13); over time, its molten surface solidified, forming a crust, through which heat has leaked ever since. As it cooled, the crust contracted, causing cracks to spread, ridges to rise, and some areas to sink under others. Lunar Reconnaissance Orbiter (LRO) has found spectacular evidence of this in Mare Frigoris (the Sea of Cold), near the Moon's north pole. Its images revealed thousands of tectonic features, including wrinkle ridges, graben, and lobate scarps (see opposite). Some of these features could be a billion years old, but the ongoing detection of "moonquakes" (see pp. 168–169) suggests that the Moon's tectonic life may be far from over.

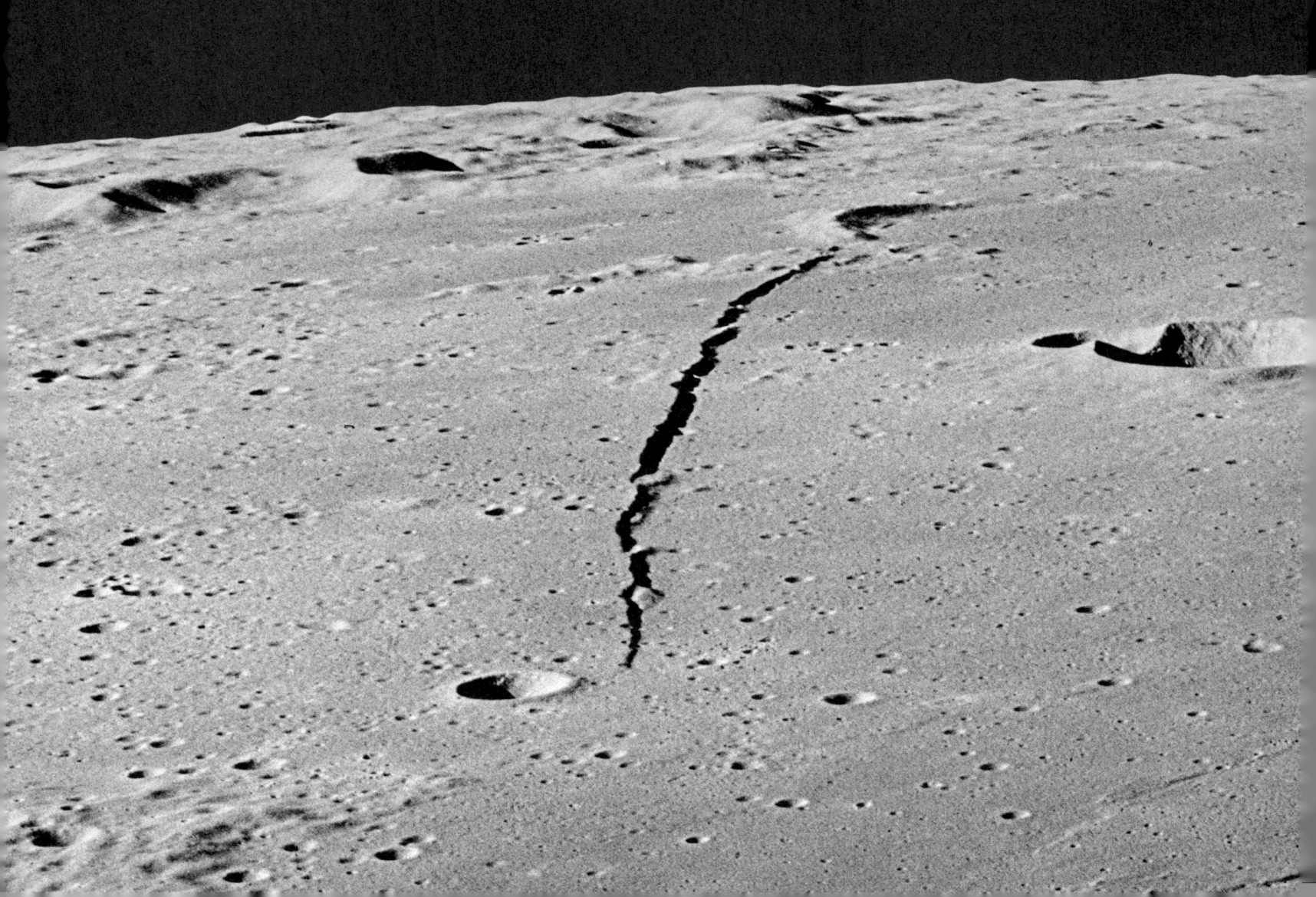

WRINKLE RIDGES

A common feature of the Moon's maria (see pp.118–119), wrinkle ridges formed when the lunar surface sank under the weight of outpourings of lava. Typically, the lava—which on the Moon is molten basalt—filled a crater- or basin-sized depression, forming a lake that then solidified. As it did so, the lunar surface sank, causing the outer edges of the depression to slump inward, raising the ground at its center. Many of these ridges extend for hundreds of miles.

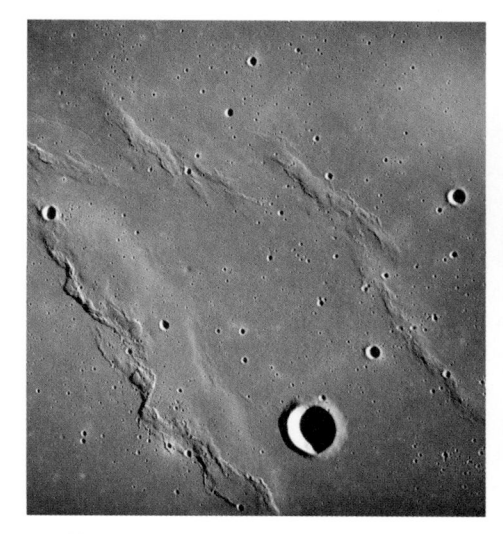

Wrinkle ridges in Flamsteed crater

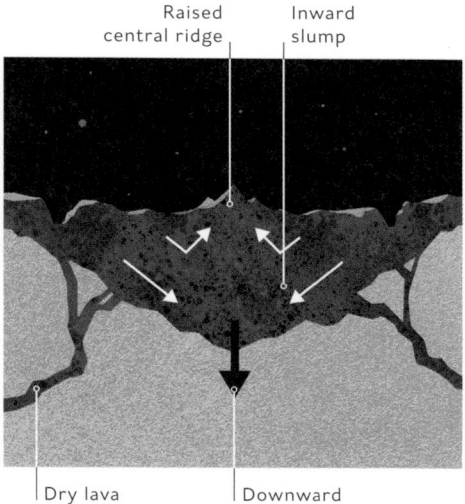

Raised central ridge

Inward slump

Dry lava dyke (intrusion)

Downward pressure

GRABEN

A graben is a trench or depression created when an area of lunar surface slumps downward due to forces pulling the surrounding terrain in opposite directions. Larger ones are often ancient and tend to be associated with the edges of lunar seas—they probably formed due to the same inward "slump" of material that created wrinkle ridges. Smaller grabens tend to be much younger (about a billion years old) and are probably responses to local tectonic movements related to the Moon's overall contraction.

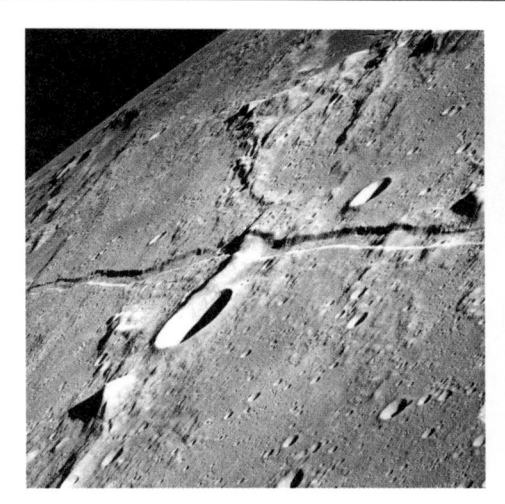

Rima Ariadaeus graben

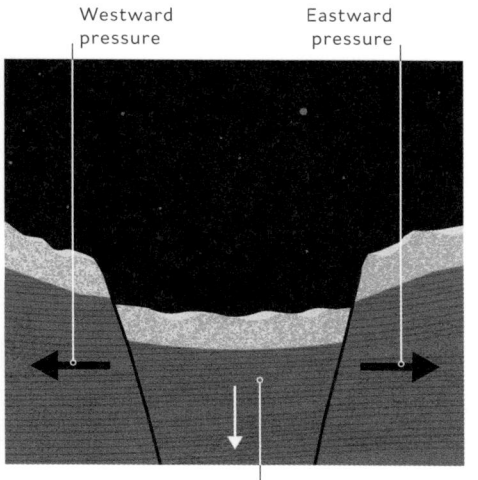

Westward pressure

Eastward pressure

Slumped central ground

LOBATE SCARPS

Unlike wrinkle ridges, which formed when lava caused the lunar surface to sink, lobate scarps are believed to have formed as the Moon shrank as a result of its cooling. Although it may only have shrunk by 328 ft (100 m), this was enough to cause its crust to compress, creating fault lines where parts of the crust are driven beneath others. Since scarps show relatively little erosion, scientists believe that they are a feature of the Moon's more recent past.

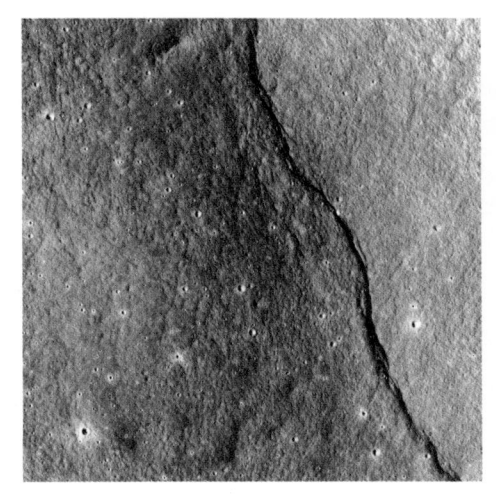

Lobate scarp near Gregory crater

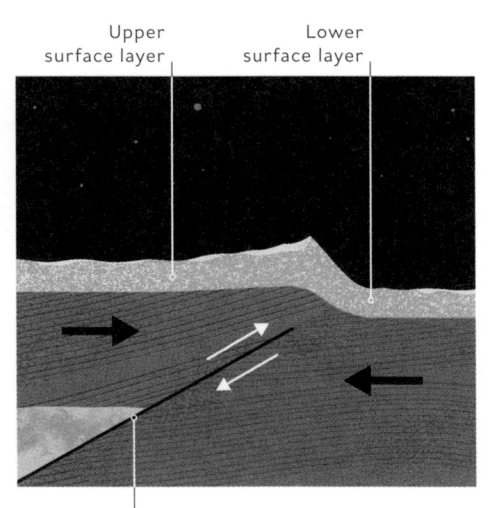

Upper surface layer

Lower surface layer

Thrust fault in contracting surface

LUNAR EXPERIMENTS

Much of the scientific bounty from the Apollo missions was provided by instruments the astronauts set up on the Moon. Typically designed to operate for a year, and left running after the astronauts' departure, in many cases the instruments continued to send back data until they were shut down in 1977.

Apollo 11 carried a limited package (an earthquake detector and a Laser Ranging Retroreflector) that could be set up rapidly in case any extra-vehicular activity needed to be cut short. Later landings deployed a full suite of instruments known collectively as the Apollo Lunar Surface Experiments Package (ALSEP), linked to a power supply that would allow them to operate through the long, cold lunar night (almost 15 Earth days). The ALSEP included seismometers and detectors to measure the gas and dust of the lunar atmosphere and the effects of material striking the Moon's surface, from micrometeorites to the subatomic particles of the solar wind.

Apollo **seismometers** detected around **10,000** moonquakes and **2,000** meteorite impacts.

1. Central Station and Dust Detector
This 55-lb (25-kg) box routed power to all the instruments, relayed instructions from Earth, and returned data. Most versions also carried a device to measure the accumulation of dust on the lunar surface.

2. Generator
Power for the ALSEP came from a Radioisotope Thermal Generator, which used heat from the radioactive decay of plutonium-238 to generate electricity. Prior to landing, the plutonium was safely stored on the Lunar Module's exterior.

3. Suprathermal Ion Detector Experiment (SIDE)
The SIDE, deployed by Apollo missions 12, 14, and 15, measured the mass/charge ratio of positively charged ions (atomic nuclei from the solar wind and other sources) near the lunar surface.

4. Solar Wind Spectrometer (SWS)
Carried on Apollos 12 and 15, this equipment measured how the solar wind interacts with the Moon, including the intensity and direction of charged solar-wind particles striking the lunar surface.

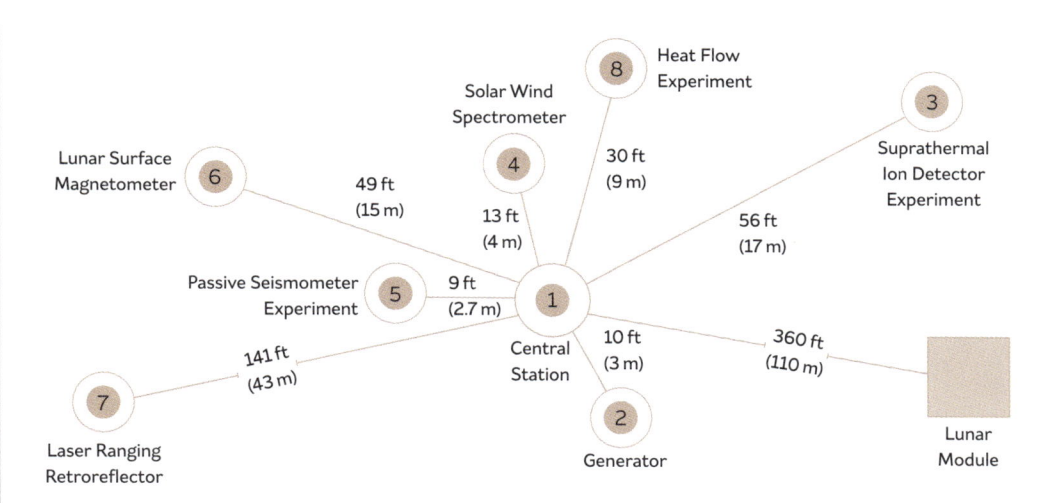

ALSEP diagram labels:

- 8 — Heat Flow Experiment
- Solar Wind Spectrometer — 4
- 6 — Lunar Surface Magnetometer
- 3 — Suprathermal Ion Detector Experiment
- 49 ft (15 m)
- 30 ft (9 m)
- 13 ft (4 m)
- 56 ft (17 m)
- 5 — Passive Seismometer Experiment
- 9 ft (2.7 m)
- 1 — Central Station
- 10 ft (3 m)
- 360 ft (110 m)
- Lunar Module
- 141 ft (43 m)
- 7 — Laser Ranging Retroreflector
- 2 — Generator

ALSEP LAYOUT

The ideal site for the ALSEP experiments was on flat ground at least 295 ft (90 m) west of the Lunar Module (minimizing any disturbance from the ascent stage blast-off). Some seismic experiments required long, straight lines for earthquake-measuring cables or for firing mortar chargers into the ground. Heat flow experiments needed areas of soil that could be drilled for inserting temperature-sensitive probes.

Perfect placement
This schematic shows the layout of key ALSEP components deployed by Apollo 15. Most instruments had to be placed on level ground, and many required alignment at specific angles to the Sun or Earth.

5. Passive Seismic Experiment (PSE)
Apollos 12 and 14–16 carried a drum-shaped unit housing three seismometers to record surface vibrations. Shockwaves from targeted impacts of spent Apollo hardware were used to calibrate the devices.

6. Lunar Surface Magnetometer (LSM)
Placed at each of the landing sites of Apollos 12, 14, and 16, this instrument measured the Moon's weak magnetic field and investigated how the field interacts with the solar wind's own magnetism.

7. Laser Ranging Retroreflector (LRRR)
Apollos 11, 14, and 15 set glassy reflectors at which laser beams were fired from Earth; their return time gave precise measurements of the Moon's position and also revealed surface distortions.

8. Heat Flow Experiment (HFE)
Apollos 15–17 deployed the HFE. Using two probes inserted into the surface, it measured the rate of heat flow and loss from the lunar regolith and the amount of heat released by internal radioactive rocks.

RESCUE IN SPACE

After a catastrophic explosion just 56 hours into its flight, Apollo 13 became a battle for survival as trapped astronauts worked with colleagues on Earth to bring their crippled spacecraft home.

APOLLO 13 MISSION DATA

Mission:	Crewed lunar landing (aborted)
Crew:	James Lovell
	John "Jack" Swigert
	Fred Haise
Date:	April 11–17, 1970

At 03:08 UTC on April 14, 1970, Apollo 13's *Odyssey* spacecraft was roughly 210,000 miles (330,000 km) from Earth when a routine procedure to stir its liquid oxygen tanks sparked an explosion in its Service Module (SM). While engineers at mission control investigated the problem, the most pressing concerns for astronauts James Lovell, Jack Swigert, and Fred Haise were dwindling oxygen levels and a loss of power from fuel cells that shared the same oxygen supply.

Crisis mode

When the astronauts reported that gas was escaping from the SM and that parts of its hull were missing, mission control decided that their best chance for survival was to abandon *Odyssey* entirely—including its Command Module (CM)—and to use its undamaged Lunar Module (LM), *Aquarius*, as a lifeboat.

"Houston, we've had a problem."

Jack Swigert, Apollo 13 Command Module Pilot

This meant that the crew would have to live for nearly four days in a vehicle that was designed to support just two astronauts for two days—the time they would have spent on the Moon.

Problem-solving

With the SM's main engines out of action, engineers came up with an ingenious idea that enabled Apollo 13 to make a safe return using only the LM's Descent Propulsion System (DPS) motor. This involved hurriedly

Stricken spacecraft
A NASA illustration depicts the crippled state of Apollo 13's SM. Despite its proximity to the explosion, the adjoining CM was not damaged.

Undamaged CM, docked with LM *Aquarius*

Damaged wiring in one of SM's oxygen tanks caused explosion

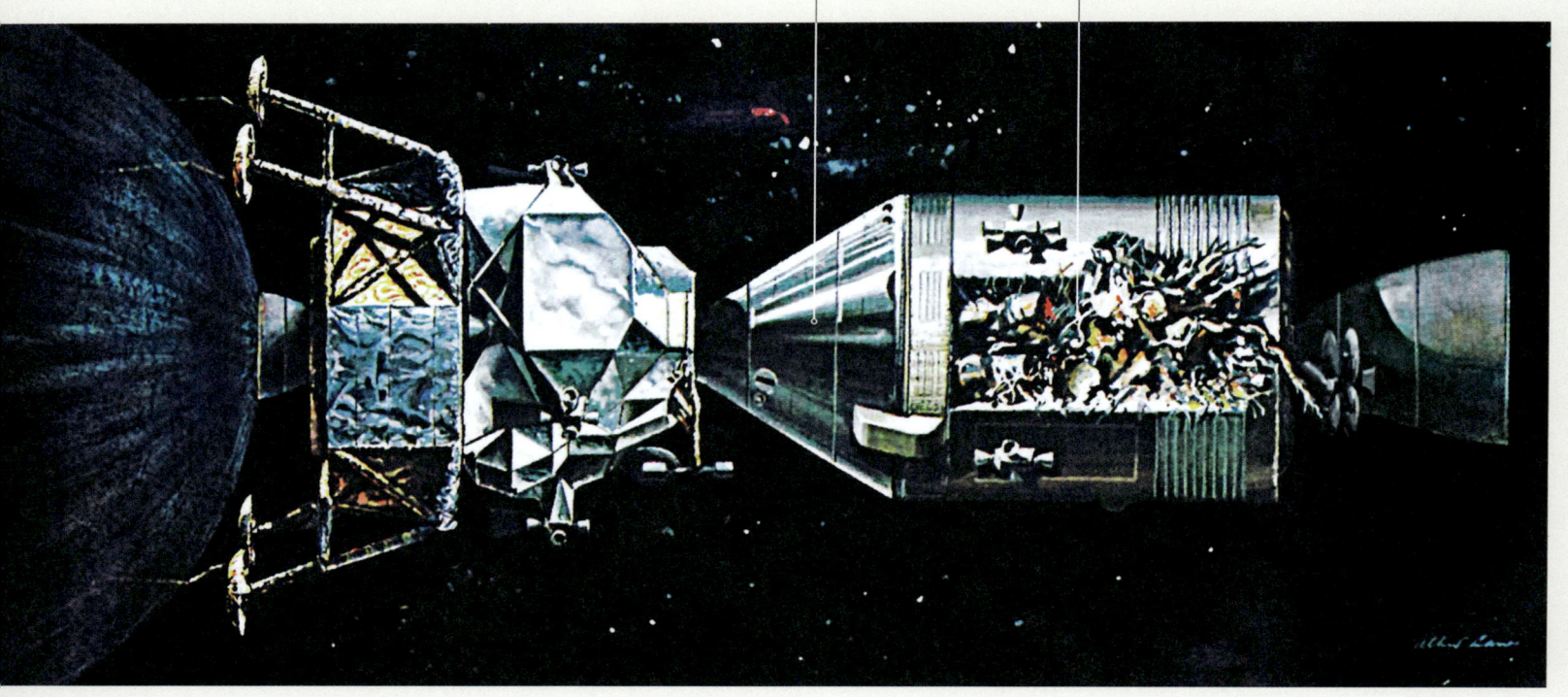

reprogramming the LM's guidance computer to slingshot the vehicle around the Moon (see below). The astronauts transferred supplies from *Odyssey*, but water and power had to be strictly rationed. To avoid dying of asphyxiation, they also had to construct a new carbon dioxide absorber, since the existing one was designed for two passengers. They did so using pieces of spare equipment.

As Apollo 13 barreled back toward Earth, mission controllers devised a procedure to reactivate *Odyssey* with limited power. Two final burns of the DPS engine, calculated using measurements of Earth's day/night boundary, made final adjustments to the spacecraft's trajectory before the crew returned to the CM and jettisoned the SM. Finally, back in Earth orbit, they abandoned *Aquarius* itself. Less than an hour later, the CM brought the astronauts home in a protracted, fiery reentry into Earth's atmosphere.

Safely home

Staff at NASA's mission control in Houston celebrate Apollo 13's return while watching live pictures of James Lovell aboard the recovery ship USS *Iwo Jima*.

Improvised lifesaver

Following instructions from mission control, the astronauts used duct tape, plastic manual covers, and a spacesuit hose to build a makeshift carbon dioxide absorber.

SLINGSHOT AROUND THE MOON

Less than six hours after the initial explosion, the Apollo 13 crew used a 34-second burn of the LM's Descent Propulsion System (DPS) to put them on a "free return" trajectory—a flightpath that swung them around the Moon and gave them enough speed to hurl them back to Earth. As they sped over the lunar far side, they reached a distance of 248,655 miles (400,171 km) from Earth—a crewed spaceflight record that remains unbeaten. Two hours after pericynthion (when they were closest to the Moon), an engine burn of 4 minutes 23 seconds boosted their velocity, shortening their return journey by about 12 hours.

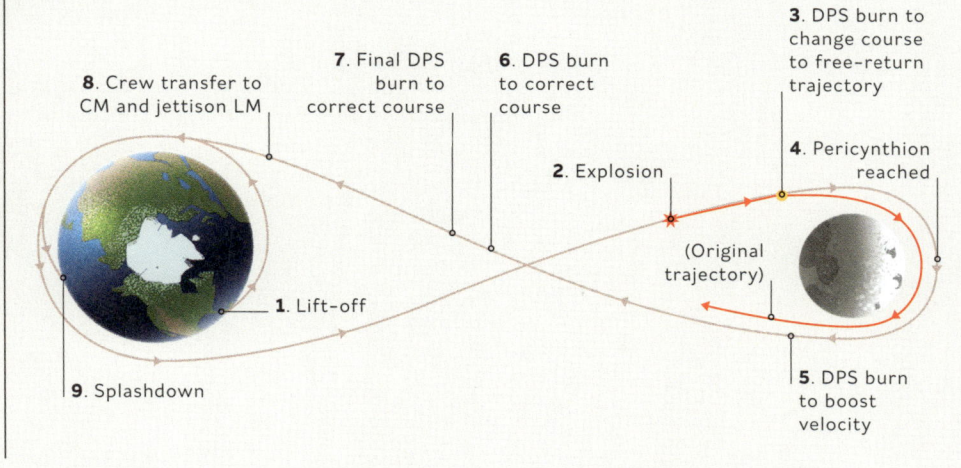

3. DPS burn to change course to free-return trajectory

8. Crew transfer to CM and jettison LM

7. Final DPS burn to correct course

6. DPS burn to correct course

4. Pericynthion reached

2. Explosion

(Original trajectory)

1. Lift-off

5. DPS burn to boost velocity

9. Splashdown

IN THE SHADOW OF MOUNT HADLEY

Significantly better equipped for lunar exploration than the previous Apollo missions, Apollo 15 was the first of the three "J-class" missions that were designed for extended stays on the Moon.

APOLLO 15 MISSION DATA

Mission: Fourth crewed lunar landing
Crew: David R. Scott
James B. Irwin
Alfred M. Worden
Date: July 26–August 7, 1971

NASA had always planned to follow its first exploratory lunar landings (the G- and H-class missions) with a series of longer, science-based expeditions. Initially, these were scheduled for Apollos 16 through to 20, but as the agency faced budget cuts—and a loss of political backing now that the space race had been won—it decided to cancel its last three flights and make Apollo 15 the first of three extended J-class missions.

Upgrades for the J missions included a modified Lunar Module capable of supporting the astronauts for 75 hours on the Moon, the addition of an electric Lunar Roving Vehicle (LRV; see pp.178–179) known as the "Moon buggy," and the addition of a Scientific Instrument Module (SIM) in a side bay of the Service Module. Extra-vehicular activity (EVA) spacesuits (see pp.162–163) were also upgraded

Mountains and valleys

They had trained to explore the area around Censorinus crater, near Mare Tranquillitatis (the Sea of Tranquility), but due to the revised Apollo program, Commander David Scott and Lunar Module (LM) Pilot James Irwin received a new landing site—the more scientifically interesting Hadley-Apennine region. At 22:16 UTC on July 30, 1971, the LM *Falcon* touched down, landing at a distinct angle with one of its feet resting in a small crater.

Scott began the surface mission by surveying the surrounding terrain from the top hatch. The astronauts then performed three EVAs over the next three days, totaling over 18 hours of surface activity. During their first EVA, Scott and Irwin deployed the LRV, making trips to the nearby Hadley Rille and the flanks of Mons Hadley Delta (to the

Working on the edge
This panoramic view of David Scott and the LRV at the edge of Hadley Rille combines several images captured by James Irwin during Apollo 15's first surface EVA.

Hadley Rille, carved by ancient lava flow

south of Mons Hadley) before setting up various scientific experiments. During their second and third EVAs, they revisited the mountain and the valley, set up more experiments, and struggled to drill into the rock to extract geological samples. Despite some glitches, the LRV was invaluable and greatly extended their range of exploration.

Orbital studies

In orbit, Command Module Pilot Al Worden kept busy operating the new SIM bay instruments, which included cameras, a laser altimeter, and spectrometers designed to analyze the chemistry of surface minerals. The LM returned from the surface on August 2; once crew and cargo were transferred, it was jettisoned and deliberately crashed onto the Moon. The astronauts continued performing experiments in orbit for another two days and then departed for Earth. One final landmark event for the mission came during the return trip, when Worden made the first ever "deep space" EVA, some 197,000 miles (317,000 km) from Earth, in order to retrieve film canisters from the SIM bay.

View from orbit
The astronauts' first view of their landing site (at bottom center) in the mountainous eastern region of Mare Imbrium (the Sea of Showers).

"There's a **fundamental truth** to our **nature. Man must explore** ... And this is **exploration at its greatest**."
David Scott, after stepping onto the lunar surface

Mons Hadley, rising 2.8 miles (4.5 km) above plain

Distant ridge, appearing clear and close due to Moon's lack of atmosphere

DRIVING ON THE MOON

Carried aboard the last three Apollo missions, the Lunar Roving Vehicle (LRV) was a versatile and robust Moon buggy that extended the exploration range and capabilities of its astronaut drivers.

NASA contractors investigated several Moon rover designs during the 1960s, but these were mostly sealed, pressurized vehicles for long-duration trips of up to 14 days. However, Wernher von Braun (see pp.134–135) also promoted studies of short-range, open buggies called "Local Science Survey Modules." Once the potential of these vehicles had been demonstrated, NASA decided to use them in the later Apollo missions.

The parameters for the rover included a strict weight limit, the ability to carry two astronauts plus other material on several lengthy trips using single-use batteries (thus avoiding the complications of solar-powered recharging), and a communications system for keeping in touch with Earth. Boeing was awarded a contract to build the LRV in October 1969, meeting a tight deadline of 17 months to have the first vehicle aboard Apollo 15.

All-terrain vehicle

The completed LRV was built to operate for at least 78 hours during lunar daytime. Four large wheels powered by independent electric motors, combined with a soft suspension system, allowed it to drive over obstacles up to 1 ft (30 cm) high, through 28-in- (70-cm-) wide crevasses, and up or down slopes of 25°. It vastly increased the area that could be explored; ultimately, however, drive distances were limited to ensure that astronauts could return to the LM on foot in an emergency.

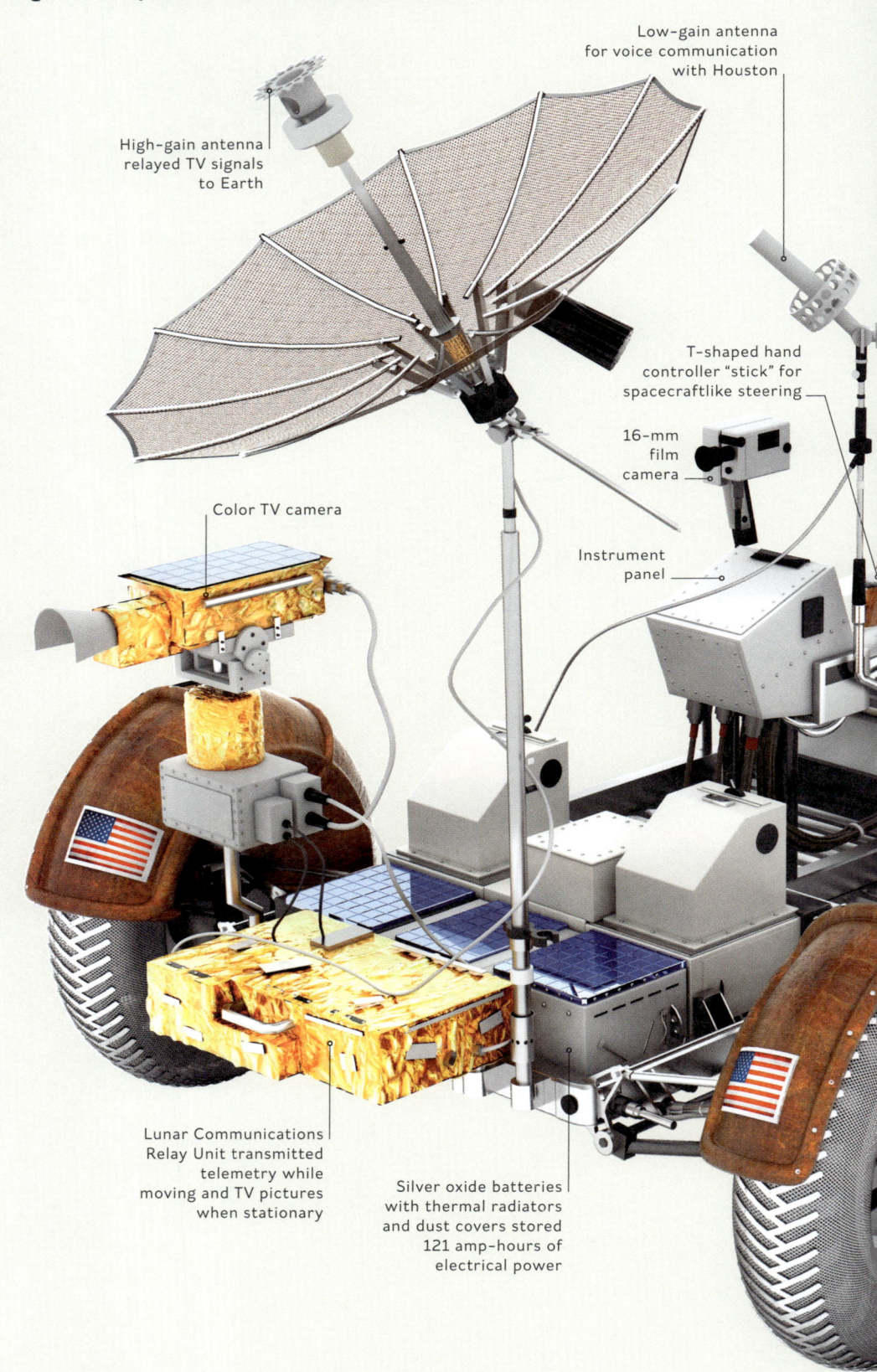

High-gain antenna relayed TV signals to Earth

Low-gain antenna for voice communication with Houston

T-shaped hand controller "stick" for spacecraftlike steering

16-mm film camera

Instrument panel

Color TV camera

Lunar Communications Relay Unit transmitted telemetry while moving and TV pictures when stationary

Silver oxide batteries with thermal radiators and dust covers stored 121 amp-hours of electrical power

Lunar Roving Vehicle
With an unladen mass of 460 lb (210 kg), the LRV was designed for a top speed of 6 mph (9.7 km/h). In use, however, it achieved an actual top speed of 11.2 mph (18 km/h).

DEPLOYMENT SEQUENCE

The LRV was stored on one side of the Lunar Module, its rear section facing outward. As the astronauts used pulleys to lower the LRV's rear end, its back wheels automatically unfolded. The front section and wheels were then manually unfolded before the LRV was lowered to the ground and assembly completed.

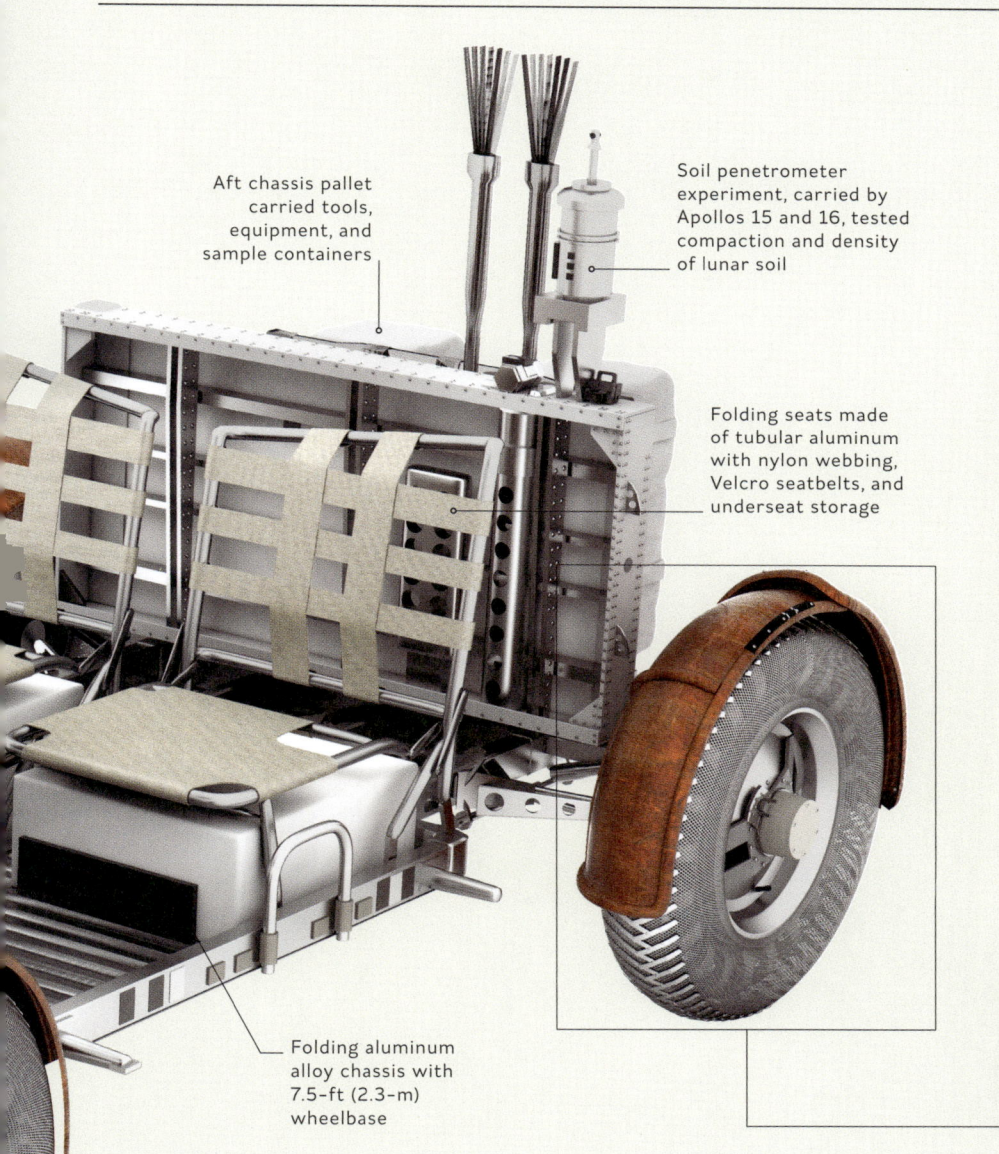

Aft chassis pallet carried tools, equipment, and sample containers

Soil penetrometer experiment, carried by Apollos 15 and 16, tested compaction and density of lunar soil

Folding seats made of tubular aluminum with nylon webbing, Velcro seatbelts, and underseat storage

Folding aluminum alloy chassis with 7.5-ft (2.3-m) wheelbase

Front wheels turned in opposite direction to rear wheels, giving rover tight turning circle

TRAINING ON EARTH

Astronauts practiced LRV operations on Earth in the 1G trainer, a rubber-wheeled simulator designed to mimic the rover's lunar performance in Earth gravity. Several other models and dummies were used to refine the vehicle's ergonomics for spacesuited drivers and to rehearse the crucial deployment procedures.

Test drive
Apollo 17 astronauts Gene Cernan and Harrison Schmidt conduct spacesuited trials aboard the 1G trainer at Kennedy Space Center in Florida.

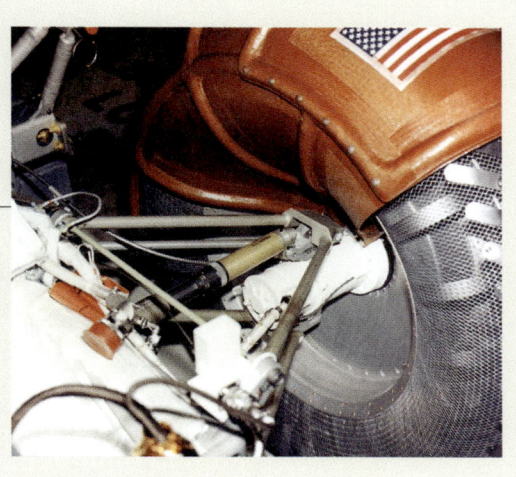

Resilient wheels
Designed to support the vehicle's weight while not becoming clogged with dust, each wheel was made of woven zinc-coated steel with a tread of chevron-shaped titanium plates. An inner frame shielded the electric motor in the wheel hub.

THE LAST MOONWALKERS

The final Moon landings of the 1970s—Apollos 16 and 17—targeted previously unexplored regions of the lunar highlands, deploying complex experiments and making increasingly ambitious surface trips.

Midair salute
John Young, Apollo 16 commander, took advantage of low lunar gravity to leap off the ground while saluting the US flag.

Tracy's Rock
Apollo 17's Harrison Schmitt inspects a boulder in the Taurus–Littrow Valley. The rock later became known as Tracy's Rock, after commander Gene Cernan's daughter.

Upon arrival at the Moon in April 1972, Apollo 16's surface expedition was almost abandoned due to problems with the engine of the Command and Service Module (CSM), which was vital for orbital maneuvers. After mission controllers had finally given the go-ahead, the *Orion* Lunar Module (LM) carrying astronauts John Young and Charlie Duke landed in the Descartes region on April 21, while Ken Mattingly remained on the CSM.

Apollo 16 was tasked with investigating whether the Descartes Highlands had been formed by volcanic activity. During three long extra-vehicular activities (EVAs) on their Lunar Roving Vehicle (LRV), Young and Duke traveled 16.6 miles (26.7 km). Among the samples they collected was "Big Muley," the largest rock returned by any Apollo mission. Young and Duke were able to conclude that, contrary to expectations, the surface was not volcanic.

A geologist on the Moon

For Apollo 17, the final Apollo mission, professional geologist Harrison Schmitt was selected as the LM Pilot. The *Challenger* LM, carrying Schmitt and commander Gene Cernan, touched down on December 11, 1972, in the Taurus-Littrow Valley, another highland region expected to show signs of relatively recent volcanism. While Cernan and Schmitt explored the surface, CSM Pilot Ronald Evans carried out orbital science, including tending to the needs of five laboratory mice.

Schmitt and Cernan's three EVAs on the rover covered 22.2 miles (35.7 km) of terrain, and their LRV carried unique experiments to measure the surface's electrical properties and changes in gravity at different locations. By the time the *Challenger* astronauts blasted off to join Evans in orbit, they had spent a record-breaking 75 hours on the Moon.

SATELLITES FROM APOLLO

Before leaving lunar orbit, Apollos 15 and 16 each deployed a small automated craft called a Particles and Fields Subsatellite (PFS). Apollo 15's PFS-1 returned data on plasma (electrically charged gas), particles, and magnetic fields for six months until its electronics failed. Due to Apollo 16's orbital difficulties, PFS-2 entered a lower lunar orbit and crashed after 34 days.

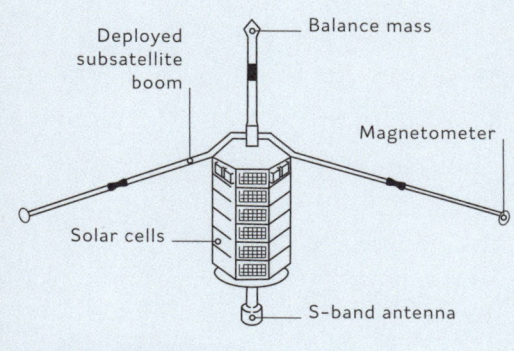

Balance mass

Deployed subsatellite boom

Magnetometer

Solar cells

S-band antenna

Tiny satellite
The PFS was a hexagonal cylinder 31 in (78 cm) long, with three 5-ft (1.5-m) booms and a mass of 80 lb (36.3 kg).

APOLLO 16 MISSION DATA

Mission: Fifth crewed lunar landing
Crew: John W. Young
Thomas K. Mattingly II
Charles M. Duke Jr.
Date: April 16–27, 1972

APOLLO 17 MISSION DATA

Mission: Sixth crewed lunar landing
Crew: Eugene A. Cernan
Ronald E. Evans
Harrison H. Schmitt
Date: December 7–19, 1972

MOON ROCKS

The Apollo missions collected around 840lb (380kg) of rocks. Along with samples from robot landers and lunar meteorites, they have proved crucial in uncovering the long history of Earth's satellite.

The Apollo astronauts used specialized tools to obtain lunar rocks. They picked up medium-sized surface rocks with tongs, gathered drifts of smaller pebbles with rakes, and collected tiny soil grains in bulk with the help of scoops or shovels. They hammered chips from larger boulders and rocky outcrops, while tubes and drills enabled them to take cores from below the surface that preserved vital information about stratigraphy (the order in which rocks and soils were laid down).

As well as taking representative samples of typical rocks, the astronauts collected any materials with unusual textures or colors. They also used "radial sampling"—taking specimens from various distances around the rims of

As rock slowly cooled and solidified, large plagioclase feldspar crystals developed

Clasts are embedded in matrix or fused together along their edges

Breccias
These rocks are typified by large fragments, or "clasts," in a matrix of once-molten material. Surfaces of other rock types can also be turned into breccias during their ejection.

Fine-grained structure due to rapid cooling of relatively fluid lava

Highland rocks
The rocks of the lunar highlands are mostly anorthosites—stones dominated by the low-density mineral plagioclase feldspar, which was buoyant in the Moon's early magma ocean. Some contain larger proportions of denser, iron-rich olivine.

Mare basalts
Lunar basaltic rocks contain higher levels of the titanium-rich mineral limenite than basalts on Earth do. The absence of water and oxygen when the maria formed also gave rise to unusual minerals.

impact craters to obtain ejecta that had come from different depths when the craters were formed. (The deepest deposits landed nearer the rims than material from shallower depths.) Each sample was bagged and numbered, while on-site photographs and measurements provided context for later interpretation back on Earth.

Three types of rock

Nearly all isolated rocks found on the Moon's surface were excavated and scattered by impact cratering, and this traumatic history must be taken into account when analyzing them. Most of this material falls into three broad classes: highland rocks, basalts, and breccias. Highland rocks, which dominate the Moon's brightest and most heavily cratered areas, are thought to have originated with the slow cooling of oceans of molten magma early in lunar history (see pp. 14–15). Basaltic rocks dominate the dark lunar seas, or maria. Like highland rocks, they are "igneous" (formed by the solidification of hot molten rock), but they have different physical and chemical compositions owing to the conditions under which they formed.

The third broad class, breccias, have been more radically transformed by heat and pressure from impact shockwaves melting and fusing jumbled debris. Breccias nevertheless still preserve evidence about the surfaces on which the original impacts occurred.

Moon glass
Glassy droplets found in regolith mostly formed when molten impact ejecta rapidly cooled and solidified without first crystallizing. However, orange glass beads—such as those in this Apollo 17 sample—formed during explosive volcanic eruptions. They owe their color to a high titanium content.

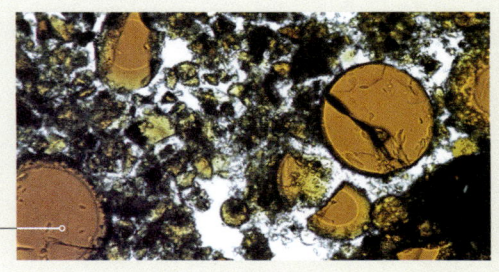

Orange spherules preserve material that originated deep beneath lunar surface

Heat from high-energy impacts may stick particles together as clumps called "agglutinates"

Lunar soil
The upper layer of the Moon's surface is dominated by regolith soil—a jumbled mix of small, loose particles and glass droplets produced by the steady bombardment of micrometeorites.

Microscope view of regolith

DATING ROCKS

Moon rocks can be dated by radiometric dating. Radioactive isotopes—unstable forms of elements—"decay" (transform) into new isotopes over very long periods of time. The isotope that decays is called the "parent"; the new isotope that it becomes is the "daughter." Absolute ages can be deduced from the proportions of parent and daughter isotopes in the rock.

● Parent isotope with a half-life of 1 billion years   ● Daughter isotope

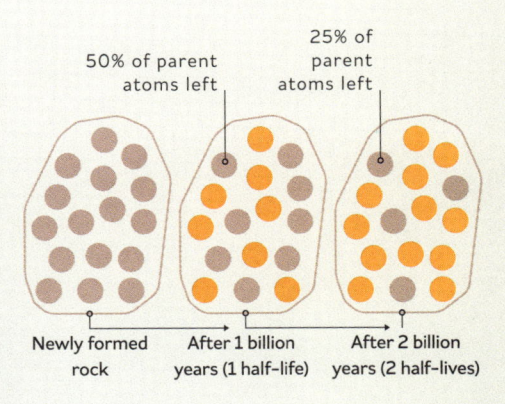

50% of parent atoms left

25% of parent atoms left

Newly formed rock

After 1 billion years (1 half-life)

After 2 billion years (2 half-lives)

Radioactive decay and half-life
The time it takes for half the parent isotope's atoms to decay is its "half-life," which is unique to that isotope. Knowing the half-life and the ratio of parent to daughter atoms in the rock enables scientists to calculate its age.

METEORITES

Powerful impacts on the Moon can eject rocks with sufficient force for them to overcome the weak lunar gravity and enter independent orbits around Earth or the Sun. If an escaped rock enters Earth's atmosphere (after anything from a few years to many millions of years), it may survive the fiery journey to become a meteorite. While knowledge of their origins is limited, lunar meteorites are scientifically valuable, since they expand our range of samples from different parts of the Moon.

Lunar meteorite DAG 262, found in the Sahara Desert

REWRITING HISTORY

As the first era of lunar exploration drew to a close in the 1970s, scientists had a wealth of new data to give fresh insight into the history not only of the Moon, but also the wider solar system.

Radiometric dating (see pp.182–183) of rock samples brought back to Earth by the early Apollo missions soon revealed the true age of the lunar maria and their place in the Moon's story. Mare Tranquillitatis (the Sea of Tranquility) and Oceanus Procellarum (the Ocean of Storms) formed from lava 3.7 and 3.2 billion years ago (BYA), but later impacts blanketed them with rock fragments up to 4.6 billion years old. The maria, it became clear, formed during an extended period of widespread volcanic eruptions long after the birth of the Moon itself. Since the Moon's near side has no major impact basins unfilled by lava "seas," geologists were also able to conclude that basin-forming impacts occurred before the maria eruptions, mostly in an intense phase from 4.1 to 3.8 BYA, now known as the "Late Heavy Bombardment" (LHB, see pp.14–15)

Geology and origins

The variety of rocks at the Apollo landing sites showed how ejecta from major impacts spread across the lunar surface over time. Core samples and other dated rocks revealed how surface materials age under exposure to micrometeorites, solar particles, and radiation, enabling scientists to establish a lunar geological timescale (see pp.14–15). The absence of complicating factors, such as a thick atmosphere or active geology, made the Moon an ideal "laboratory" for understanding how planetary surfaces are shaped by the space environment; the patterns revealed not only aid the dating of other areas on the Moon, but also of planets and moons with a similar history of events such as the LHB.

Analysis of lunar rocks showed that the Moon's raw materials, while related to Earth's, had some key differences, especially an absence of volatile chemicals due to intense past heating in a global magma ocean (see pp.14–15). Along with the discovery that the Moon is slowly spiraling away from Earth, this data inspired a revival of the long-neglected "giant impact hypothesis" (see pp.12–13).

Crater dating
The radiometric dating of layered lunar rocks helped turn broad principles (such as the degradation of older craters by later impacts) into useful tools. Planetary geologists can now date craters, such as the 3.9-billion-year-old Vendelinus (above center), with some precision—not only on the Moon, but on other worlds, too.

Modeling impact
Scientists investigating the giant impact hypothesis use computer simulations to model impact scenarios and compare their outcomes with the known properties of Earth and the Moon.

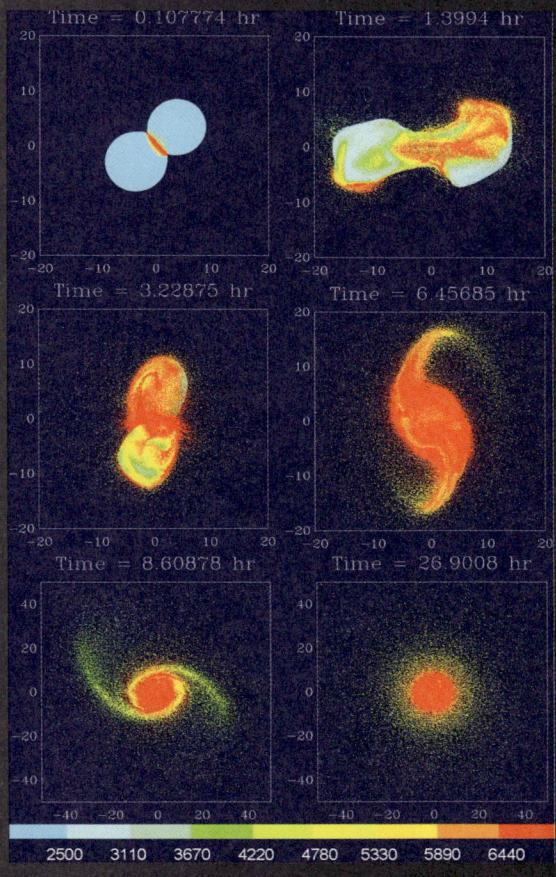

> ## "Apollo ... opened up our **understanding** of the entire **solar system**."

Samuel Lawrence, planetary scientist, NASA podcast, 2019

Late Heavy Bombardment
Around 4 BYA, the evolving orbits of the giant outer planets disrupted the paths of nearby asteroids and comets, causing them to bombard the inner planets and their moons.

CONSPIRACY THEORIES

As the Apollo landings faded into history, theories that they were hoaxes began to emerge. Such claims probably originate from a mix of honest misunderstanding, willful cynicism, and malicious falsehood. Photographs of the Moon's surface can be difficult for nonexperts to interpret, and NASA images of astronauts practicing in simulated lunar environments have been used to reframe Apollo as a vast, Hollywood-like production. The sheer scale of cover-up required, the impossibility of fooling the world's scientific community, and images of the landing sites from orbit all undermine the conspiracy idea—but still the claims persist.

HUMAN TRACES

Artifacts left by people lie scattered over the lunar surface, including probes and rocket stages purposely crashed onto the Moon—the first of these being Luna 2, a Soviet probe that impacted in 1959. Other uncrewed craft have subsequently landed and remained on the surface. Each of the six crewed Apollo missions left behind its Lunar Module's descent stage; the ascent stages fell back to the surface once the astronauts had rejoined the Command and Service Module. Three Lunar Roving Vehicles were also discarded, while two subsatellites (see pp.180–181) and the third stages of some Saturn V rockets collided with the Moon postmission.

The human artifacts on the Moon also include commemorative items, such as flags and plaques, as well as personal effects that the astronauts left in honor of friends and families on Earth.

1. Goodwill messages
In Mare Tranquillitatis (the Sea of Tranquility) on July 24, 1969, the Apollo 11 astronauts left a silicon disk etched with microscopic goodwill messages from the leaders of 73 nations around the world. Here, a US 50-cent coin illustrates the relative size of the disk.

2. Apollo plaques
Each Apollo Lunar Module had a commemorative plaque mounted on its descent stage, which remained on the Moon when the ascent stage blasted off. The Apollo 11 plaque is signed by the three astronauts and the then-president of the United States, Richard Nixon.

3. Golf balls
Using a makeshift club, Apollo 14's Alan Shepherd hit two golf balls on the Moon on February 6, 1971. Although the balls were never recovered, decades later, they were spotted in the astronauts' original photographs when the images were enhanced.

4. Falcon feather
A falcon feather remains on the Moon after it was used by Apollo 15's David Scott to demonstrate that all objects fall at the same speed in a vacuum. He performed the experiment, which featured the feather and a hammer, on August 2, 1971.

Since Luna 2 crashed into the Moon in 1959, nearly 70 more items of space hardware—from artificial satellites to entire spacecraft—have impacted on the lunar surface. These collisions have created the Moon's human-made, or "anthropogenic," craters.

The crater made by Apollo 16's third-stage booster

5. Commemorative sculpture
Fallen Astronaut is an aluminum figurine that commemorates the 14 astronauts and cosmonauts who were killed in the early days of their space programs. It was left near Hadley Rille by the Apollo 15 astronauts on August 2, 1971.

6. Family portrait
Apollo 16's Charles Duke left a portrait of himself with his wife and two children in the Descartes Highlands. On the back of the photograph, he wrote: "This is the family of Astronaut Duke from Planet Earth. Landed on the Moon, April 1972."

7. Flags
On December 12, 1972, Apollo 17's Harrison Schmitt became the last astronaut of the Apollo era to plant a flag on the lunar surface. Of the other five flags left on the Moon, Apollo 11's no longer stands—it was blown over when the mission's ascent stage took off.

8. Eugene Shoemaker
On July 31, 1999, NASA's Lunar Prospector was deliberately crashed onto the Moon at the end of its mission. It carried some of the ashes of Apollo geologist Eugene Shoemaker (1928–1997), who thus became the first person to be buried on the Moon.

Lunar Module (LM) Pilot James Irwin salutes the United States flag on August 2, 1971. Apollo 15 was the first mission to use a Lunar Roving Vehicle (right), which enabled the astronauts to explore areas within 3 miles (5 km) of the LM landing site.

DUST AND GAS

Various missions and observations have shown that although the Moon lacks a substantial atmosphere, its gravity is strong enough to trap sparse clouds of gas and dust into its orbit.

Although the Moon's lack of atmosphere was confirmed by the 19th century, early space missions observed a faint light above the lunar surface. This "lunar horizon glow," as it became known, was first spotted by NASA's Surveyor probes in the 1960s and was later sketched by astronauts in lunar orbit. The glowing material is most likely to be dust that has fountained above the lunar surface, lifted by electrostatic repulsion as it is alternately bombarded by radiation during the day and by solar wind (charged particles) at night.

Lunar gases

During the Apollo era, a number of gases were discovered around the Moon. Apollos 15, 16, and 17 each carried mass spectrometers that analyzed the space they traveled through, Apollo 17 deploying a similar device on the lunar surface, as well as equipment designed to measure solar wind particles. Together, these instruments showed that the Moon's dominant gases are argon and helium, although it also has traces of more complex molecular gases, such as carbon monoxide, carbon dioxide, methane, oxygen, and nitrogen.

In the 1980s, Earth-based researchers detected further elements of the tenuous lunar atmosphere, observing emissions from sodium and potassium vapor. The ratio of these gases matched that of the sodium and potassium found on the lunar surface, suggesting that their origin was "sputtering"—a process by which solar wind particles collide with and eject surface atoms. Sputtering, along with micrometeorite bombardment, constantly injects new material into near-lunar space, explaining how the Moon can have developed a thin, surface-layer atmosphere long after the gases from its formation were lost to space.

Lunar horizon glow
Although most of the glow in this backlit image of the Moon comes from the Sun's corona (atmosphere), part of it is light being reflected from high-altitude lunar dust.

Lunar Prospector and LADEE

Other lunar gases were discovered in the 1990s, when NASA's Lunar Prospector satellite identified radon and polonium close to the Moon's surface. These gases are produced when radioactive uranium in the lunar crust decays, transforming into gaseous radon that escapes in a process known as "outgassing." More recently, in 2013–2014, the Lunar Atmosphere and Dust Environment Explorer (LADEE) satellite discovered lunar neon, originating from the solar wind, and found that the Moon's argon is outgassed during the decay of radioactive potassium.

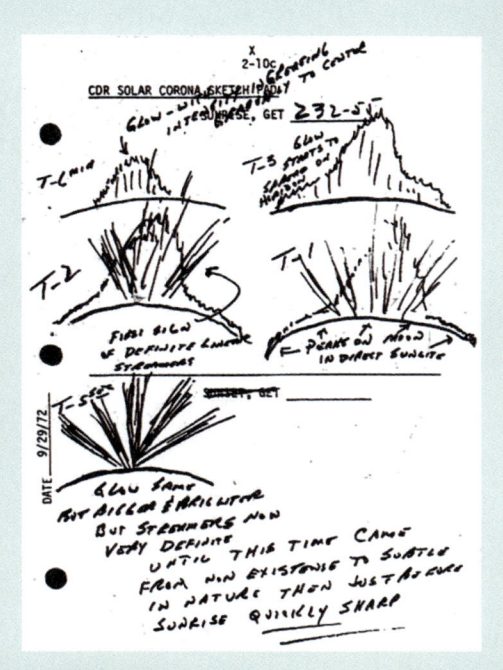

Crepuscular rays
Apollo 17's Gene Cernan sketched crepuscular rays rising above the Moon's horizon before sunrise, caused by light scattering in the thin lunar atmosphere.

The Moon's atmosphere exerts a pressure just **3 million billionths** of that of Earth.

KORDYLEWSKI CLOUDS

There are five places in the Moon's orbital path where the gravitational fields of the Moon and Earth are balanced, creating areas where objects could orbit Earth without being disturbed by the Moon. Scientists have long predicted that escaped lunar dust should gather in clouds at two of these "Lagrangian points," as they are known—L4 and L5, which lie ahead of and behind the Moon. Polish astronomer Kazimierz Kordylewski discovered such clouds in 1961, but their existence was only conclusively proved in 2018.

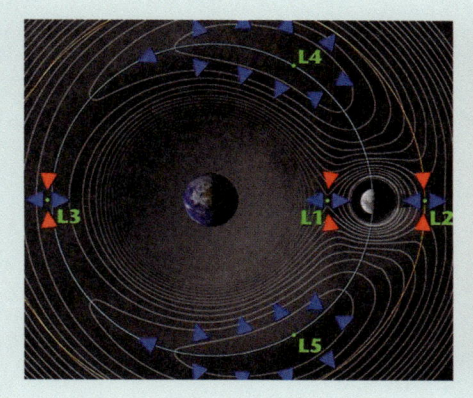
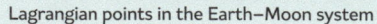

Lagrangian points in the Earth–Moon system

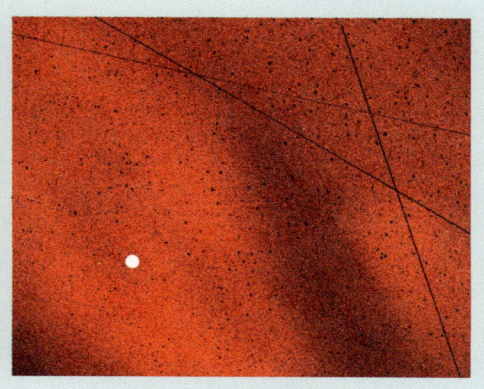

Dust at L5, reflecting distinctive polarized light

TRANSIENT LUNAR PHENOMENA

For centuries, astronomers have reported seeing brief glows and sparkles on the lunar surface. Known as Transient Lunar Phenomena (TLPs), these are often seen in places where the lunar crust is under stress, suggesting that at least some TLPs may be caused by outgassing. Other instances may be caused by asteroid strikes or light scattering in clouds of lunar dust.

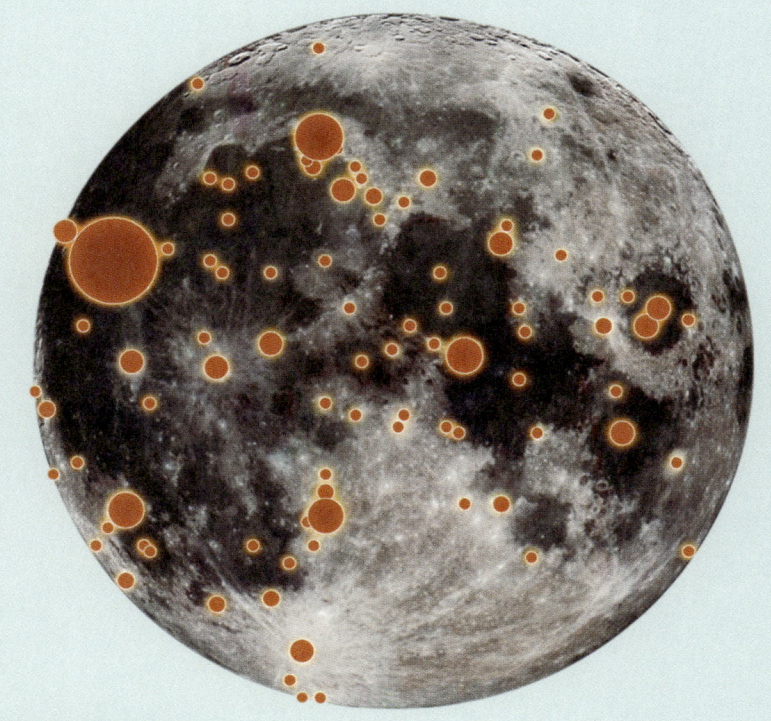

Locations of recorded TLPs

REMAPPING THE MOON

Since the 1990s, spacecraft and lunar probes have employed innovative technologies developed for studying Earth and other planets to discover new aspects of our satellite.

Gravity map
GRAIL's map of the Moon's gravity identified 74 ring-shaped areas (colored red and blue) where gravity rises and falls steeply due to concentrations of mass known as "mascons." They tend to coincide with basins filled with dense basalt rocks (see pp.112–113).

An unexpected benefit of the space race was the fresh perspective it offered on our own planet. Astronaut reports of the surprising detail visible from orbit soon inspired engineers to develop specialized "remote sensing" equipment for studying Earth from above. Since the 1990s, the same technology has been used to study the Moon.

One method, called multispectral imaging, involves photographing the lunar surface through narrow-wavelength filters that correspond to specific ranges of visible,

ultraviolet, and infrared light. Different parts of the surface reflect or absorb the Sun's light in different ways, depending on their physical properties. This enabled spacecraft such as Clementine, which launched in 1994, to produce detailed maps of the Moon's surface composition. Lunar Prospector, launched in 1998, achieved greater precision by measuring gamma-ray and particle emissions—discharges produced by the radioactive decay of unstable nuclei and the interactions of particles with the solar wind (see pp.190–191).

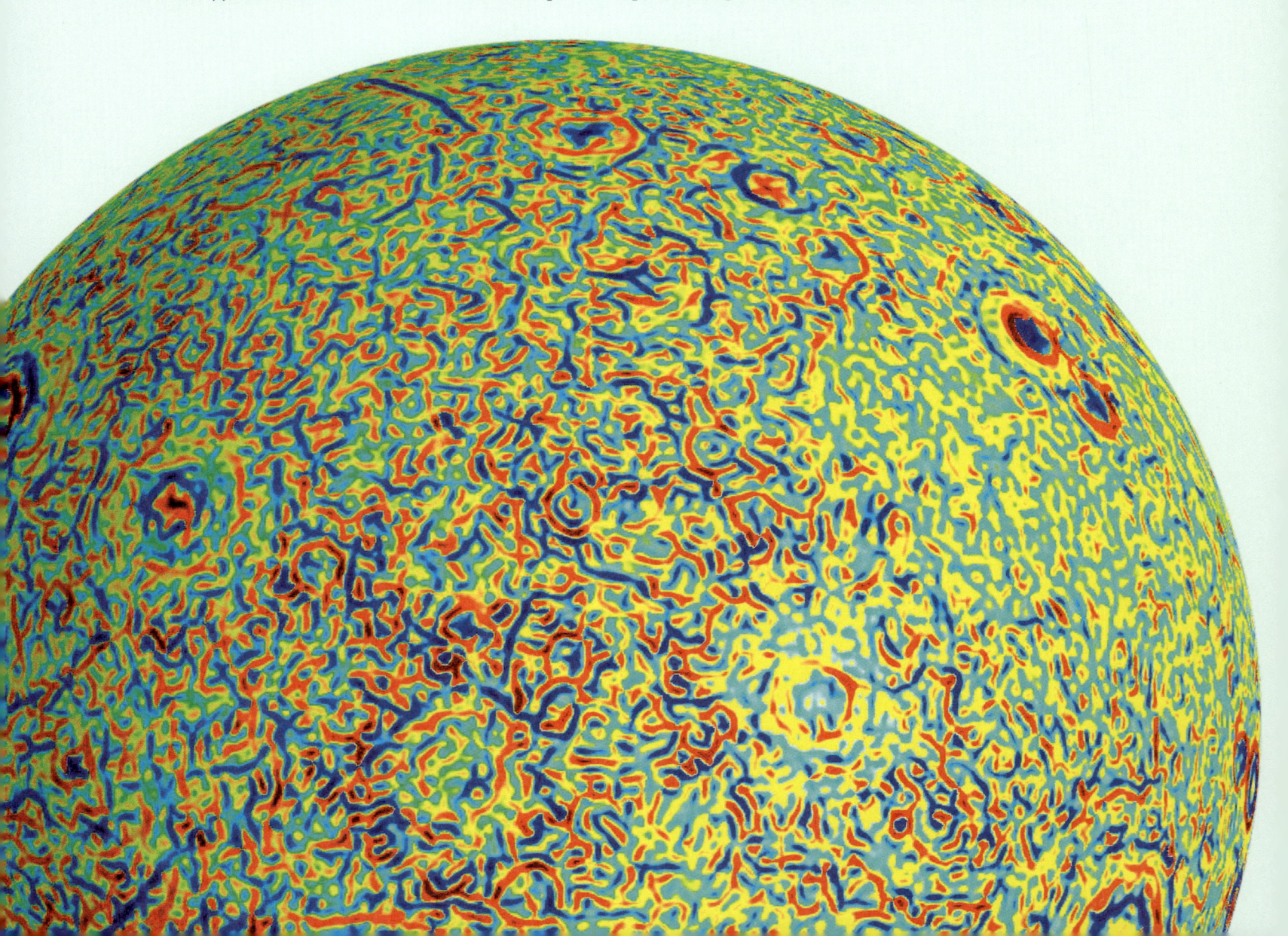

Three dimensions and beyond

Clementine was the first mission to map the Moon's landscape in three dimensions using a Laser Image Detection and Ranging (LIDAR) system, which measured the precise distance between Clementine and the lunar surface based on the return time of laser-light pulses. Launched in 2009, Lunar Reconnaissance Orbiter (LRO) employed a similar technology, Synthetic Aperture Radar (see below), to map a range of surface properties. Japan's SELENE, launched in 2007, used two telescopic cameras to image the surface from slightly different angles, creating a stereoscopic (3D) image.

Using other techniques, orbiting probes can map hidden properties of the Moon's surface and interior. The Moon lacks a global magnetic field, but Lunar Prospector detected regional magnetic fields and trace magnetic minerals on the surface (see right). Tracking the speed of probes along their orbit allowed engineers to map variations in the Moon's gravity due to internal mass concentrations. Initially, this could only be done via direct radio link to Earth and so was restricted to the lunar near side, but in 2007, SELENE bounced signals back from the far side by relay satellite. In 2012, the twin GRAIL (Gravity Recovery and Interior Laboratory) satellites produced an even more accurate gravity map.

Mapping minerals
NASA's Galileo mission took multispectral images of the Moon in 1992. Anorthosite-rich highlands are red, while titanium-rich lowlands range from orange to blue.

MAGNETIC ROCKS

Lunar Prospector carried an Electron Reflectometer to measure the magnetism of lunar rocks, based on the angles at which electrons from the solar wind deflected as they bounced back into space. It showed that magnetic areas often lie directly on the opposite side of the Moon from its major impact basins.

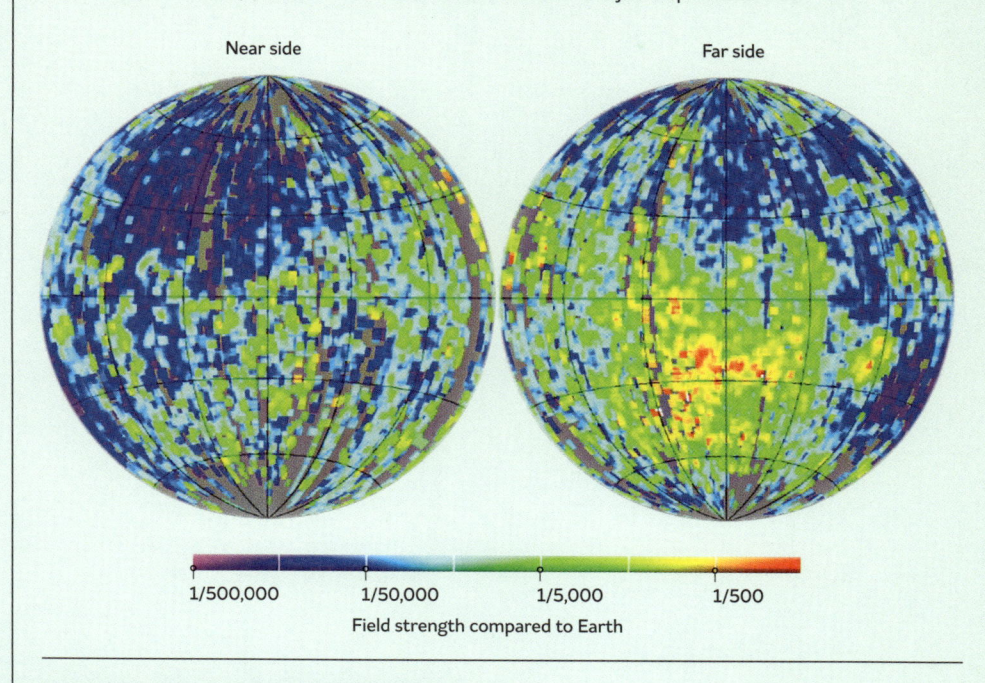

Near side   Far side

| 1/500,000 | 1/50,000 | 1/5,000 | 1/500 |

Field strength compared to Earth

SYNTHETIC APERTURE RADAR (SAR)

LRO's Miniature Radio Frequency (Mini–RF) device uses synthetic aperture radar, which beams radio waves at the lunar surface and measures their return as the satellite moves along its orbit. Properties such as elevation, slope, roughness, and reflectivity can be calculated by analyzing the returning signal.

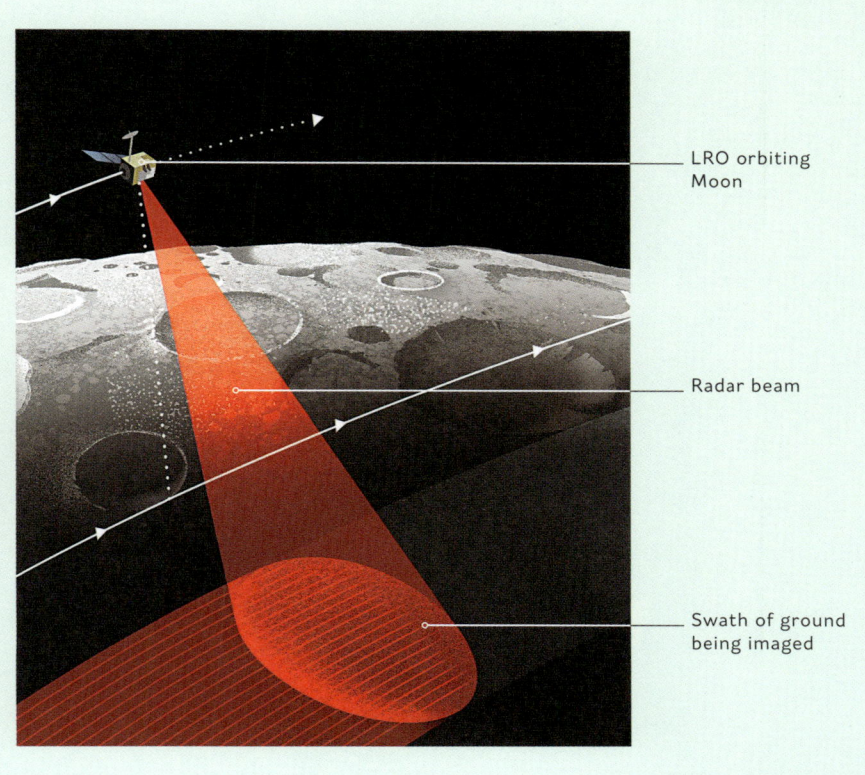

LRO orbiting Moon

Radar beam

Swath of ground being imaged

A NEW GENERATION
OF PROBES

Since the 1990s, numerous uncrewed automated space probes have been sent to the Moon, dispatched by a variety of nations and even by commercial firms (see pp.218–219).

Chinese rover
The Chang'e-3 lander snapped this picture of its Yutu ("Jade Rabbit") rover in December 2013. Although difficult surface conditions soon limited the rover's operation, it was able to continue transmitting data.

In 1992, with its Hiten probe, Japan became the third country to put a spacecraft in orbit around the Moon. Originally planned to follow an elongated orbit of Earth and study dust in near-Earth space, its mission was changed after the failure of the small Hagoromo probe that Hiten was meant to release into lunar orbit.

Over the next 12 years, only NASA's Clementine and Lunar Prospector probes went to the Moon (in 1994 and 1998–1999 respectively), followed by the European Space Agency's SMART-1, which entered lunar orbit in November 2004. SMART-1 used a variety

Low-angle sunrise
NASA's Lunar Reconnaissance Orbiter took this image of Tycho crater (left). The boulder on top of the 6,700-ft (2,000-m) central peak is about 330 ft (100 m) wide.

of new techniques to investigate the lunar surface, utilizing an ion engine (a high-efficiency rocket) to get there.

A month after Japan returned to the Moon in October 2007 with its SELENE orbiter, China followed with Chang'e 1, beginning a series of ambitious Chinese lunar missions (see pp.200–201). India's first foray into lunar exploration, Chandrayaan-1, reached orbit in November 2008 and found strong evidence for ice in the regolith (see pp.196–197).

Over the next five years, NASA launched a series of probes intended to pave the way for new crewed Moon landings. In June 2009, Lunar Reconnaissance Orbiter launched alongside the ingenious LCROSS impactor (see pp.196–197), while in January 2012, the twin GRAIL (Gravity Recovery and Interior Laboratory) satellites slipped into orbit for a year-long mission to map the Moon's gravity field. Finally, from September 2013, LADEE made studies of the lunar atmosphere and dust halo (see pp.190–191). In December 2013, meanwhile, China claimed another impressive first with the landing of its Chang'e 3 rover.

The next wave

In 2019, a further wave of probes began with the privately funded Israeli Beresheet, but it crashed into the Moon in April instead of making its intended soft landing. India's Chandrayaan-2 reached orbit in July of the same year, although its Vikram lander and rover also crashed. In 2022, however, South Korea's first Moon probe, Danuri, began a successful mission in lunar orbit.

Rashid, a rover developed by the United Arab Emirates, was destroyed along with its parent Japanese lander, Hakuto-R, during its descent in April 2023. India finally reached the surface when the orbiting Chandrayaan-3 deployed a new Vikram lander in August 2023 (shortly after the crash of Luna 25, Russia's attempt at a lunar return). In early 2024, Japan's SLIM and the privately developed US *Odysseus* (see above) both reached the surface intact, only to encounter problems due to uneven landings that hampered the illumination of their solar panels.

Odysseus flies free
Cameras aboard a SpaceX Falcon 9 rocket snapped the IM-1 Nova-C lander—named *Odysseus* and developed by US company Intuitive Machines—at the beginning of its lunar mission in February 2024.

The **Lunar Reconnaissance Orbiter camera** can reveal details as small as **20 in (50 cm)** across from a height of **31 miles (50 km)**.

Chandrayaan-3 landing
In 2023, India became the first nation to land near the Moon's south pole. Its Vikram probe touched down at latitude 69°S. Pragyan, a 57-lb (26-kg) rover, explored the surface for 12 days until lunar nightfall.

Lunar near–side ice
This map shows frozen water and hydroxyl ions (blue/purple), pyroxene (orange), and sunlight intensity (green).

IN SEARCH OF ICE

For decades, the Moon was presumed to be arid and waterless.
In the 21st century, however, probes found signs of ice near its poles,
sparking a search that could transform future exploration plans.

The rise of spectroscopy (the study of the absorption and emission of light and other radiation by matter) in the 19th century gave scientists a better understanding of the Moon's rocky surface. By 1892, astronomer William Pickering had concluded that the Moon must be entirely dry owing to its lack of atmosphere and great temperature variations. Apollo rock samples later confirmed the absence of "hydrated" water-containing minerals.

Breakthrough investigations

As early as 1961, however, some scientists proposed that near the poles, craters with deep floors that lie in permanent shadow might preserve frozen water delivered by ice-rich comets. The possibility had major implications for the practicality of future crewed lunar exploration, and its investigation became a major goal of lunar probes from the 1990s on.

Decisive evidence had to wait until India's Chandrayaan-1 mission (2008). Its Moon Mineralogy Mapper experiment (M3; see below) detected features linked solely to icy material. Meanwhile, its Moon Impact Probe (deliberately crashed into an icy crater) found water in the lunar atmosphere prior to impact.

In 2009, NASA resolved any lingering doubts with its Lunar Crater Observation and Sensing Satellite (LCROSS) mission. Launched along with the Lunar Reconnaissance Orbiter, it consisted of a Shepherding Spacecraft trailing behind a Centaur rocket booster, both

Ice detection
NASA's LCROSS analyzed icy material ejected from the crater Cabeus after it was hit by an empty 5,000-lb (2.3-tonne) Centaur rocket stage (on the right).

targeting the southern crater Cabeus. In the six minutes between the booster's impact and its own demise, the Shepherding Spacecraft confirmed escaping ice in the Centaur's ejecta cloud. The quantity of icy material present and its accessibility remain questions to be answered by future probes and landers.

MAPPING THE MOON'S ICE

One of two NASA science experiments on India's Chandrayaan-1 orbiter, the M3 was an imaging spectrometer that captured a high-resolution map of the Moon's surface in 260 different spectral "channels." This enabled scientists to analyze reflected and emitted radiation across a range of wavelengths, identifying water, related molecules, and minerals on the lunar surface from the way they absorb infrared radiation (see opposite).

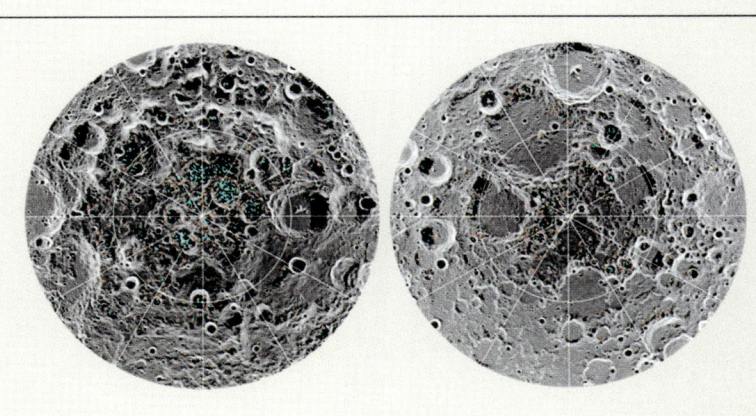

Polar deposits
M3 maps of the Moon's south (far left) and north poles reveal surface ice (highlighted in blue) in permanently shadowed craters.

THE CRUST AND THE CORE

The rocky layers that make up the Moon's internal structure have been shaped not only by events early in its history, but also by our satellite's intimate relationship with Earth.

The Moon's gravity and the residual fluidity of its rocks due to heat from its early molten period have caused its interior, like those of other large solar system bodies, to separate into distinct layers with different densities and compositions: the mantle and core. The layers, although inaccessible beneath the solid crust, can be probed using data from seismometers deployed during the Apollo Moon landings, which operated for several years until they were switched off in 1977. Lunar seismology makes use of the fact that shockwaves from moonquakes (see pp.168–169) change speed and direction as they cross between layers. Gravitational measurements from orbiting satellites also enable scientists to calculate the layers' density and map concentrations of mass closer to the Moon's surface. Analysis of lunar rock samples and comparison with minerals on Earth, meanwhile, reveals their probable composition. Combined with simulations of events in early lunar history, this data provides us with our most detailed understanding of the interior of another world.

Cutaway Moon
Under the crust, most of the interior is rocky mantle. The iron–rich core is relatively small, probably because heavier elements from Earth and the colliding planet Theia (see pp.12–13) were mostly absorbed by Earth during the impact that created the Moon.

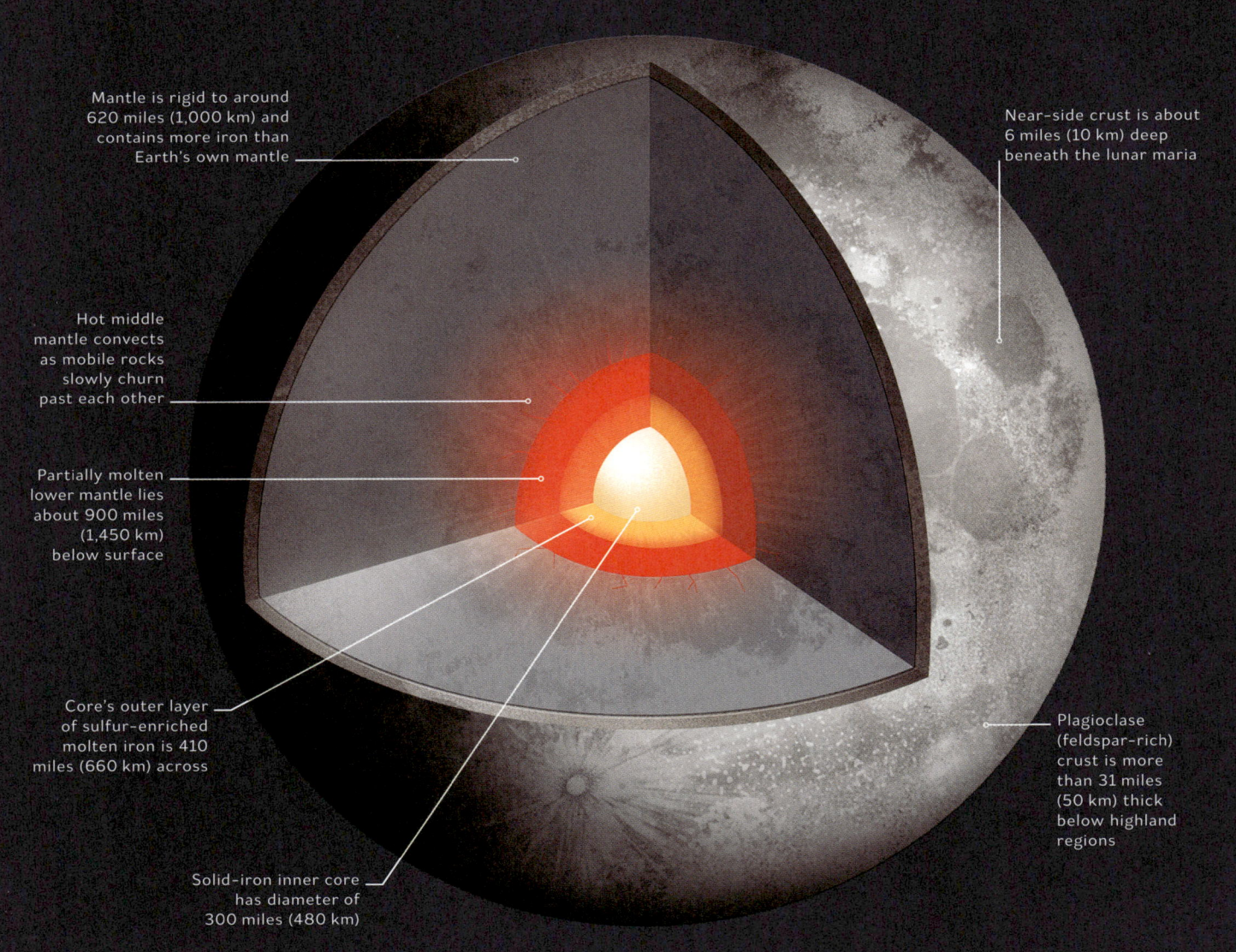

Mantle is rigid to around 620 miles (1,000 km) and contains more iron than Earth's own mantle

Near-side crust is about 6 miles (10 km) deep beneath the lunar maria

Hot middle mantle convects as mobile rocks slowly churn past each other

Partially molten lower mantle lies about 900 miles (1,450 km) below surface

Core's outer layer of sulfur–enriched molten iron is 410 miles (660 km) across

Plagioclase (feldspar–rich) crust is more than 31 miles (50 km) thick below highland regions

Solid–iron inner core has diameter of 300 miles (480 km)

OFFSET STRUCTURE

Measurements from orbit reveal that the Moon's center of mass is offset by about 1.2 miles (2 km) from its geometric center, and almost directly in line with the direction of Earth. The offset was once attributed to Earth's gravity pulling the lunar core toward it. Today, the core's offset position is thought to be a result of variations in the crustal thickness, which themselves intensified the tidal forces that slowed the Moon's rotation in its early history (see pp.16–17).

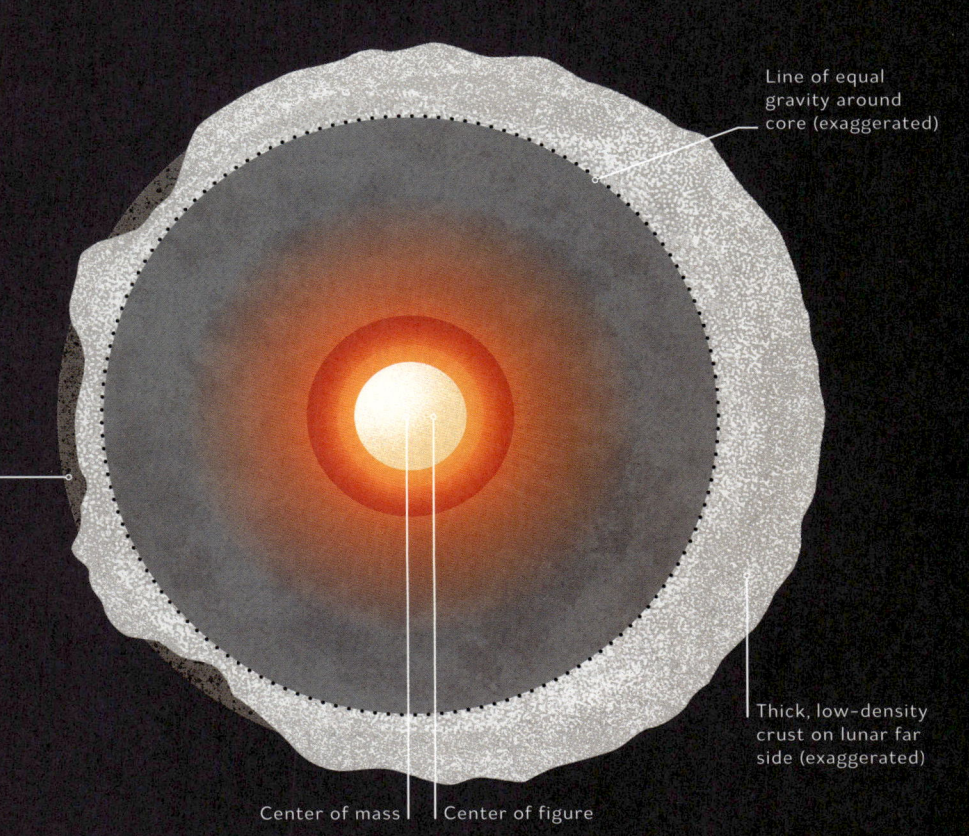

Line of equal gravity around core (exaggerated)

Thin, dense crust on lunar near side

Thick, low-density crust on lunar far side (exaggerated)

Center of mass   Center of figure

Thick crust, thin crust
Overall, the crust on the far side is much thicker than that beneath the lunar near-side maria. This is probably the result of repercussions from the impact that formed the South Pole–Aitken basin (see pp.136–137).

LAYER FORMATION

After the giant impact that formed it, the young Moon stayed molten long enough for dense materials to sink to its center and buoyant ones to rise to the surface. Different elements and compounds solidified into crystalline minerals under varying temperatures and pressures, leaving the remaining material in a liquid "melt" that continued to transport solid materials due to convection currents until virtually the entire interior had solidified.

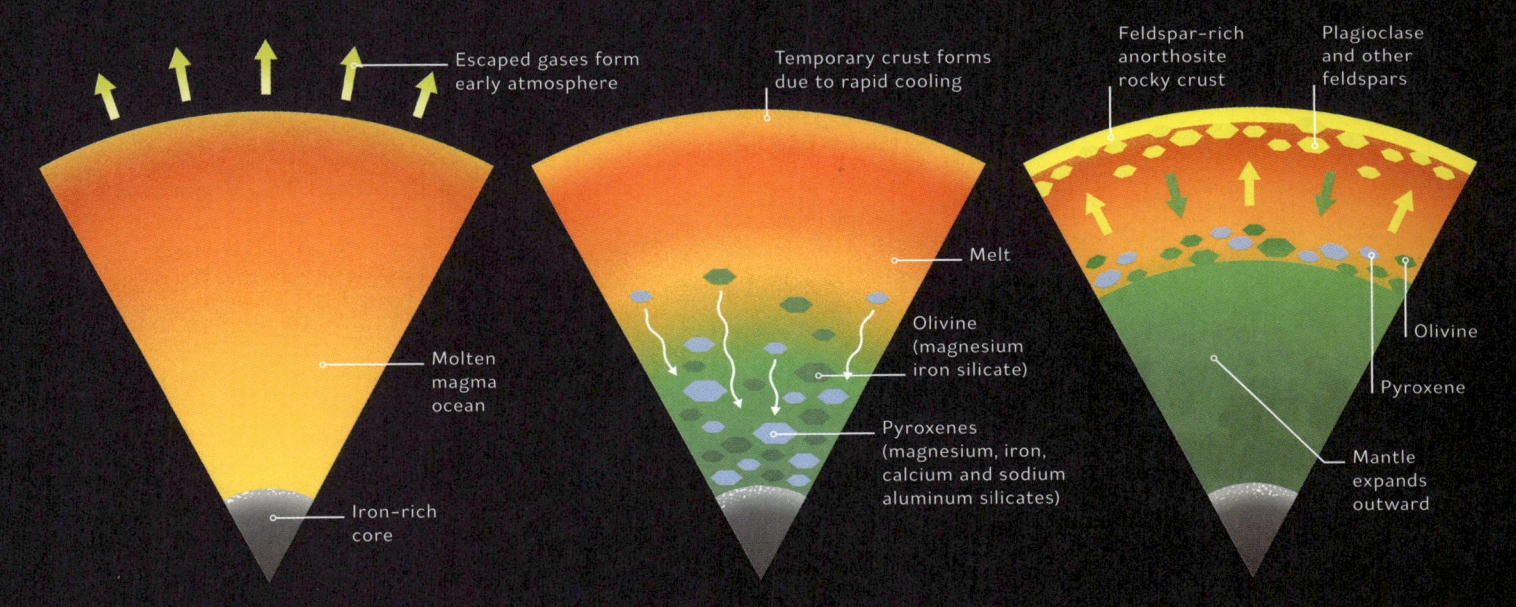

Escaped gases form early atmosphere

Molten magma ocean

Iron-rich core

Temporary crust forms due to rapid cooling

Melt

Olivine (magnesium iron silicate)

Pyroxenes (magnesium, iron, calcium and sodium aluminum silicates)

Feldspar-rich anorthosite rocky crust

Plagioclase and other feldspars

Olivine

Pyroxene

Mantle expands outward

1. Core formation
Heavy metals sink to the core within 1,000 years of the impact. Gases escaping from the interior form a thick early atmosphere.

2. Mantle accumulation
A temporary crust forms. As olivine and pyroxene minerals solidify and fall inward, a mantle accumulates around the core.

3. Stabilized crust
Further mantle minerals fall inward, while feldspars crystallize and rise upward, forming a thicker and more stable crust.

EXPLORING THE FAR SIDE

Always out of sight and beyond direct communication with Earth, the Moon's far side presents a huge exploration challenge. Since 2019, Chinese robot probes have pioneered its investigation.

Yutu-2 rover
The 310-lb (140-kg) "Jade Rabbit" rover can reach a top speed of 656 ft (200 m) per hour. It has six-wheeled suspension and the ability to climb slopes of up to 30°.

Named after Chinese Moon goddess Chang'e (see pp.66–67), China's Moon program began in 2007 with the first of two lunar orbiters. Its second phase commenced with Chang'e 3's soft landing on the near side in 2013; the mission also deployed a small, six-wheeled rover named Yutu ("Jade Rabbit"), after the goddess's traditional companion.

An identical follow-up mission was already in development, but the success of Chang'e 3 encouraged a more ambitious plan: The Chang'e 4 mission would use a relay satellite (see opposite page) to send back data from the far side of the Moon. The Queqiao-1 ("Magpie Bridge") satellite was duly launched in May 2018, and on January 3, 2019, the Chang'e 4 lander touched down in Von Kármán crater, within the far side's vast South Pole-Aitken basin.

Scientific goals

Both Chang'e 4 and its Yutu-2 rover used solar power for daytime operations and radioisotope heaters (see pp.172–173) to survive the cold lunar night in standby mode. In addition to cameras, the lander carried instruments to study solar radiation, test its possible effects on simple organisms, and measure doses that might be received by astronauts. The rover, meanwhile, carried a ground-penetrating radar, an imaging spectrometer to view the surface at different light wavelengths, and a Swedish experiment to measure the effects of the solar wind on exposed materials. While the original Yutu's mission was curtailed by mechanical failure within weeks of landing, its successor was able to far exceed the planned three-month mission. By May 2024, it had explored almost 1 mile (1.6 km) of lunar terrain, providing valuable data about lunar mantle rocks exposed in the formation of the deep Von Kármán crater, and also finding traces of smaller and more recent impacts.

New rock samples

In late 2020 and mid-2024, a third phase of Chinese plans unfolded as the Chang'e 5 and 6 missions landed on the near and far sides respectively, collecting the first lunar rock samples to be returned to Earth since 1976. Future missions are expected to explore the potential of making useful resources from lunar raw materials (see pp.224–225), paving the way for future crewed landings.

Far side panorama

Chang'e 4's first view from its landing site shows the freshly deployed Yutu-2 rover nearby. Lava flooding of Von Kármán crater at some point after its formation is responsible for the mostly flat landscape.

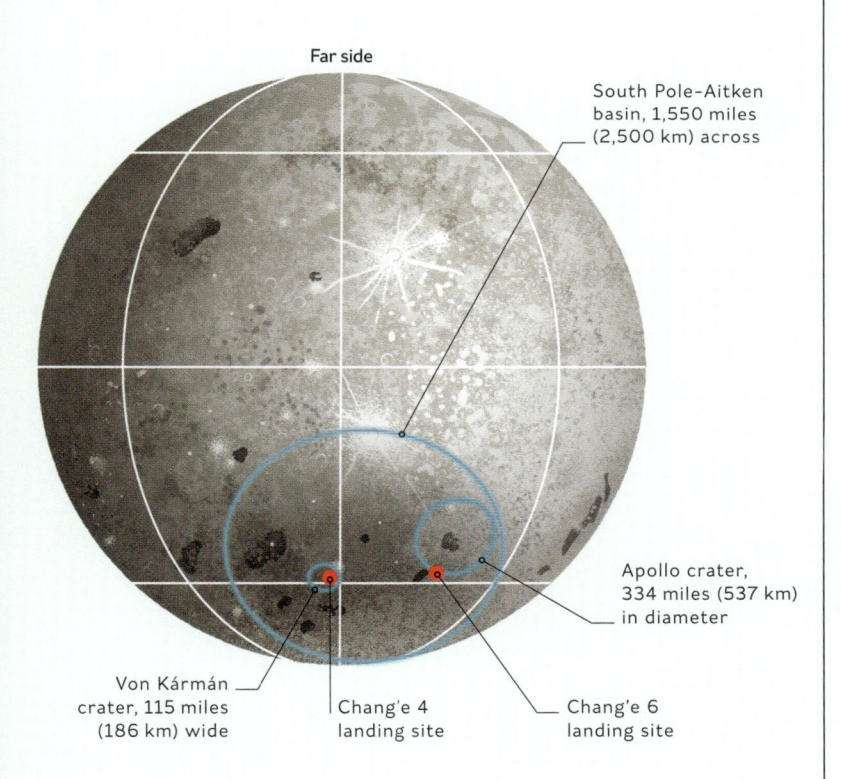

Far side

South Pole-Aitken basin, 1,550 miles (2,500 km) across

Apollo crater, 334 miles (537 km) in diameter

Von Kármán crater, 115 miles (186 km) wide

Chang'e 4 landing site

Chang'e 6 landing site

Landing sites

Despite appearances on a flat lunar map, both Chang'e 4 and Chang'e 6 landed at relatively midlatitude sites (45.4°S and 41.6°S respectively). This enabled them to maintain clear contact with the Queqiao-1 relay satellite while exploring their surroundings in the South-Pole-Aitken basin.

KEEPING IN TOUCH

The Queqiao-1 relay satellite was placed into an orbit that maintains more or less the same orientation to both Earth and the Moon, keeping both in sight and ensuring constant communication. However, sustaining this orbit requires intermittent engine burns, limiting the satellite's operational lifetime. For this reason, Queqiao-2, launched in March 2024, instead follows a "frozen orbit" around the Moon itself—an inclined elliptical path that ensures gravitational disturbances from Earth are minimized. No engine burns are needed to stabilize this orbit, although contact with the far side is sometimes lost when the satellite passes behind the Moon.

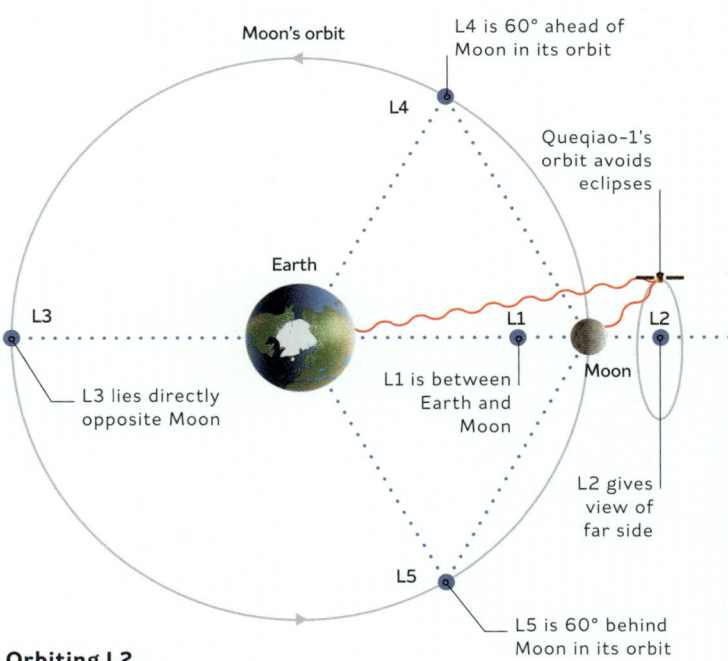

Moon's orbit

L4 is 60° ahead of Moon in its orbit

Queqiao-1's orbit avoids eclipses

Earth

L3

L3 lies directly opposite Moon

L1 is between Earth and Moon

Moon

L2

L2 gives view of far side

L5 is 60° behind Moon in its orbit

Orbiting L2

Queqiao-1 orbits the L2 Lagrange point, 38,100 miles (61,300 km) beyond the Moon. Lagrange points are five locations in the Earth–Moon system where the influences of the two bodies balance so that a third small object can follow a stable orbit.

INTO THE FUTURE

In the current decade, NASA plans to return crewed missions to the Moon. These will explore the lunar surface and set up the Gateway station, which will serve as a hub for astronauts working in lunar orbit. Russia and China also have crewed expeditions planned, and private companies have similar goals for the future. Collectively, their aims include establishing a permanent base on the Moon and using the Moon as a stepping stone to Mars.

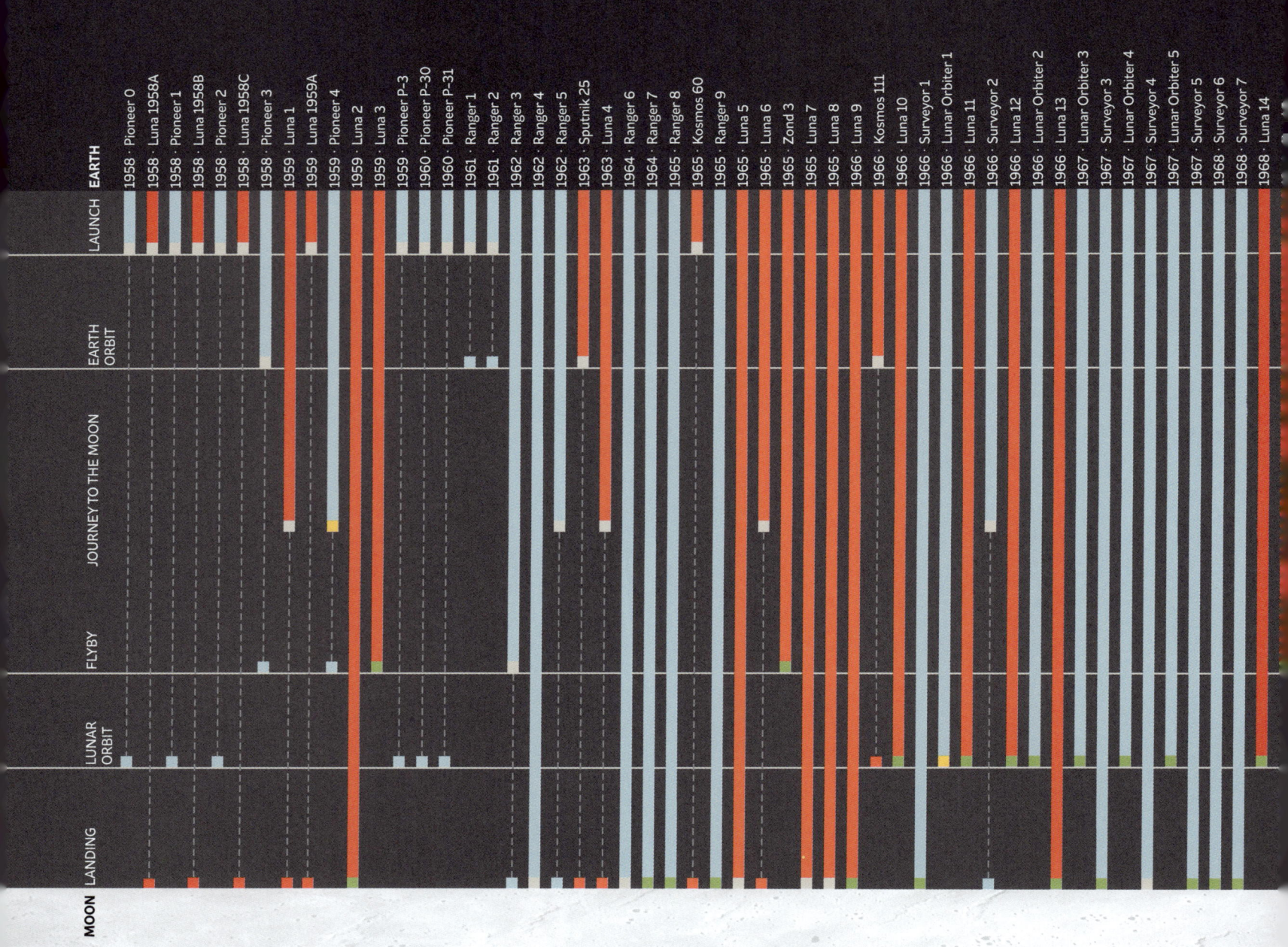

MISSIONS TO THE MOON

There have been dozens of missions to the Moon, during almost 70 years of efforts to reach our nearest astronomical neighbor, using both robotic and crewed spacecraft.

Starting in the 1950s, most of the early Moon missions were unsuccessful. For example, the Soviet Luna 1 probe—the first spacecraft to travel beyond Earth's orbit—missed the Moon on its approach owing to a technical error. However, spacecraft continued to head moonward, including those of NASA's six successful Apollo missions, which carried 12 astronauts to the lunar surface. Traditionally, lunar exploration was dominated by the US and the Soviet Union (USSR), but in recent years, other countries have launched Moon missions, including China, India, and Japan, as well as the European Space Agency (ESA).

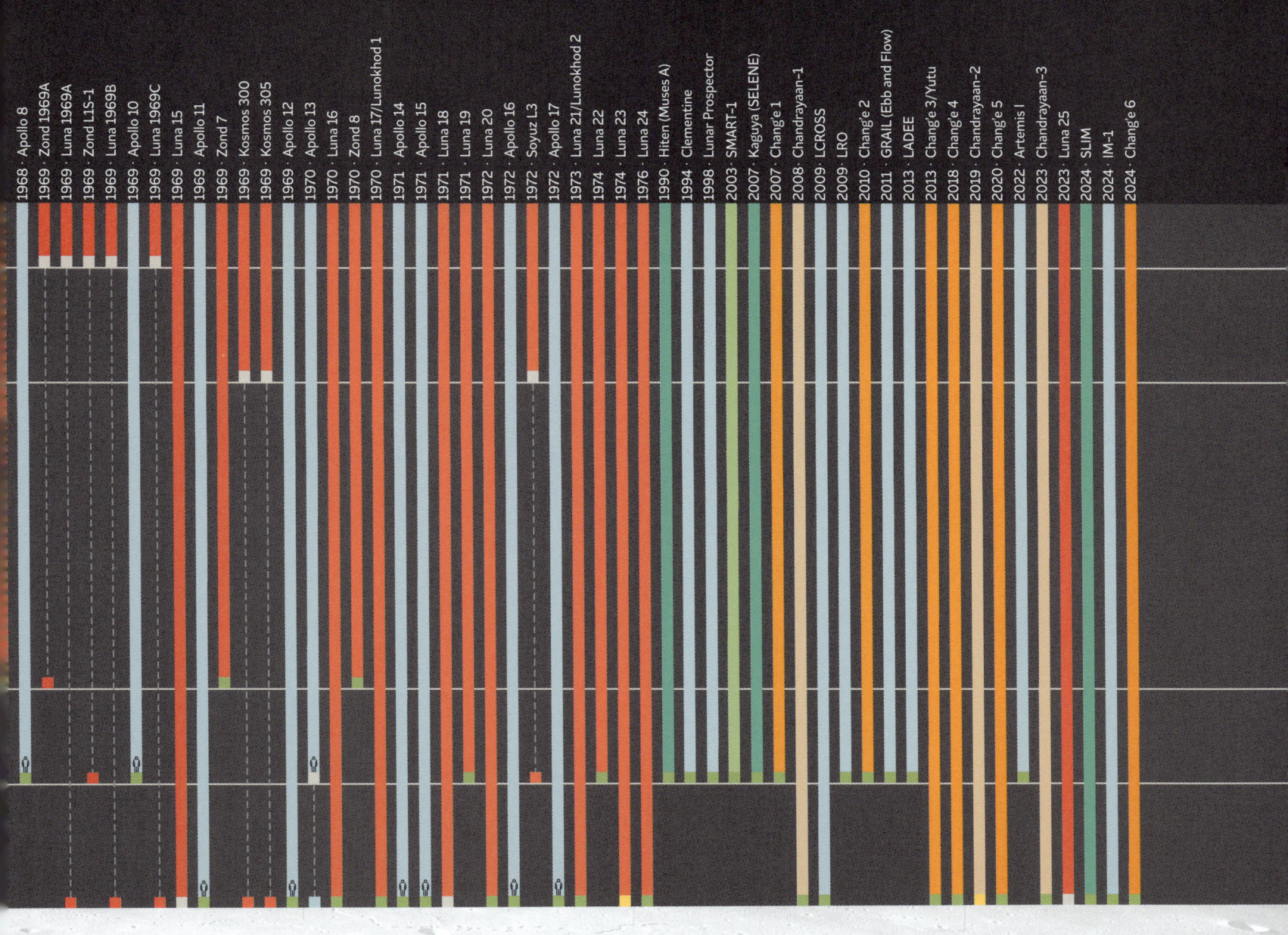

Mission timeline (top, left to right): 1968 · Apollo 8 | 1969 · Zond 1969A | 1969 · Luna 1969A | 1969 · Zond L1S-1 | 1969 · Luna 1969B | 1969 · Apollo 10 | 1969 · Luna 1969C | 1969 · Luna 15 | 1969 · Apollo 11 | 1969 · Zond 7 | 1969 · Kosmos 300 | 1969 · Kosmos 305 | 1969 · Apollo 12 | 1970 · Apollo 13 | 1970 · Luna 16 | 1970 · Zond 8 | 1970 · Luna 17/Lunokhod 1 | 1971 · Apollo 14 | 1971 · Apollo 15 | 1971 · Luna 18 | 1971 · Luna 19 | 1972 · Luna 20 | 1972 · Apollo 16 | 1972 · Soyuz L3 | 1972 · Apollo 17 | 1973 · Luna 21/Lunokhod 2 | 1974 · Luna 22 | 1974 · Luna 23 | 1976 · Luna 24 | 1990 · Hiten (Muses A) | 1994 · Clementine | 1998 · Lunar Prospector | 2003 · SMART-1 | 2007 · Kaguya (SELENE) | 2007 · Chang'e 1 | 2008 · Chandrayaan-1 | 2009 · LCROSS | 2009 · LRO | 2010 · Chang'e 2 | 2011 · GRAIL (Ebb and Flow) | 2013 · LADEE | 2013 · Chang'e 3/Yutu | 2018 · Chang'e 4 | 2019 · Chandrayaan-2 | 2020 · Chang'e 5 | 2022 · Artemis I | 2023 · Chandrayaan-3 | 2023 · Luna 25 | 2024 · SLIM | 2024 · IM-1 | 2024 · Chang'e 6

LANDING SITES

Many factors are considered when choosing where to land a spacecraft on the Moon (see pp.160–161). These include a suitable landscape (there are deep craters and peaks taller than Mount Everest on the Moon), scientific interest, and ease of communications. Apollo 11 aimed for Mare Tranquillitatis because it is relatively smooth and level. Shown here are a selection of notable touchdown points.

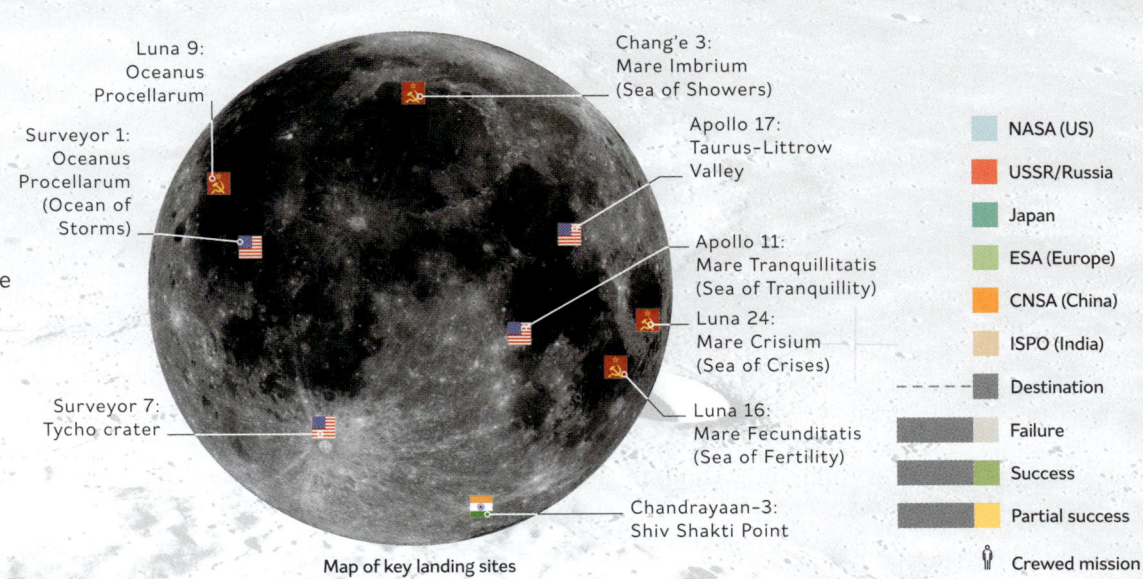

Luna 9:
Oceanus
Procellarum

Surveyor 1:
Oceanus
Procellarum
(Ocean of
Storms)

Surveyor 7:
Tycho crater

Chang'e 3:
Mare Imbrium
(Sea of Showers)

Apollo 17:
Taurus-Littrow
Valley

Apollo 11:
Mare Tranquillitatis
(Sea of Tranquillity)

Luna 24:
Mare Crisium
(Sea of Crises)

Luna 16:
Mare Fecunditatis
(Sea of Fertility)

Chandrayaan-3:
Shiv Shakti Point

Map of key landing sites

Legend:
- NASA (US)
- USSR/Russia
- Japan
- ESA (Europe)
- CNSA (China)
- ISPO (India)
- Destination
- Failure
- Success
- Partial success
-  Crewed mission

India's Chandrayaan-3 launch, from the Satish Dhawan Space Centre in Sriharikota, Andhra Pradesh, resulted in the first soft landing near the lunar south pole.

THE MOON RACE 2.0

Efforts to return astronauts to the Moon and establish a permanent lunar base are being led by the US and China. But this new "Moon Race" also involves many other nations and even commercial companies.

The 20th century's contest to put humans on the Moon was a two-horse race between the US and the Soviet Union. A second Moon race ("Moon Race 2.0") is now underway, with the two largest geopolitical powers in the world, the US and China, each carrying out uncrewed lunar landings in preparation for delivering astronauts to the surface. However, Moon Race 2.0 is more ambitious than its predecessor.

Moon Race 2.0 is often described as being between the US and its allies (signatories to the US-led Artemis Accords) and China and its allies (participants in the China-led International Lunar Research Station). The reality is more complex. In the past, the huge budgets required for space programs could only be met by government funding, which excluded all but the wealthiest countries. Over time, however, the costs of space travel have reduced considerably, not least because of investment from private companies and innovations such as SpaceX's reusable rockets (see pp.218–219). More than 80 countries now have space programs (see pp.194–195).

The competitors in Moon Race 2.0 aim to set new milestones in space exploration, including building permanent bases on the lunar surface and/or in lunar orbit. These would support scientific research and even act as a springboard for future missions to Mars (see pp.228–229).

The discovery of valuable resources on the Moon (see pp.224–225) could make lunar mining a lucrative prospect. Several countries have expressed an interest in future human landings on the Moon, with their attendant commercial potential and enhanced global prestige. This opens up the prospect of intense competition between states, as well as mutually beneficial cooperation, although codes of conduct and binding regulations for this new frontier remain a work in progress.

ASTRO-POLITICAL ALLIANCES

Broad geopolitical alliances are reflected in the way countries are engaging with the US-led Artemis Accords (agreements laying out norms for space exploration) and the International Lunar Research Station led by China and Russia. However, most nations are currently not involved with either project.

ARTEMIS

Lead nation: United States
Other participants: Currently more than 40 countries, many of them in Europe and the Americas, but also including India, Japan, South Korea, and Israel.
Mission: To return astronauts to the surface of the Moon, establish a permanent base on the lunar surface, and a space station in lunar orbit.

Artemis logo

SPACE LAUNCH SYSTEM (SLS)

The backbone of NASA's Artemis program, the SLS rocket (see pp.210–211) comes in a range of configurations, or "blocks," suited to different mission types. Block 1 can deliver 30 tons (27 tonnes) to the Moon; Block 2 will be able to carry up to 51 tons (46 tonnes) into deep space.

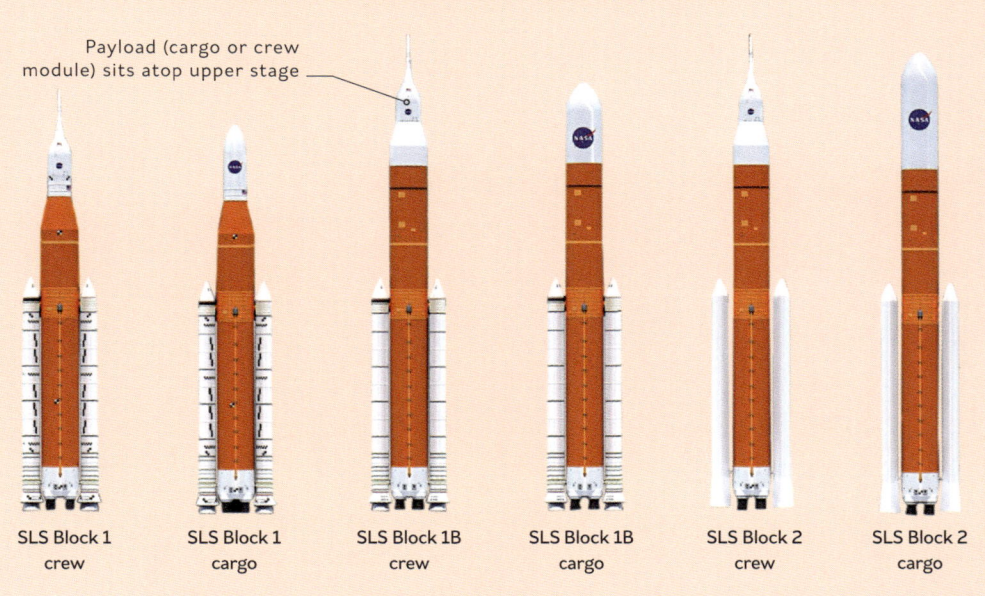

Payload (cargo or crew module) sits atop upper stage

SLS Block 1 crew   SLS Block 1 cargo   SLS Block 1B crew   SLS Block 1B cargo   SLS Block 2 crew   SLS Block 2 cargo

ILRS

Lead nations: China, Russia
Other participants: Currently around a dozen countries, including Pakistan, Venezuela, South Africa, Egypt, and Thailand.
Mission: To build a research station either on the lunar surface or in lunar orbit.

Chinese National Space Administration logo   Roscosmos (Russian space agency) logo

THE ARTEMIS ERA

Spearheaded by the US, a new Moon-exploration program called Artemis aims to return humans to the lunar surface and build a permanent base that could be a staging post on voyages to Mars.

Artemis logo
A red trajectory (symbolizing the path to Mars) links the crescent Earth and half Moon. Shaped like an arrowhead from Artemis's quiver, the A represents launch and points beyond the Moon.

Artemis, named for the ancient Greek goddess of the hunt and twin sister of Apollo, will explore the Moon for scientific research, technological development, and economic benefit, with a view to making it possible to live and work on the Moon. The long-term goal is to establish a permanent lunar presence, making the Moon a stepping stone for future crewed missions to Mars.

Artemis will use a very large and powerful rocket, the Space Launch System (SLS), to propel the Orion spacecraft toward the Moon (see pp. 210–211). The first mission, the uncrewed Artemis I, took place in 2022. The SLS successfully launched Orion, which orbited the Moon for several days before returning to Earth and splashing down safely in the Pacific Ocean. Artemis II, the next major

mission, will test the SLS and Orion with a crew aboard. Artemis III (see pp. 212–213) will land astronauts on the Moon for the first time in over half a century, using the new Human Landing System (HLS) to touch down near the Moon's south pole. Astronauts on later missions will journey to and from the surface via the Lunar Gateway—a space station that NASA plans to build in orbit around the Moon. Support missions will carry components of the Lunar Gateway for assembly in lunar orbit and survey the Moon for useful resources.

While NASA is leading the Artemis program, it is working alongside private companies—such as SpaceX and BlueOrigin (through the Commercial Lunar Payload Services initiative)—in addition to the national space agencies of several US allies.

FUTURE ROVERS

Artemis will use robots, as well as astronauts, to explore the Moon, including a new generation of lunar rovers. These vehicles will need to be tough enough to cope with the uneven terrain, have the resilience to function in extreme heat and cold, and possess the dexterity to perform a range of operations (such as constructing a communications array).

THE FIRST ARTEMIS CREW

The Artemis II crew was announced in April 2023. Clockwise from left: Christina Koch, Victor J. Glover, Jeremy Hansen, and Reid Wiseman. They will be the first people to travel beyond low Earth orbit since the Apollo 17 astronauts in 1972.

Moon bus
Astronauts would be able to travel long distances in a pressurized vehicle, such as is shown in this NASA concept. NASA plans for these rovers to utilize natural resources—for example, converting water ice into oxygen and fuel (see pp.220–221).

"We go for all and by all"
The crews will be more diverse than those of Apollo. Artemis aims to land the first woman and first person of color on the Moon.

On the surface
This NASA-commissioned artwork imagines Artemis astronauts on the lunar surface collecting samples and conducting experiments.

SPACE LAUNCH SYSTEM

NASA's Space Launch System (SLS) is a powerful new rocket designed for crewed missions to the Moon, and is an integral part of the agency's plans for deep-space exploration.

After completing the Apollo project, NASA developed the Space Shuttle, which it used as a vehicle for 135 missions over the next four decades. These included helping to construct the International Space Station in low-Earth orbit. Then, in 2011, NASA began work on its successor: SLS.

Engineers incorporated some technology in the new vehicle that had been used in the Space Shuttle, including the RS-25 engine. They also designed it to have different configurations, known as "blocks," each of which utilize the same 213-ft- (65-m-) long rocket core, with upper stages and boosters suited to the increasingly ambitious missions of the Artemis program. For example, the Exploration Upper Stage allows for greater payload mass (the ability to carry heavier cargo) and has enhanced safety features that will be useful during the later Artemis missions.

Returning to the Moon

SLS is the world's most powerful rocket and is designed to be able to launch heavy and large payloads into deep space. Critically, it can send the Orion Module, four astronauts, and an unprecedented amount of cargo to the Moon with a single launch. Its first mission was Artemis I, which sent Orion on a crewless test flight around the Moon in November 2022. It was launched from the Kennedy Space Center in Florida, where all SLS launches will take place.

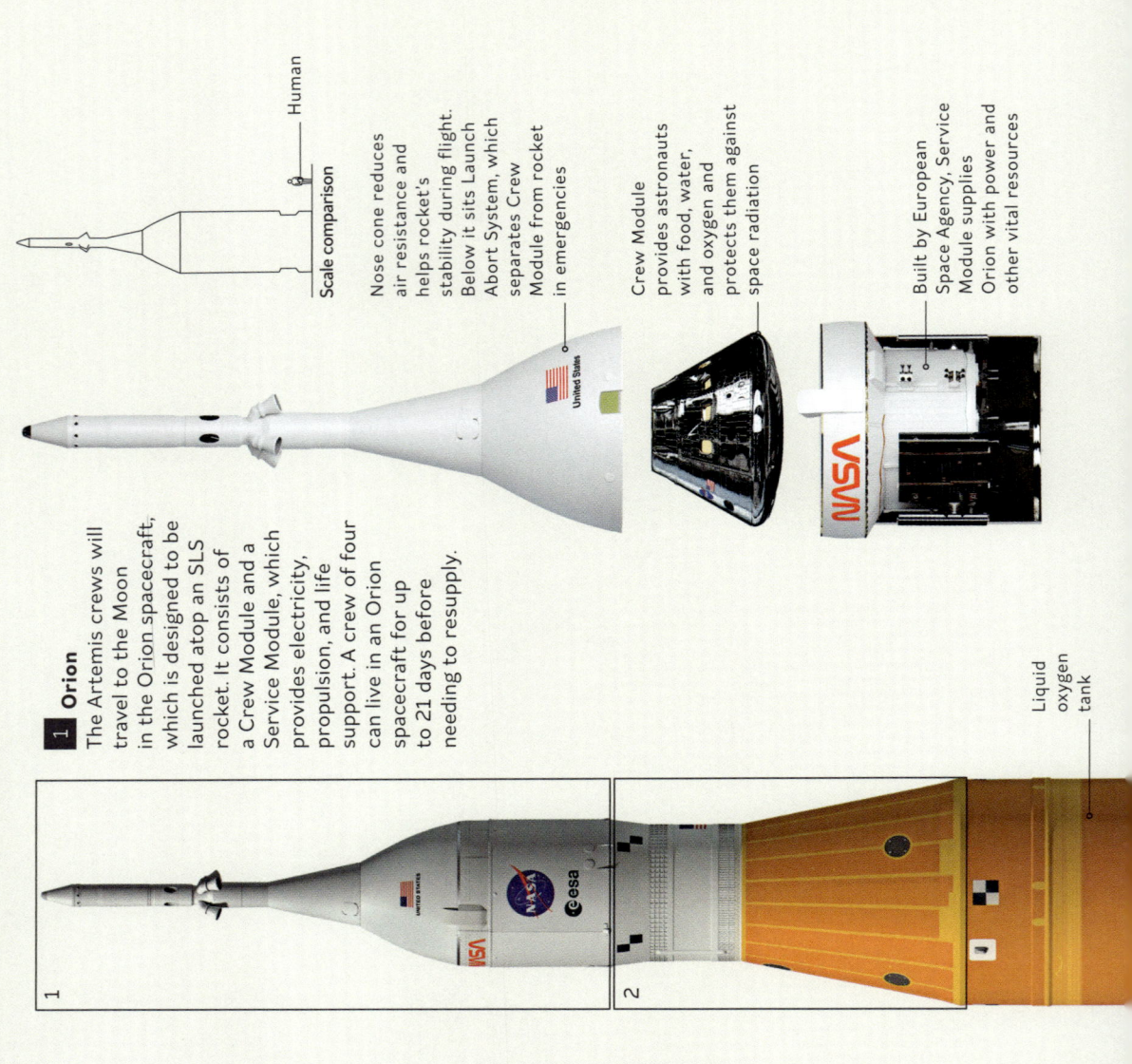

Human

Scale comparison

Nose cone reduces air resistance and helps rocket's stability during flight. Below it sits Launch Abort System, which separates Crew Module from rocket in emergencies

Crew Module provides astronauts with food, water, and oxygen and protects them against space radiation

Built by European Space Agency, Service Module supplies Orion with power and other vital resources

1 Orion

The Artemis crews will travel to the Moon in the Orion spacecraft, which is designed to be launched atop an SLS rocket. It consists of a Crew Module and a Service Module, which provides electricity, propulsion, and life support. A crew of four can live in an Orion spacecraft for up to 21 days before needing to resupply.

Liquid oxygen tank

Intertank section

Liquid oxygen feedline

Liquid hydrogen tank

Solid rocket booster

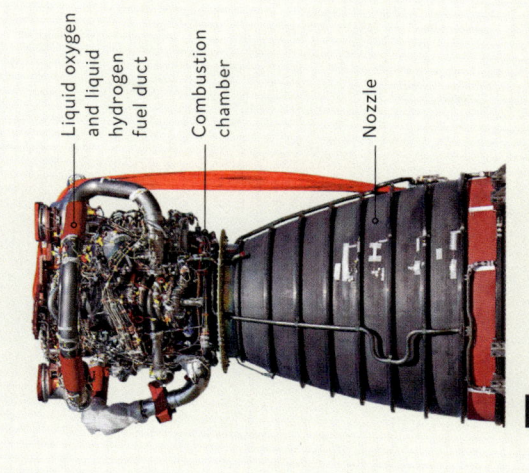

Liquid oxygen and liquid hydrogen fuel duct

Combustion chamber

Nozzle

3 RS-25 engine

Utilizing cryogenic (extremely low-temperature) liquefied gas, the RS-25 engine partly powered the Space Shuttle, and has been adapted for use in SLS. To meet the more extreme demands of the Artemis missions, it has been modified—for example, to provide greater thrust. While the Space Shuttle used three RS-25s, SLS uses four.

2 The ICPS

The first few Artemis missions will carry Orion to the Moon with an upper stage known as the Interim Cryogenic Propulsion Stage (ICPS). This has two tanks that feed the engine—one containing liquid hydrogen and one containing liquid oxygen. With Artemis IV, the ICPS will be replaced by the Exploration Upper Stage.

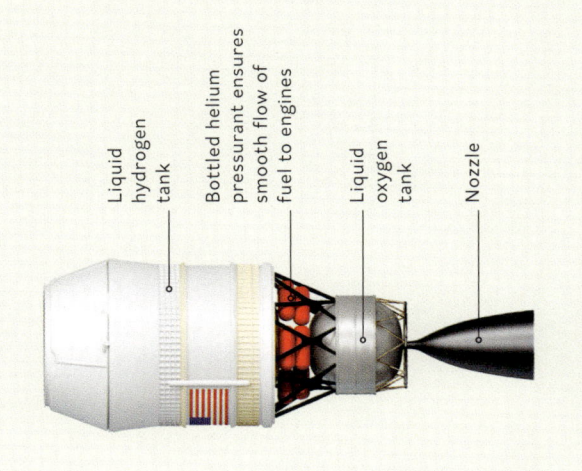

Liquid hydrogen tank

Bottled helium pressurant ensures smooth flow of fuel to engines

Liquid oxygen tank

Nozzle

SLS DATA

Function	Super heavy-lift launch vehicle
Country of origin	US
Height	Block 1 (crew and cargo to Moon): Up to 322 ft (98 m) Block 2 (crew and cargo to Moon, Mars, and deep space): Up to 365 ft (111 m)
Diameter	Core stage: 27.6 ft (8.4 m) ICPS: 16.7 ft (5.1 m)
First Flight	November 16, 2022, 1:47.44 A.M. EST

BACK TO THE MOON

The next humans to walk on the surface of the Moon are very likely to do so as part of NASA's Artemis III mission. Their stay on the Moon will be the longest ever.

Artemis III aims to put the first astronauts at the lunar south pole, where they will conduct experiments that will enrich our knowledge of the Moon and the wider solar system. The area is particularly interesting geologically and may even contain water ice. The astronauts' week-long stay will set a new record, beating the nearly 75 hours spent on the Moon by the Apollo 17 astronauts in 1972.

Exploring the south pole

The crew of Artemis III will travel in an Orion multipurpose crew vehicle, which will be launched atop a Space Launch System rocket (SLS; see pp. 210–211). The uppermost section of the SLS—the Interim Cryogenic Propulsion Stage (ICPS)—will propel Orion to the Moon, where, in a Near-Rectilinear Halo Orbit (NRHO; see pp. 214–215), it will rendezvous with a Starship Human Landing System (HLS).

Made by SpaceX, the HLS will transport two of the four astronauts from Orion to the lunar surface and will then serve both as a habitat and base of operations. The astronauts will spend a week or so on the Moon—putting on spacesuits each day and exiting the HLS via an airlock, descending to the surface via an elevator to survey the area and collect samples and data. When they return to Orion, they will spend a few days working in lunar orbit before Orion's engines reignite and return the craft to Earth.

Despite careful planning, uncertainties remain, including lingering safety concerns about the Orion spacecraft. Artemis III has already been delayed—from an original 2024 launch date to a revised launch now no earlier than September 2026—and it could be delayed even further.

Space Launch System (SLS)

1. SLS takes off from Launch Pad 39B at Kennedy Space Center, Florida

15. Crew module reenters Earth's atmosphere

2. SLS enters Earth orbit

14. Crew module separates from Service Module

16. Crew are recovered from Pacific Ocean

3. Orion separates from SLS Core Stage

HUMAN LANDING SYSTEMS

An HLS is a reusable vehicle for ferrying astronauts to the lunar surface. The commercial companies SpaceX and Blue Origin have won contracts to design and build HLS vehicles for the Artemis program. On missions after Artemis III, astronauts from Earth will transfer to such landers via the Gateway space station (see pp. 214–215).

Blue Moon

The Blue Moon lander was designed for the Artemis V mission. The crew will be housed in the bottom of the lander so they can easily access the Moon.

Starship HLS

The HLS will take the astronauts of Artemis III and Artemis IV to the lunar surface and back. The crew cabin is located at the top of the vehicle.

On **reentry** into Earth's atmosphere, the **Artemis III crew** will travel at about **24,855 mph** (40,000 km/h).

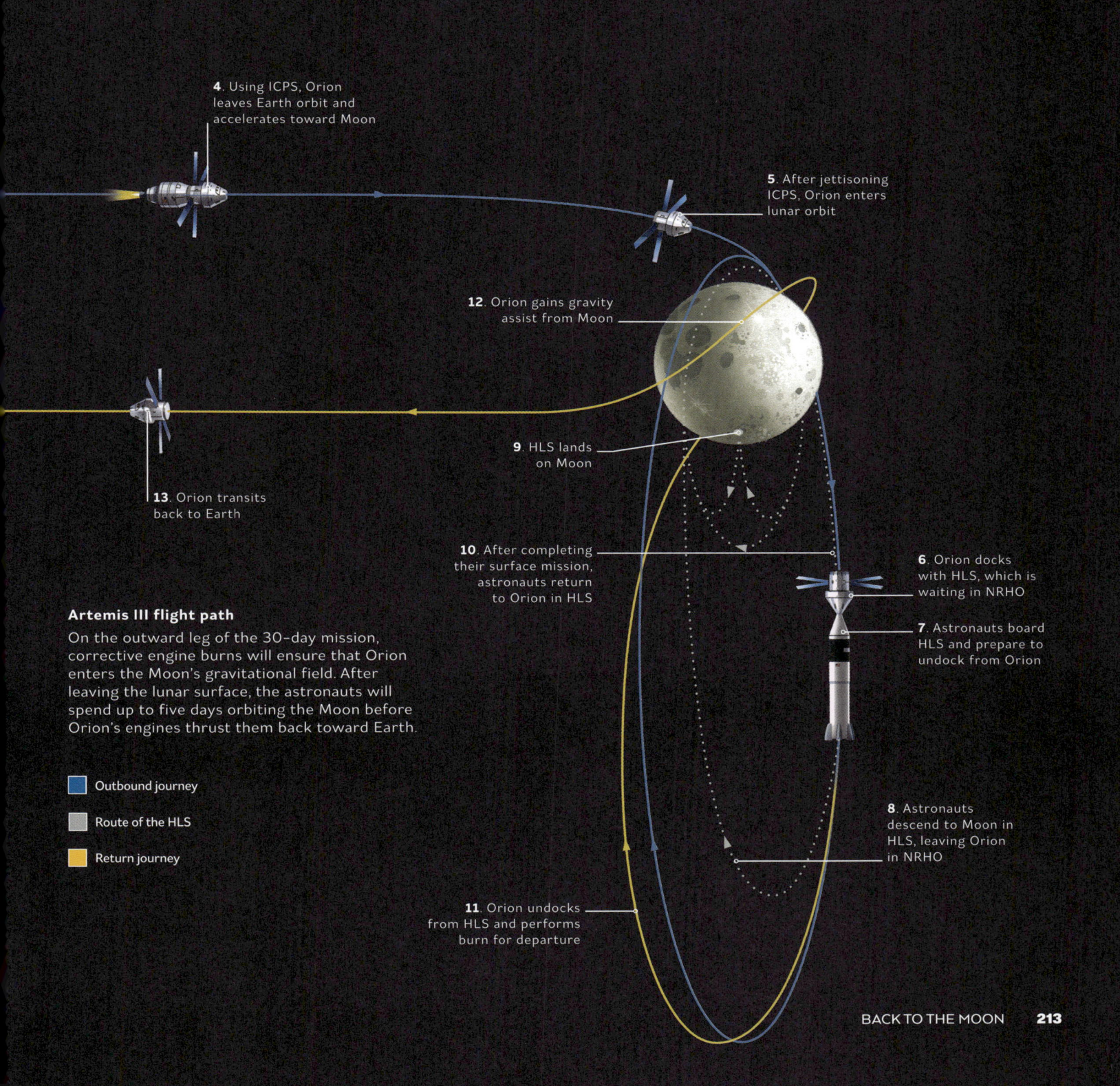

4. Using ICPS, Orion leaves Earth orbit and accelerates toward Moon

5. After jettisoning ICPS, Orion enters lunar orbit

12. Orion gains gravity assist from Moon

9. HLS lands on Moon

13. Orion transits back to Earth

10. After completing their surface mission, astronauts return to Orion in HLS

6. Orion docks with HLS, which is waiting in NRHO

7. Astronauts board HLS and prepare to undock from Orion

Artemis III flight path

On the outward leg of the 30-day mission, corrective engine burns will ensure that Orion enters the Moon's gravitational field. After leaving the lunar surface, the astronauts will spend up to five days orbiting the Moon before Orion's engines thrust them back toward Earth.

■ Outbound journey

■ Route of the HLS

■ Return journey

8. Astronauts descend to Moon in HLS, leaving Orion in NRHO

11. Orion undocks from HLS and performs burn for departure

THE LUNAR GATEWAY

The Lunar Gateway—or Gateway—will be the first space station orbiting the Moon and will support the most-distant crewed space missions ever attempted.

A space station is an orbital spacecraft that can dock and undock with other craft to receive supplies and personnel. So far, all space stations have orbited Earth and have served, among other things, as laboratories in which astronauts have performed all kinds of scientific experiments. Gateway will be the first such station to orbit the Moon, where it will function as a base for lunar exploration and a research laboratory that will be uniquely free of Earth's gravity and magnetic field. In the longer term, it will serve as a staging point for trips to Mars (see pp.228–229).

When completed, Gateway will be 66 ft (20 m) long, which will make it less than half the size of the International Space Station (the first modular space station). Astronauts will be able to live and work in it for up to three months at a time, but for most of its life, Gateway will operate without a crew.

HALO ORBIT

Gateway will be placed in what is known as a "near-rectilinear halo orbit" around the Moon. At its closest and farthest points, it will pass within 932 miles (1,500 km) and 43,000 miles (70,000 km) of the Moon. This unique orbit will be stable, give the space station access to the entire lunar surface, and enable constant line-of-sight communications with Earth.

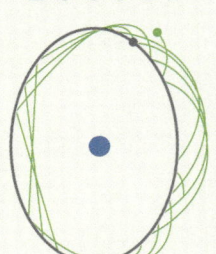

Lunar orbit
Gateway's one-week orbit will ensure that the station is never blocked by the Moon. It will also bring Gateway close both to the Moon and to rendezvous points with supply ships from Earth.

Earth orbit
Gateway will also circle Earth, its orbit rotating once per month with the Moon.

- Gateway
- Moon
- Earth

The Moon and beyond
Gateway—here, depicted by a NASA-commissioned artist—is designed to support both lunar surface missions and excursions into deep space.

ANATOMY OF A SPACE STATION

Like most space stations, Gateway will have a modular structure, which means that it will have a central core (the Habitation and Logistics Outpost) to which other modules can be docked. These modules will be constructed separately on Earth and then assembled in space, where the final structure will offer multiple docking ports for a variety of spacecraft, including Orion (see pp.210–211). Its Human Landing System will take astronauts down to the lunar surface and back.

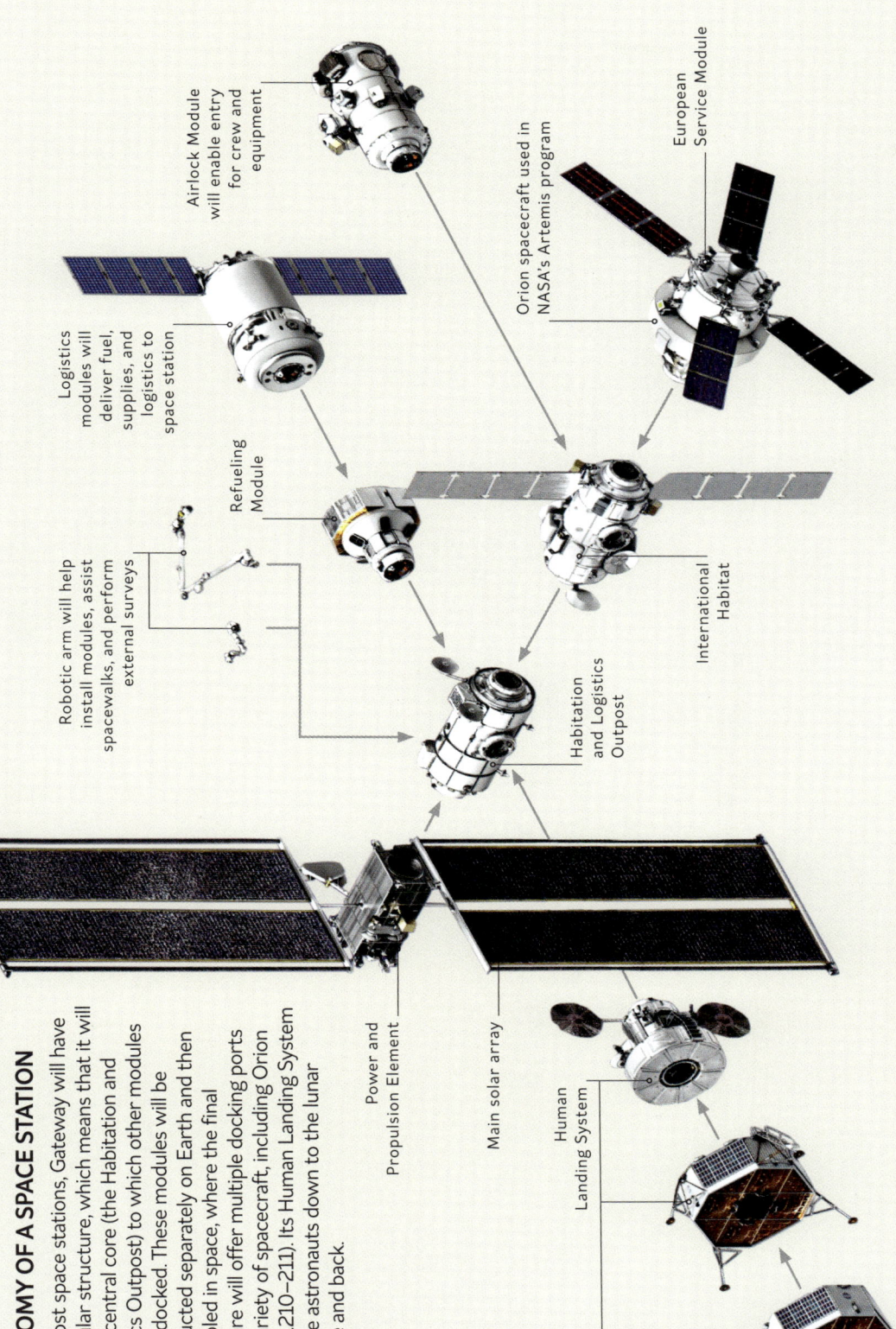

Airlock Module will enable entry for crew and equipment

Logistics modules will deliver fuel, supplies, and logistics to space station

Refueling Module

Robotic arm will help install modules, assist spacewalks, and perform external surveys

Orion spacecraft used in NASA's Artemis program

European Service Module

International Habitat

Habitation and Logistics Outpost

Power and Propulsion Element

Main solar array

Human Landing System

Piece by piece
Due to its modular structure, Gateway can be assembled in space and then added to in the future if necessary.

Ion thruster
Gateway's propulsion system will feature three Hall-effect thrusters (engines), which will have a characteristic blue glow. These produce thrust by ejecting streams of ionized gas.

PROPULSION

Gateway's Power and Propulsion Element will both power the station and keep it in orbit. It will feature two huge solar arrays, which will be capable of generating 50 kW of solar electric power. It will also have a chemical propulsion system, which will use xenon gas for fuel. Together, these will minimize the amount of fuel that will have to be launched with the PPE.

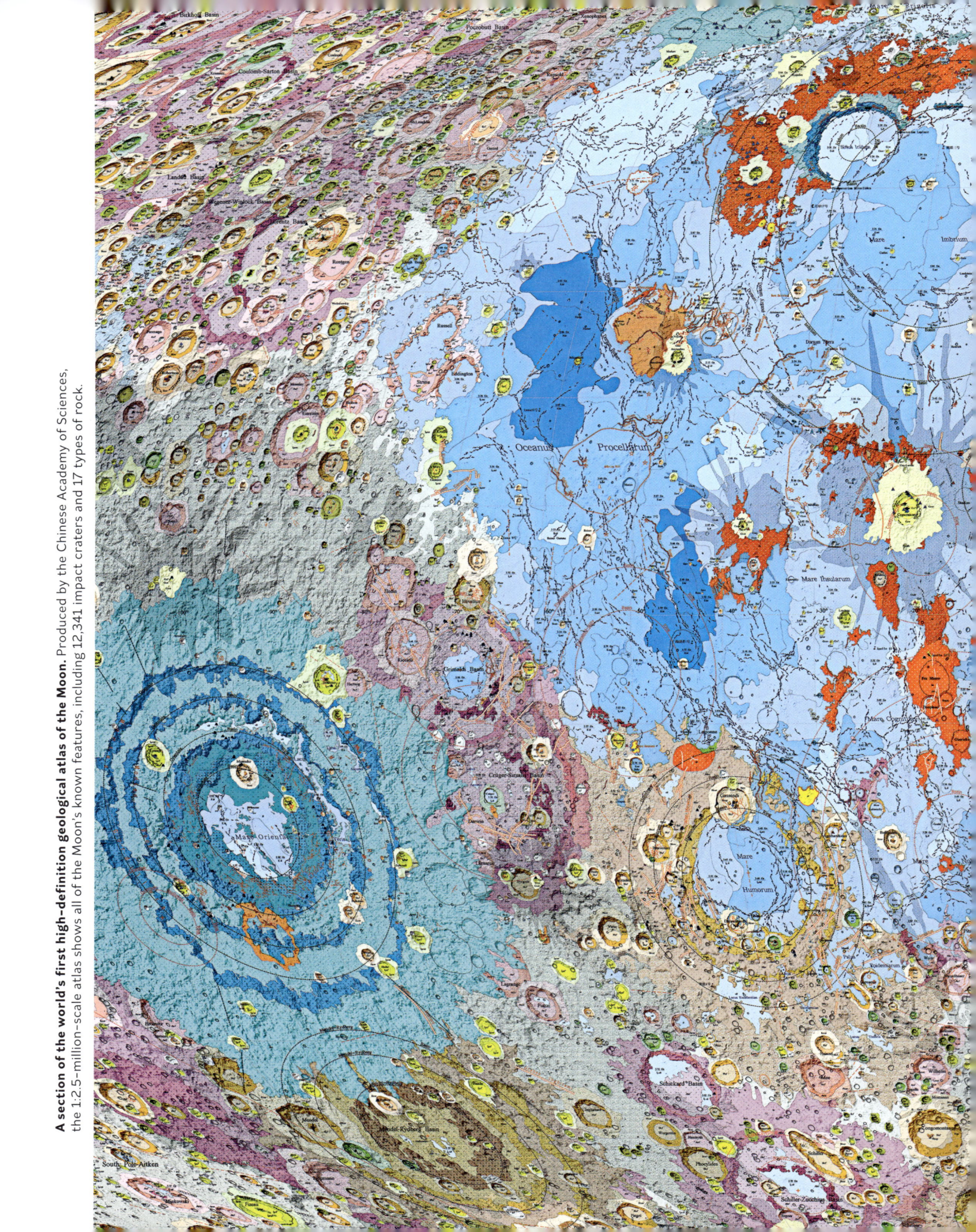

A section of the world's first high-definition geological atlas of the Moon. Produced by the Chinese Academy of Sciences, the 1:2.5-million-scale atlas shows all of the Moon's known features, including 12,341 impact craters and 17 types of rock.

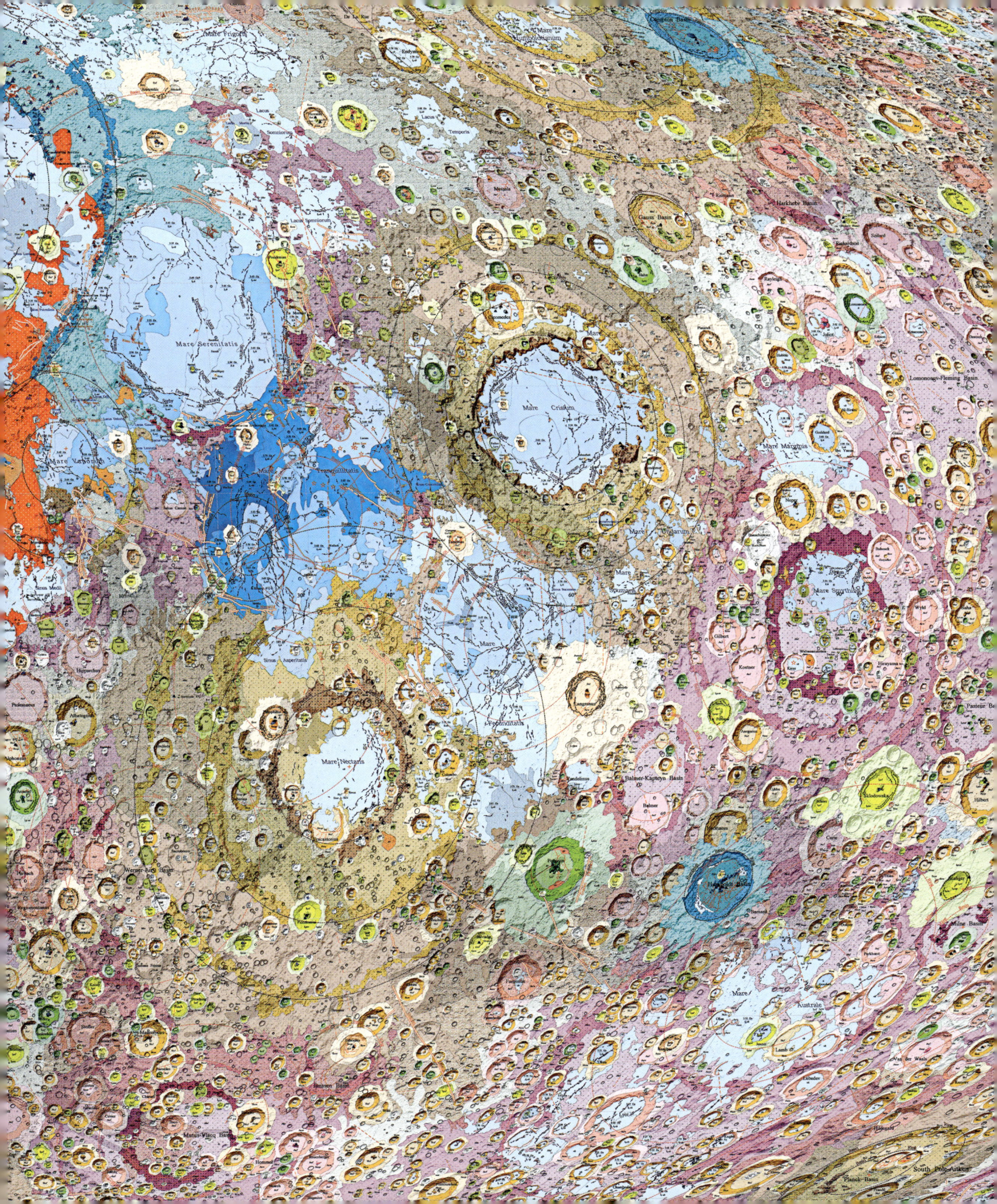

COMMERCIAL SPACEFLIGHT

NASA will not return to the Moon on its own. Project Artemis depends on the contributions of private companies, which are building spacecraft and other equipment for the space agency.

Traditionally, only government-funded organizations—such as NASA in the US, Russia's Roscosmos, and the European Space Agency (ESA)—had the resources to pursue spaceflight. The early 21st century, however, has seen a rapid rise in commercial agencies providing satellite and launch services and exclusive tourism opportunities, as well as carrying out their own increasingly ambitious space programs. Many of these companies are based in the US, which has supported the space industry with a series of laws and programs intended to incentivize the private sector to offer space services. In 2012, soon

Falcon 9 landing
SpaceX's Falcon 9 was the first commercial rocket used to launch a spacecraft (*Crew Dragon*) to supply the International Space Station. On its return to Earth, Falcon 9 lands vertically, which enables the rocket to be refueled and reused many times.

The **space tourism market** is projected to be worth around **$16.3 billion** by 2032.

after the retirement of the Space Shuttle, SpaceX became the first private company to transport cargo to and from the International Space Station (ISS). Since 2020, SpaceX has also carried astronauts to and from the ISS.

Expanding partnerships

NASA intends for private companies to play a significant part in the Artemis program. SpaceX will provide a human landing system for Artemis III and Artemis IV, while Blue Origin will provide a lander for Artemis V. In addition, a number of companies will be contracted to send small robots to the lunar surface as part of the Commercial Lunar Payload Services (CLPS) initiative (see right).

Space tourism

For more than 20 years, commercial firms have been designing and building vehicles specifically for space-going tourists, including Virgin Galactic's SpaceShipTwo spaceplane and Blue Origin's New Shepard capsule. There

is a small but growing interest among wealthy individuals in paying to fly into orbit around Earth or potentially make a flyby of the Moon.

Tourists have been visiting the ISS since 2001, although such sojourns were put on hold after the retirement of the Space Shuttle program in 2011. In 2019, however, NASA announced it would again allow tourists to visit the ISS at a cost of $35,000 per night and around $60 million for the round trip, with transport provided by SpaceX or Boeing.

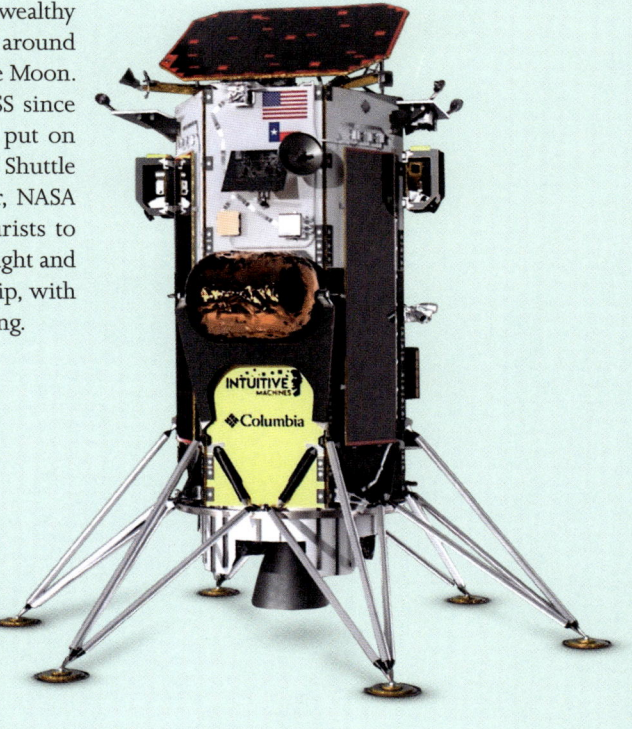

Collaboration with NASA

Intuitive Machines is one of the companies working with NASA on Artemis under the CLPS scheme. Its Nova-C landers will deliver small science payloads to the Moon. Intuitive Machines became the first private operator to make a successful soft landing on the lunar surface when its IM-1 *Odysseus* lander (right) touched down in February 2024.

HOW REUSABLE ROCKETS WORK

The new generation of reusable rockets, including SpaceX's Falcon 9, are designed so that their most important parts (such as engines and boosters) can be recovered, refurbished, and reused. They greatly reduce launch costs, allowing smaller companies to, for instance, put their own satellites in orbit. Reusable rockets have been critical in ushering in the "New Space Age." By reducing launch costs, they have enabled national space agencies, companies, and researchers with fewer resources to access space.

Typical Falcon 9 flight profile
Falcon 9's first stage carries the second stage and its payload to a predetermined altitude and orbit before separating. Thrusters then flip the first stage so that it can make a controlled, vertical landing on its extendable legs.

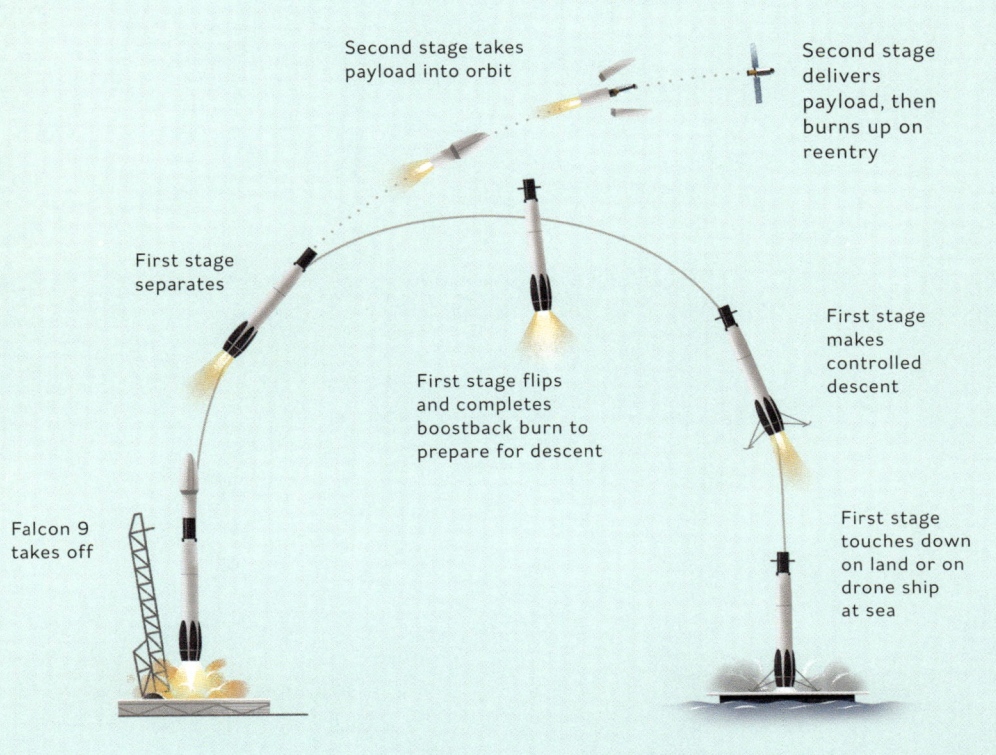

Second stage takes payload into orbit

Second stage delivers payload, then burns up on reentry

First stage separates

First stage makes controlled descent

First stage flips and completes boostback burn to prepare for descent

Falcon 9 takes off

First stage touches down on land or on drone ship at sea

BUILDING A MOON BASE

Moon bases could enable humans to make extended stays on the lunar surface. Initially, these may rely on resupply missions from Earth, but eventually they could become self-sufficient.

VISUAL TOUR

1. Habitat module
The habitat provides protective housing for a small number of astronauts.

2. Lunar rover
Astronauts can use rovers to move around the base and explore farther afield.

3. Greenhouse
Growing food frees the inhabitants from relying on food deliveries from Earth.

4. Solar array
Solar panels harness the Sun's energy to power electronic devices throughout the base.

Settling on other worlds is an age-old human dream. Artemis (see pp.208–209) aims to realize that dream by establishing a permanent base on the Moon (the Artemis Base Camp), as does China's Chang'e program, with its proposed International Lunar Research Station. Such bases will enable humans to explore much more of the lunar surface and to carry out scientific experiments. However, they must also protect their inhabitants from extreme temperatures and solar radiation and provide the resources essential for life.

ESA base camp
This Moon base proposed by the European Space Agency (ESA) could include factories that can turn regolith into building materials and water into fuel (see right).

Ultimately, Moon bases will have to achieve some form of self-sufficiency. They will need to utilize the Moon's resources to create water, oxygen, and construction materials to expand the base beyond what can be made on Earth and flown to the Moon. Available resources include sunlight and regolith (lunar material such as rock fragments and minerals created by meteorite impacts; see pp.182–183). In-Situ Resource Utilization (ISRU) processes are currently being finessed in preparation for the establishment of the Artemis Base Camp.

> "We will ... establish the **first long-term presence** on the Moon."
> **Artemis plan**

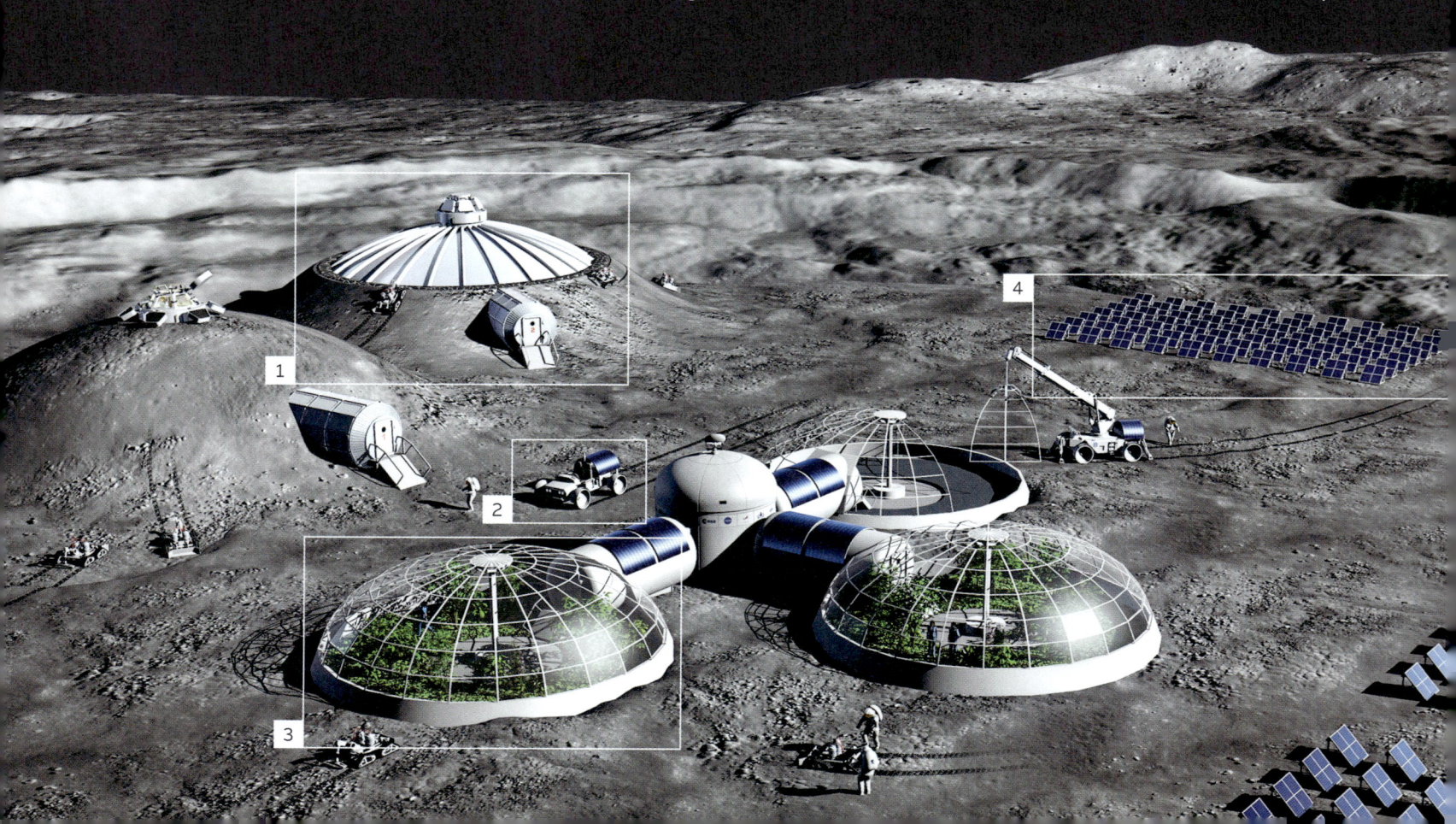

BUILDING WITH REGOLITH

Additive manufacturing (also known as 3D printing) is the process of constructing an object from a digital model. One method being considered by space agencies involves building the object up in successive stages by focusing a beam of light onto a layer of regolith. This "sinters" (solidifies) the regolith without melting it, after which further layers are added until the object is complete. The finished object can be a brick, a panel, or any part necessary for constructing buildings, roads, or even launch pads.

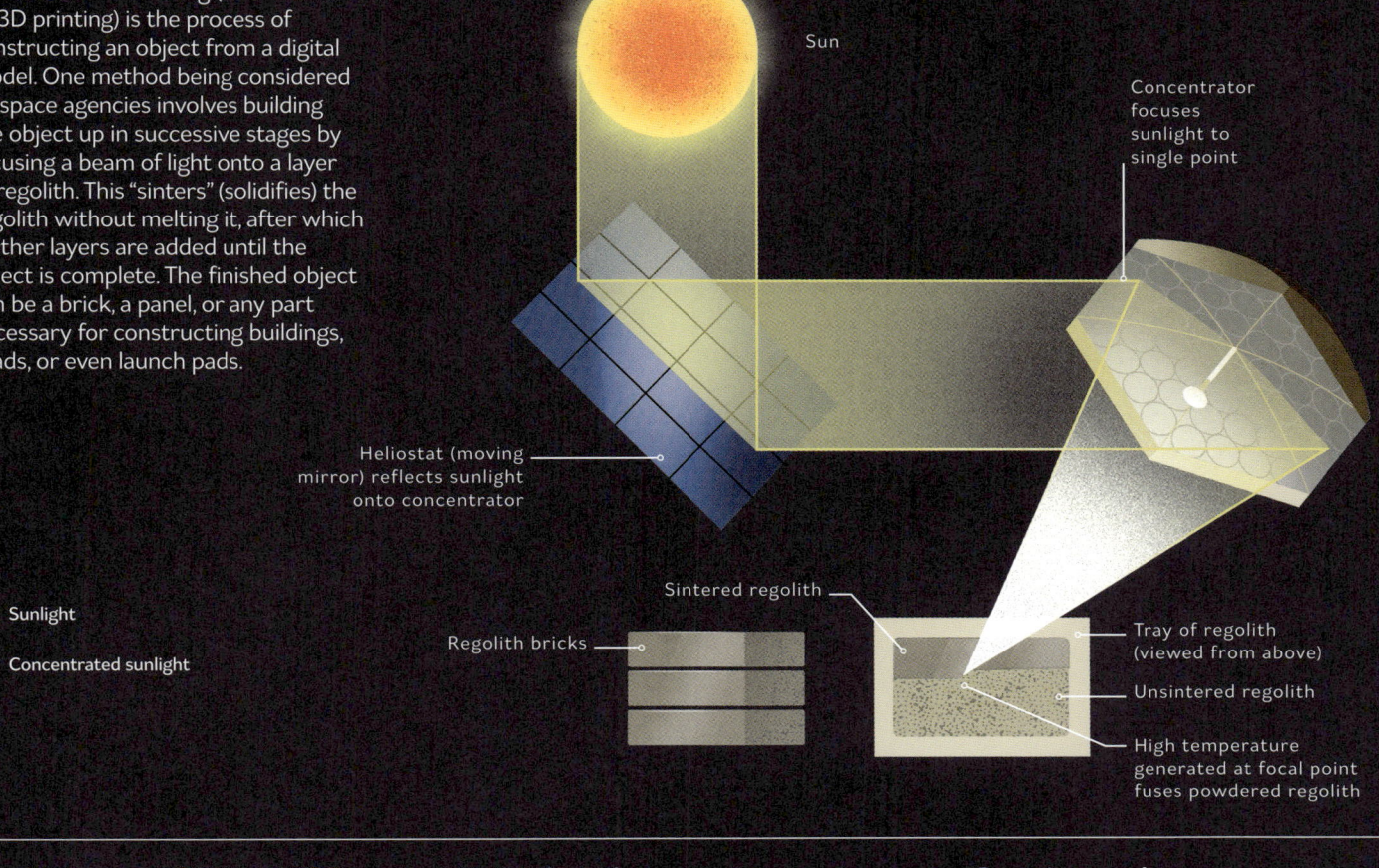

Sun

Concentrator focuses sunlight to single point

Heliostat (moving mirror) reflects sunlight onto concentrator

Sintered regolith

Regolith bricks

Tray of regolith (viewed from above)

Unsintered regolith

High temperature generated at focal point fuses powdered regolith

- Sunlight
- Concentrated sunlight

MAKING FUEL FROM WATER

Rocket fuel can be heavy, and heavy launches are expensive and risky. The inconvenience of carrying rocket fuel is particularly relevant when it comes to long-term missions to the Moon, which require supplies to sustain the astronauts' lives, as well as enough fuel for a return journey. One solution is to create fuel on site by a process known as "water electrolysis." This separates water into its component gases—hydrogen and oxygen—which can then be stored as liquids and used as rocket fuel when needed.

Water electrolysis
When water molecules are split using submerged electrodes, hydrogen gas forms at the negative electrode and oxygen gas forms at the positive electrode.

Powering a rocket
The hydrogen and oxygen can be recombined in a rocket's combustion chamber and ignited. This creates exhaust, which provides thrust that propels the rocket.

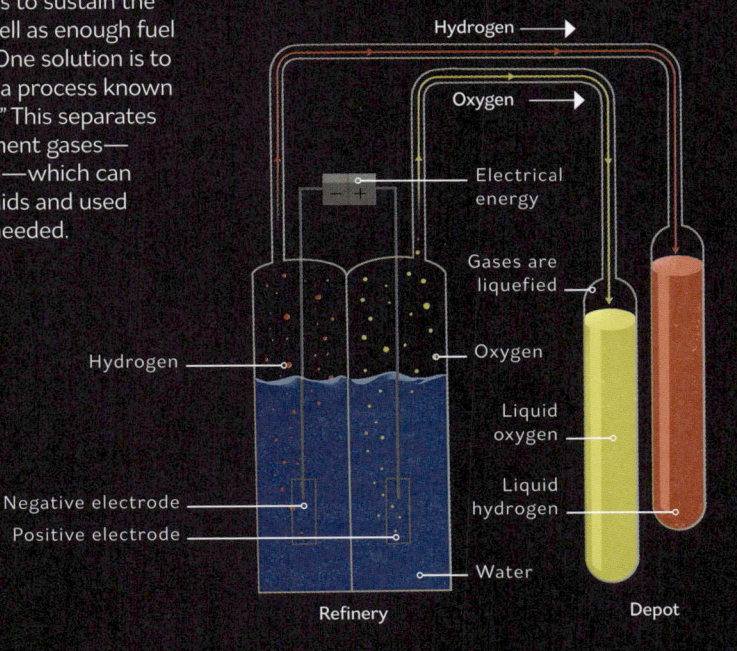

Hydrogen

Oxygen

Electrical energy

Gases are liquefied

Oxygen

Liquid oxygen

Liquid hydrogen

Water

Hydrogen

Negative electrode

Positive electrode

Refinery

Depot

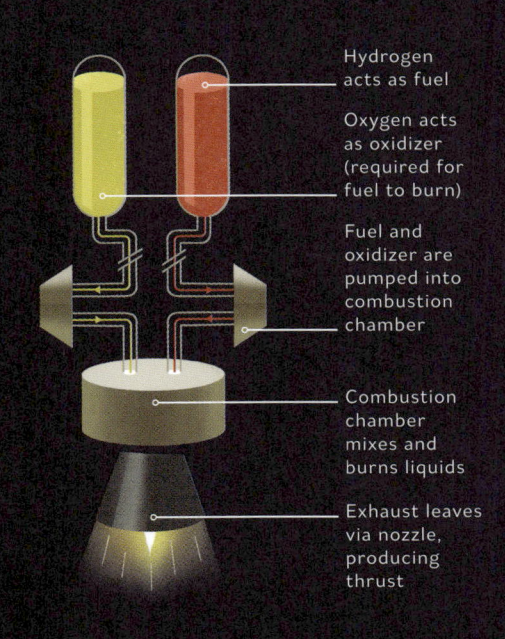

Hydrogen acts as fuel

Oxygen acts as oxidizer (required for fuel to burn)

Fuel and oxidizer are pumped into combustion chamber

Combustion chamber mixes and burns liquids

Exhaust leaves via nozzle, producing thrust

In 2022, the Orion capsule of Artemis I traveled 268,563 miles (432,210km) away from Earth, beating the previous distance record for a spacecraft designed for humans, which was set by Apollo 13.

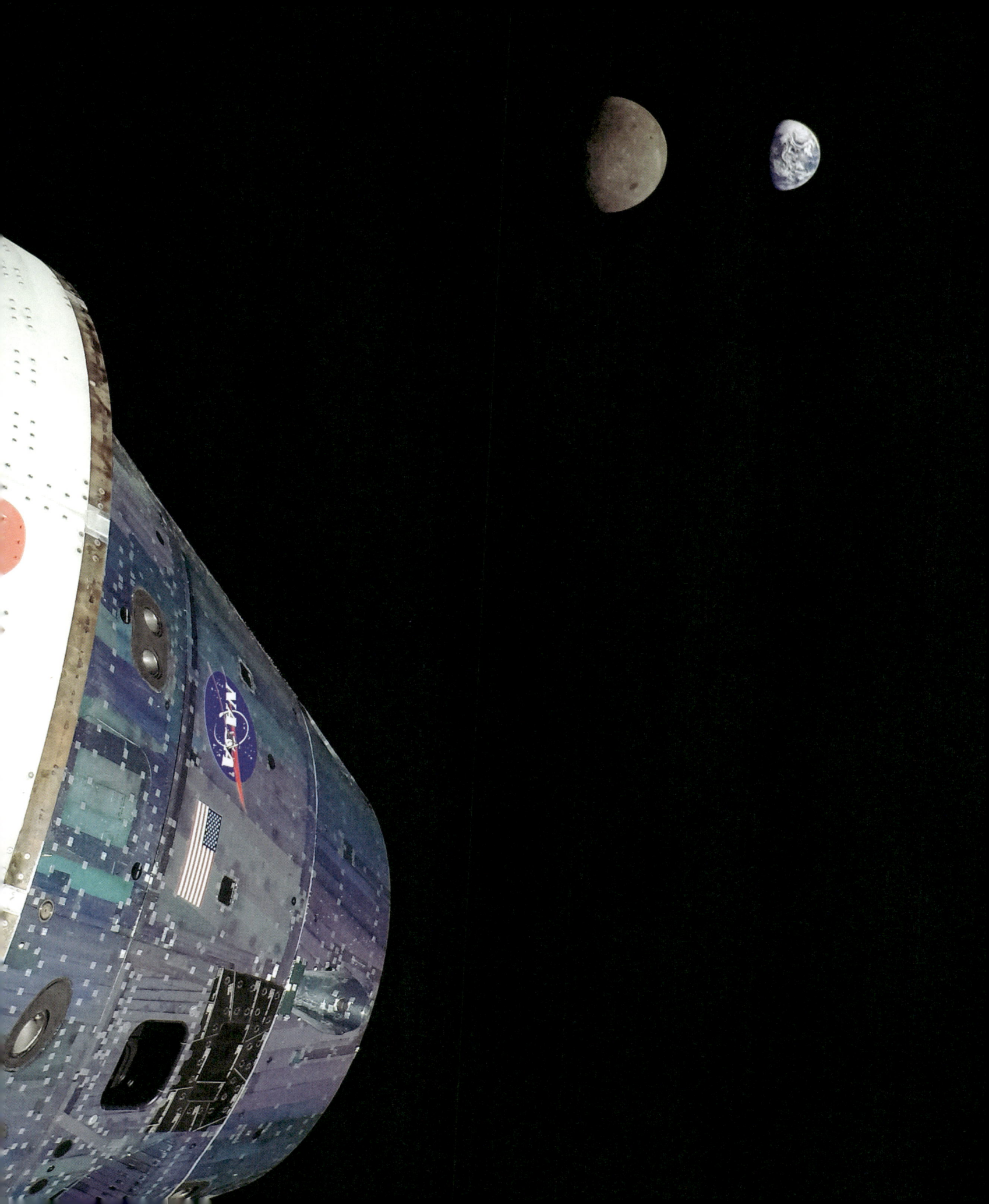

A NEW GOLD RUSH

The Moon is an untapped well of natural resources, and governments and commerce already have their eyes on the prize. Space mining, once merely the stuff of science fiction, could soon be a reality on the Moon.

Thorium map
Based on Lunar Prospector data, this map shows thorium levels in the Moon's surface. High thorium levels typically indicate the presence of rare-earth metals and other resources. Thorium itself could potentially be mined as fuel for fission reactors (in which atomic nuclei are split apart).

Lunar resources might be mined for their inherent value, or usefulness to industries back on Earth, or because they would make exploration of the Moon cheaper and easier. Among the many elements now known to be present on the Moon are oxygen, silicon, iron, magnesium, calcium, aluminum, titanium, and thorium. However, much remains to be learned about the Moon's materials and in what proportions they exist. Robotic and human missions to collect samples will help reveal more about this—and identify sites worth exploring for resource exploitation.

Key commodities
Lunar resources that might be extracted include water, helium-3, and rare-earth metals. Water could support human settlements on the Moon and be a source of rocket fuel (see pp.220–221). Helium-3, an isotope of helium

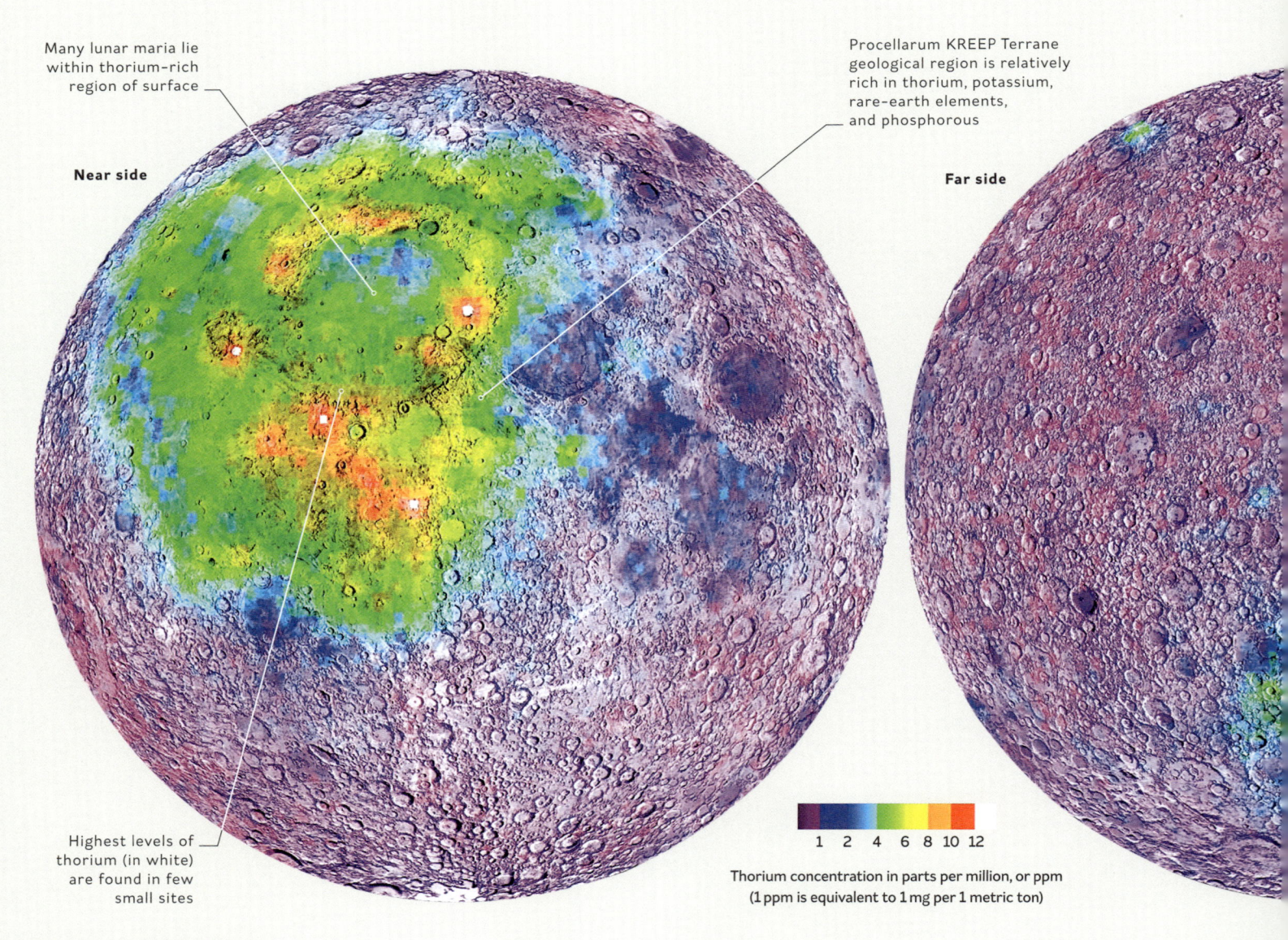

Many lunar maria lie within thorium-rich region of surface

Procellarum KREEP Terrane geological region is relatively rich in thorium, potassium, rare-earth elements, and phosphorous

Near side

Far side

Highest levels of thorium (in white) are found in few small sites

1 2 4 6 8 10 12

Thorium concentration in parts per million, or ppm
(1 ppm is equivalent to 1 mg per 1 metric ton)

uncommon on Earth, may be abundant on the Moon, prompting debate as to whether it could be a viable fuel for fusion reactors (in which atomic nuclei would be fused together; see below). Rare-earth metals, such as yttrium and scandium—vital for electronic devices—might be worth returning to Earth if plentiful enough.

With uncertainties over regulation and the technical and economic feasibility of mining on the Moon, the possibility of a "lunar gold rush" remains speculative, however. In 2020, NASA proposed the Artemis Accords: a framework governing lunar exploration and exploitation. While more than 40 countries are signatories to the accords, Russia and China are not. NASA is also encouraging commercial involvement. It has said that it will buy small amounts of lunar regolith collected by private companies for use in its Artemis program.

Felspathic Highlands Terrane covers 65 percent of lunar surface; it has low thorium levels

South Pole-Aitken Terrane is another thorium-enriched region, although its thorium levels are lower than in Procellarum KREEP Terrane

DRILLING FOR RESOURCES

NASA engineers have designed a drill known as TRIDENT (The Regolith Ice Drill for Exploring New Terrain), which will be attached to future robotic rovers. TRIDENT will drill out surface samples that will help locate viable concentrations of extractable resources.

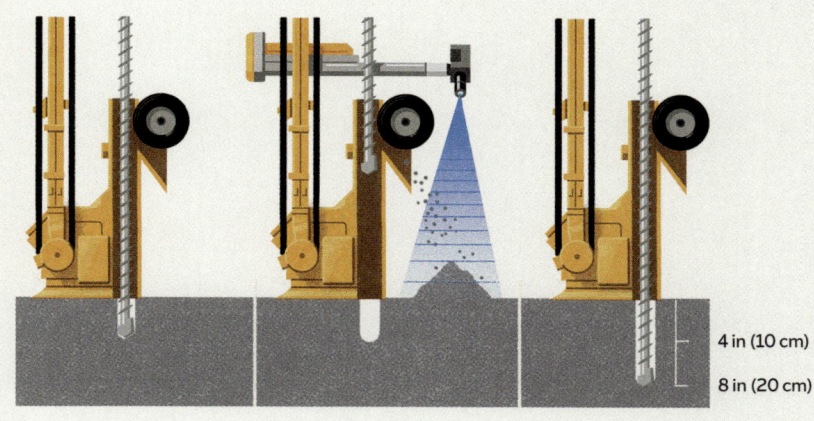

4 in (10 cm)

8 in (20 cm)

1. Cuttings are caught in the flutes on the side of the drill as it "bites" 4 in (10 cm) into the rock-hard regolith.

2. As the drill retracts, a brush deposits the cuttings in a pile to be assessed by the spectrometers.

3. Each bite goes 4 in (10 cm) deeper than the last, up to a depth of 3.28 ft (1 m), to bring up new samples.

FUSION OF HELIUM-3

Helium-3 is an exciting prospect as a nuclear fuel: It does not release the dangerous radiation that traditional nuclear fusion does. However, serious doubts exist as to whether helium-3 fusion is technically achievable.

Energetic reaction
In a fusion reactor, helium-3 would smash into deuterium, a hydrogen isotope. As the two fuse into normal helium, vast amounts of energy would be released.

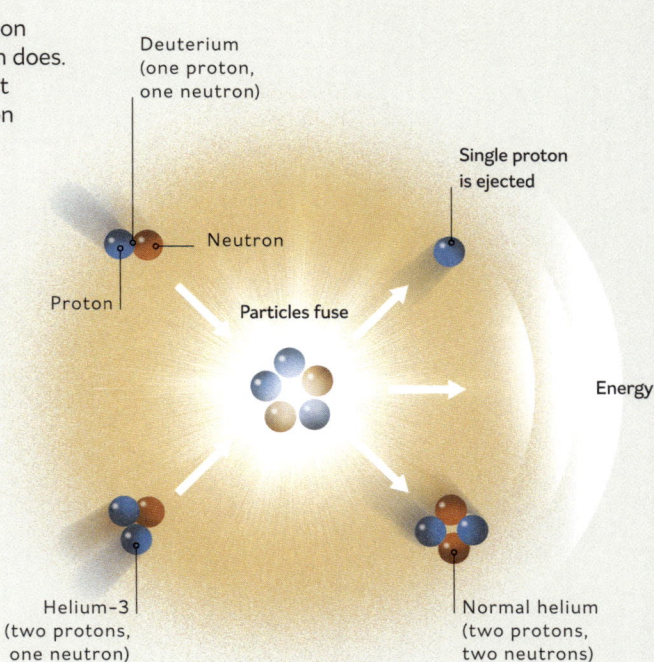

Deuterium (one proton, one neutron)

Neutron

Proton

Single proton is ejected

Particles fuse

Energy

Helium-3 (two protons, one neutron)

Normal helium (two protons, two neutrons)

Analyzing samples
Spectrometers on VIPER will identify water, minerals, and other resources in TRIDENT's drill cuttings by analyzing the light they emit or absorb.

OBSERVING THE UNIVERSE

As the Moon becomes more accessible, building and maintaining a lunar observatory—which will have advantages over both Earth- and space-based observatories—will become increasingly feasible.

A lunar observatory could accommodate large, complex telescopes that require regular maintenance (which is currently not possible in space) and offer views of the sky that would not be blocked or distorted by Earth's atmosphere. The Moon's lack of atmosphere would also enable astronomers to study all wavelengths of the electromagnetic spectrum, and its slow spin would allow for much longer observations than are possible on Earth.

In February 2024, the *Odysseus* lander—part of a mission run by the US company Intuitive Machines and backed by NASA—soft-landed near the lunar south pole. *Odysseus* performed a scientific experiment called Radio wave Observations at the Lunar Surface of the photo Electron Sheath (ROLSES). This was the first radio-astronomy experiment to work successfully on the Moon. The miniature ROLSES observatory—designed to investigate

Lunar Crater Radio Telescope
A lander delivers the LCRT to the center of a crater. Wall-climbing robot rovers, also delivered by lander, deploy and anchor the receiver and wire-mesh dish.

Suspended receiver

Wire-mesh reflector dish

Crater

Earth positioned between LCRT and Sun

Noise produced by Earth sources

LCRT located on Moon's far side

Area shielded from noise

Sun

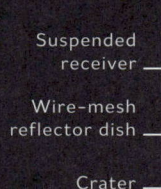

Earth

Moon

Incoming long-wavelength radio emissions

Isolated from noise
The Moon shields the LCRT from radio interference, or "noise," from Earth and the Sun. This enables astronomers to study deep-space radio waves.

how the Moon's surface scatters radio waves—represented the first step toward establishing astronomical observatories on the Moon.

On the drawing board

Proposals for future lunar observatories include NASA's Lunar Crater Radio Telescope (LCRT), a radio telescope located in a crater on the far side of the Moon (see opposite page). With a diameter of 0.8 miles (1.3 km), the LCRT would be sensitive to wavelengths longer than 33 ft (10 m). The Lunar Surface Electromagnetics Experiment (LuSEE-Night; see right), being developed by NASA and the US Department of Energy, would also be located on the far side and would observe the early, "Dark Ages" of the universe. Another Dark-Ages project, the FarView observatory, is a concept for a radio telescope array of 100,000 antennae spread over 180 sq miles (200 sq km), to be built on site using metals extracted from lunar regolith.

LUSEE-NIGHT

Consisting of four antennae on a rotating platform, LuSEE-Night is a planned radio telescope designed to investigate the feasibility of long-wavelength radio astronomy from the far side of the Moon (see below). The advantage of the far side is that it faces away from Earth, making it the one place in the solar system that is always shielded from Earth's radio interference.

Antennae for picking up radio waves

OBSERVING THE DARK AGES

The aim of the LuSEE-Night mission is to observe the sky over several lunar nights to gather data from the cosmic Dark Ages—the time between the Big Bang and the formation of the first luminous stars and galaxies. Since there was no optical light at that time, astronomers search for radio waves that can tell us about this early phase of the universe's life. These radio waves—which come from deepest space, and so from billions of years in the past—are scattered by Earth's ionosphere and therefore can only be detected from space. Since the far side of the Moon is also sheltered from Earth's radio interference, it is the ideal place to make such observations.

The early universe
During the Dark Ages—between 370,000 and 400 million years after the Big Bang—the universe was electrically neutral and transparent. This period ended when stars began to form.

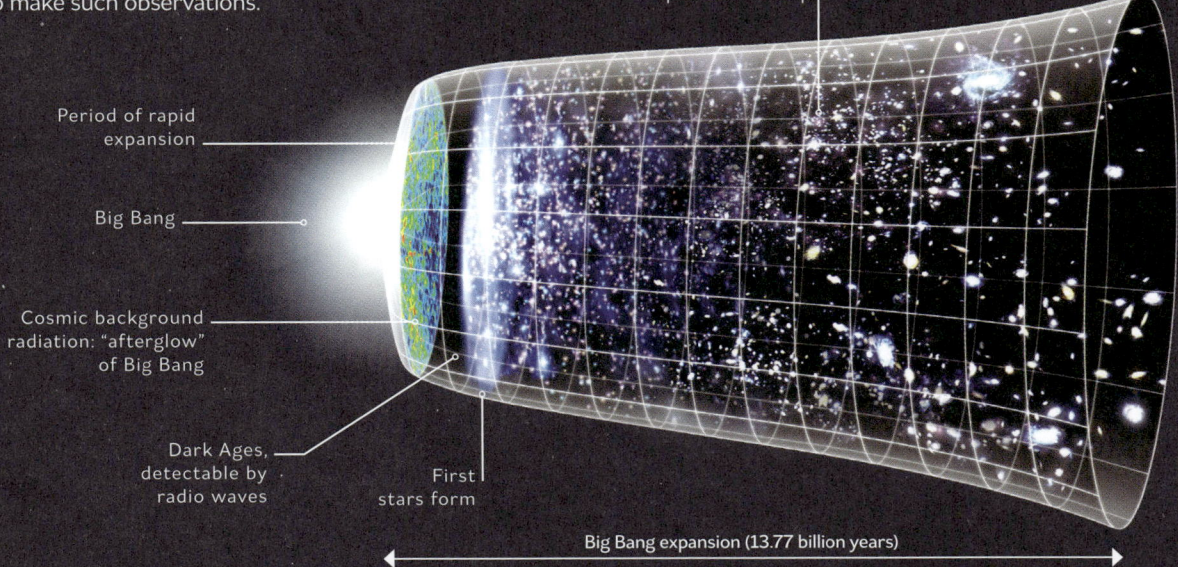

Galaxies and planets develop

Period of rapid expansion

Big Bang

Cosmic background radiation: "afterglow" of Big Bang

Dark Ages, detectable by radio waves

First stars form

Big Bang expansion (13.77 billion years)

A STEPPING STONE TO MARS

The long-term goal of the Artemis program is to prepare NASA to send humans farther into space than ever before—to Mars. This will test NASA's technical and operational capabilities to their limits.

NASA has stated: "The sooner we go to the Moon, the sooner we send astronauts to Mars." Ultimately, the Artemis program considers the Moon a stepping stone to Mars. While the journey to Mars will take seven months, the Moon can be reached in just three days, making it a convenient testing ground for crewed missions to the Red Planet.

Testing environment

Following the return of astronauts to the Moon in Artemis III, NASA plans to carry out human missions on and around the Moon in readiness for an expedition to Mars. The establishment of the Lunar Gateway (see pp.214–215) and the Artemis Base Camp at the lunar south pole (see pp.220–221) are critical elements of this preparation. The Gateway will not only provide a staging point for lunar missions, but also a simulation of how a crewed Mars expedition might be carried out, with astronauts remaining in orbit and only descending to a Martian base camp when required.

On the Moon's surface, the testing of key technologies, such as mobility systems and on-site resource utilization (including building structures from available materials), will be vital to make extended stays on Mars possible. The later Artemis missions will also allow for investigations into the effects of microgravity on the human body over long periods of time.

Spacecraft for Mars

Vehicles for Mars are already being planned. Block 2, the next phase of the Space Launch System (see pp.210–211), will have advanced boosters and is intended to enable launches of astronauts and cargo to Mars. The Deep Space Transport, currently at concept stage, is a spacecraft that would support astronauts on years-long Mars missions. Comprising an Orion capsule and a habitat module, it would depart from the Lunar Gateway.

"**Mars** remains our **horizon goal** for human **exploration**."
NASA

MAKING OXYGEN

MOXIE (Mars Oxygen In-Situ Resource Utilization Experiment) is a device for generating oxygen—essential to sustain life during long-term space missions. MOXIE uses an electrochemical process to separate oxygen atoms from carbon dioxide in the Martian atmosphere.

Loading MOXIE onto Perseverance
MOXIE was tested on the NASA Mars rover Perseverance. It successfully extracted oxygen from the Martian atmosphere in 2021, producing up to 0.4 oz (12 g) of oxygen per hour.

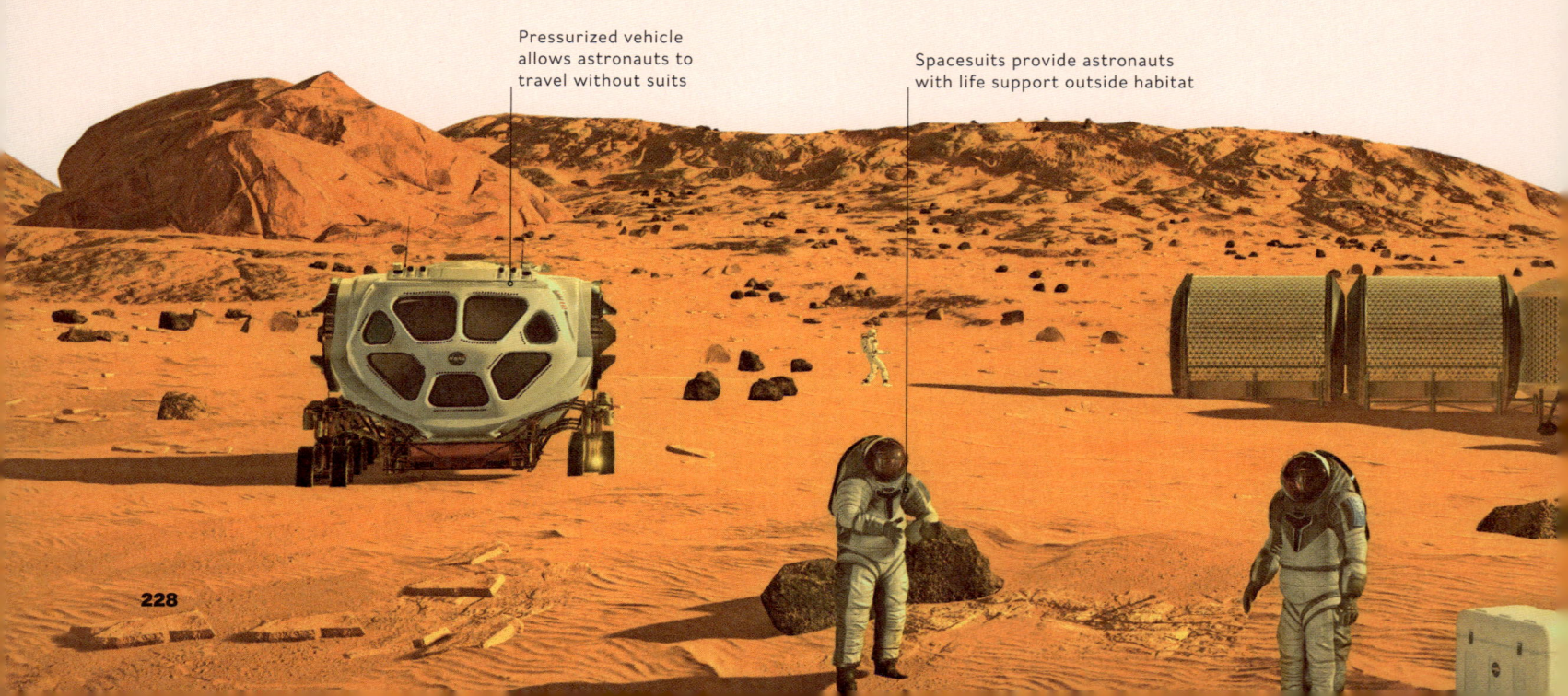

Pressurized vehicle allows astronauts to travel without suits

Spacesuits provide astronauts with life support outside habitat

PERSEVERANCE ROVER

In July 2020, NASA sent the car-sized Perseverance rover to explore the surface of Mars, successfully landing in Jezero crater in February 2021. Jezero is thought to have been flooded with water long ago, and therefore might once have harbored microbial life. Perseverance will search for signs of this ancient life and study how the region formed. It has already collected samples of rock and soil that will be returned to Earth by later missions.

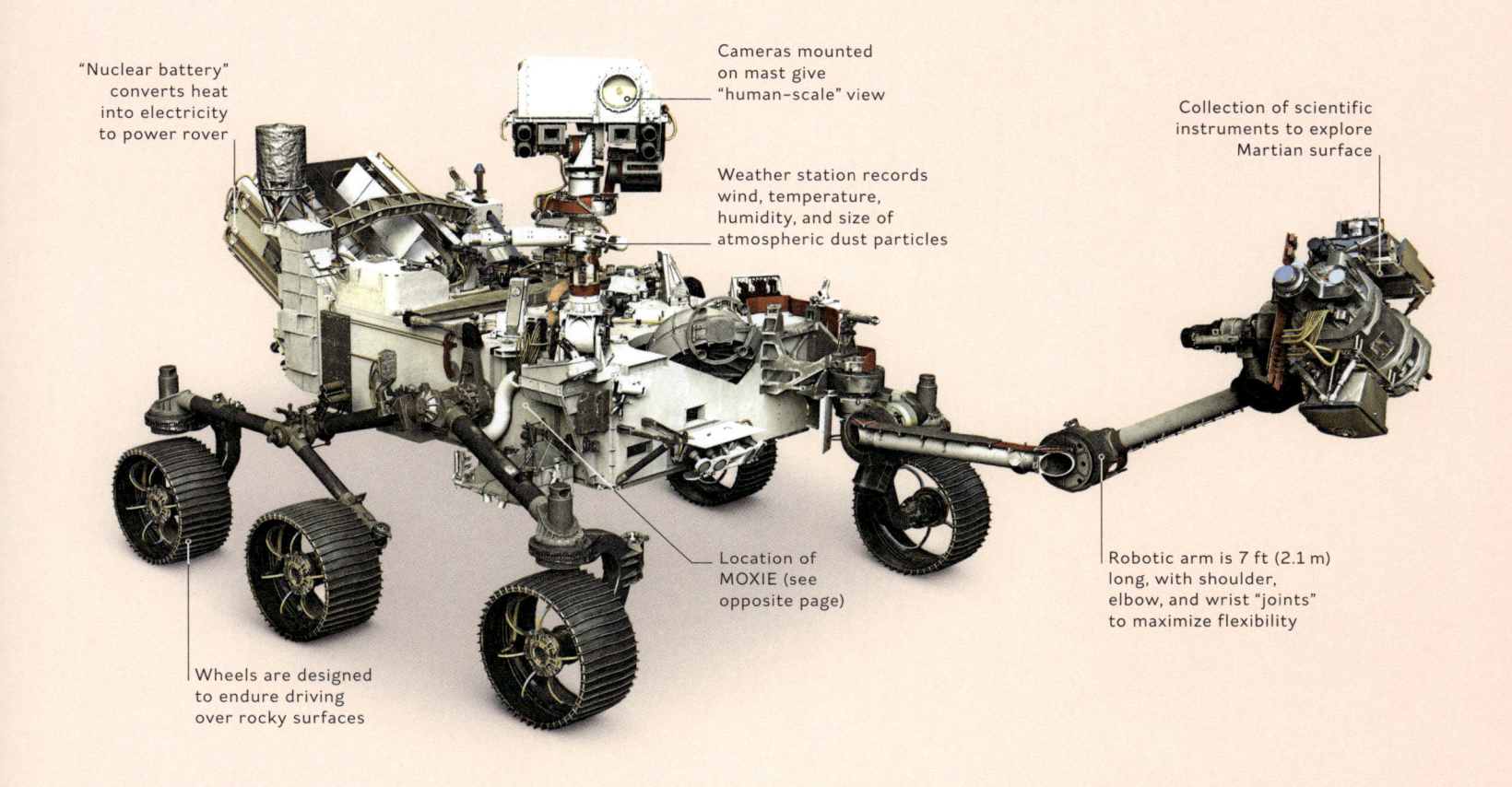

"Nuclear battery" converts heat into electricity to power rover

Cameras mounted on mast give "human-scale" view

Collection of scientific instruments to explore Martian surface

Weather station records wind, temperature, humidity, and size of atmospheric dust particles

Location of MOXIE (see opposite page)

Robotic arm is 7 ft (2.1 m) long, with shoulder, elbow, and wrist "joints" to maximize flexibility

Wheels are designed to endure driving over rocky surfaces

Martian base camp
This artist's concept depicts astronauts walking on the surface of Mars beside a protective modular habitat in which they live and do most of their work. Such a camp will embody technologies tested—and lessons learned—on the lunar surface.

Habitat comprises mix of modules designed for different purposes

Modules can exist separately or interlock

DIRECTORY

With no atmosphere and a mantle too cool to generate significant tectonic activity, the Moon is as it has been for billions of years. Its surface is scarred by meteorite impacts, which—along with volcanic eruptions and the Moon's shrinking as it cooled after its fiery birth—have given the Moon its unique appearance. The following pages show a range of features that make up the lunar landscape.

FEATURES OF THE NEAR SIDE

ARISTARCHUS CRATER

Hemisphere
Northern

Type
Impact crater

Age
450 million years

Diameter
25 miles (40 km)

Aristarchus, photographed by Lunar Orbiter 4 in 1967

One of the brightest features of the Moon, Aristarchus is a polygonal crater with terraced outer walls and a steep central peak. Its brightness is also due to its youth; at 450 million years old, it has yet to darken due to space weathering, which, by changing its chemistry, makes the lunar soil less reflective. Its brightest feature is its peak, which, like all of the Moon's pale highland features, is made of anorthosite (see pp.182–183). The peak formed when rock beneath the original impact compressed and rebounded up into the crater's center.

Aristarchus is located on a mare–highland boundary, which makes it particularly interesting geologically. It is a quarry of various preimpact crustal rocks overlaid with pyroclastic deposits, including volcanic glasses (see pp.182–183). It is also rich in ilmenite, a titanium oxide mineral that could be used in the production of lunar oxygen.

MARIUS HILLS

Hemisphere
Northern

Type
Volcanic domes

Age
3.8 billion years

Height
985 ft (300 m)

A view of the hills taken by the Lunar Reconnaissance Orbiter (LRO)

Located in Oceanus Procellarum (the Ocean of Storms), the Marius Hills are a group of volcanic domes that are believed to have been made when an unusually stiff type of lava breached the surface and failed to "pool" to form a mare (see pp.118–119). Giving the area a mottled appearance, the domes scatter a plateau that is scarred with the remnants of all kinds of volcanic activity, including rilles, channels, cones, and even a hole where a lava tube has caved in—possibly due to a meteorite strike (see pp.106–109). Estimated to be 260 ft (80 m) deep and about 985 ft (300 m) wide, this hole (like other lava-tube openings) has been suggested as a place where, sheltered from radiation, human beings could establish a colony. Although the hills have not been visited either by humans or robotic probes, they were on the shortlist of Apollo landing sites and have since been extensively surveyed by NASA's LRO, Japan's SELENE, and India's Chandrayaan-1 orbiter spacecraft (see pp.194–195).

SCHRÖTER'S VALLEY

Hemisphere
Northern

Type
Sinuous rille

Age
3.6 billion years

Length
100 miles (160 km)

Reaching a width of 6 miles (10 km), Schröter's Valley spills from a 3.7-mile- (6-km-) wide volcanic crater known as Cobra Head and snakes to the edge of Oceanus Procellarum. At 100 miles (160 km) in length, it is the largest lava tube, or sinuous rille, on the Moon (see pp.124–125). It began as a stream of lava, the roof of which cooled into solid rock, forming a tunnel, which then collapsed when the lava stopped flowing. The original stream plunged over a 3,000-ft- (915-m-) high cliff and partly filled Oceanus Procellarum below.

A mosaic of Schröter's Valley assembled from Apollo 15 photographs

Copernicus, with its three central peaks

COPERNICUS CRATER

Hemisphere Northern	**Age** 860 million years
Type Impact crater	**Diameter** 58 miles (93 km)

Located in eastern Oceanus Procellarum, Copernicus is the young crater that gives its name to the Moon's most recent "Copernican" geological period (see pp.14–15). Hexagonal in form, 58 miles (93 km) wide, and with central peaks that rise 3,900 ft (1,200 m) above its floor, Copernicus is 860 million years old and may have been created by a fragment of asteroid 495 Eulalia. It has three main terraces, caused by subsidence, and a ray system that extends 500 miles (800 km) into the surrounding maria. Scientists are particularly interested in its peaks, which, due to the way in which the surface rebounded after the impactor struck, are made of the deepest rocks to be exposed by the impact. As with all such peaks, they shed light on the composition of the Moon's interior, which, in this case—as discovered by India's Chandrayaan-1 lunar orbiter—is an olivine-rich material. Since olivine is present in igneous rocks, the Copernican peaks may have been raised from a magma chamber or even from the Moon's upper mantle.

MONTES APENNINUS

Hemisphere Northern	**Age** 3.9 billion years
Type Mountain range	**Length** 373 miles (600 km)

The 3.9-billion-year-old Montes Apenninus (Apennine Mountains) mark part of the boundary of the Imbrium basin—a colossal hole that was gouged out of the Moon by the collision of a protoplanet during the Late Heavy Bombardment (see pp.14–15). The range curves in a northeasterly direction from Eratosthenes crater to the headland of Promontorium Fresnel (Cape Fresnel), reaching a height of 3 miles (5 km). Apollo 15 landed in a valley at the northern end of the range, where astronauts David Scott and John Irwin explored Hadley Rille, one of the Moon's most spectacular volcanic features (see pp.124–125).

The southeast rim of the Imbrium basin

MARE TRANQUILLITATIS

Hemisphere Northern	**Age** 3.6 billion years
Type Sea	**Diameter** 542 miles (873 km)

One of the older impact basins, Mare Tranquillitatis (the Sea of Tranquility) is famous for being where Apollo 11 touched down and Neil Armstrong and Edwin "Buzz" Aldrin became the first humans to walk on the Moon (see pp.154–157). However, it is also the site where Ranger 8 was deliberately crashed onto the lunar surface in 1965 and where, two years later, Surveyor 5 landed and spent two weeks transmitting images back to Earth. Geologically, the area is interesting for having no mascon (central concentration of mass; see pp.112–113) and for having a slightly bluish tint due to the large amounts of titanium in its soil.

Buzz Aldrin and the Apollo 11 Lunar Module at Tranquility Base

VALLIS ALPES

Hemisphere	Age
Northern	3.8 billion years
Type	**Length**
Valley	103 miles (166 km)

A flat, lava-filled valley bisected by a sinuous rille (see pp.124–125), Vallis Alpes (Alpine Valley) is one of the Moon's most visually striking features. Reaching a maximum width of 6 miles (10 km), it extends 103 miles (166 km) from Mare Imbrium (the Sea of Showers) to Mare Frigoris (the Sea of Cold) and cuts through the Montes Alpes (the Lunar Alps), where peaks rise to 7,870 ft (2,400 m) on either side. Astronomers have speculated about its origins—whether it was caused by the asteroid that opened the Imbrium basin or whether expansions in the lunar crust elsewhere caused the surface to fall in this area. However, most agree that it is a graben—a place where fault lines have caused the land to subside (see pp.170–171). The rille that snakes through its center is a lava tube—a vein of lava that emptied, leaving a tunnel that eventually collapsed.

Vallis Alpes with its rille, as viewed by Lunar Orbiter 4

MARE SERENITATIS

Hemisphere	Age
Northern	3.9 billion years
Type	**Diameter**
Sea	400 miles (650 km)

One of the eyes of the Man in the Moon, Mare Serenitatis (the Sea of Serenity) is slightly brighter, and therefore younger, than its neighboring maria (see pp.118–119). It is a multi-ring basin with two distinct ramparts and is filled with basalts of varying ages, laid down between 3.9 and 1 billion years ago—the older, iron- and titanium-rich layers being on the edges. It is streaked with two huge ejecta rays from Tycho crater and features the Rimae Sulpicius Gallus rille system and numerous prominent wrinkle ridges (see pp.170–171).

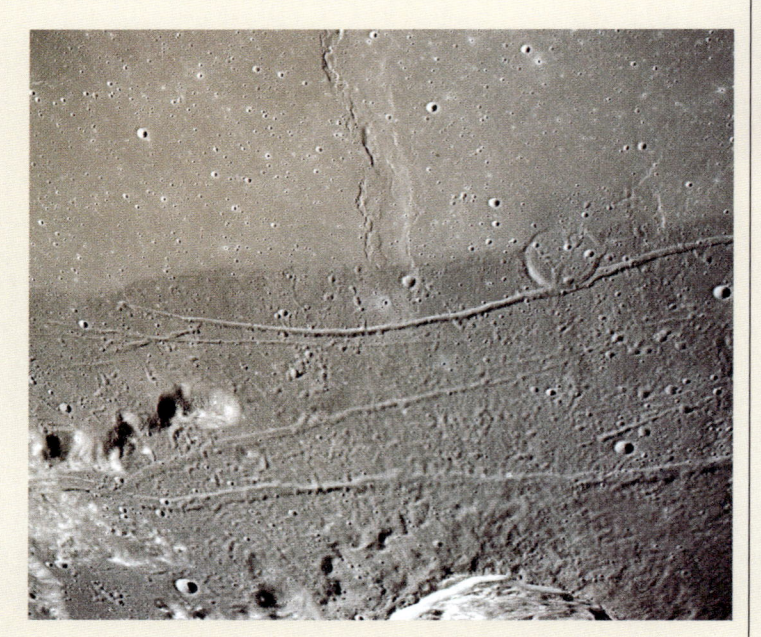

A wrinkle ridge cut by linear rilles in Mare Serenitatis

HADLEY RILLE

Hemisphere	Age
Northern	3.3 billion years
Type	**Length**
Sinuous rille	50 miles (80 km)

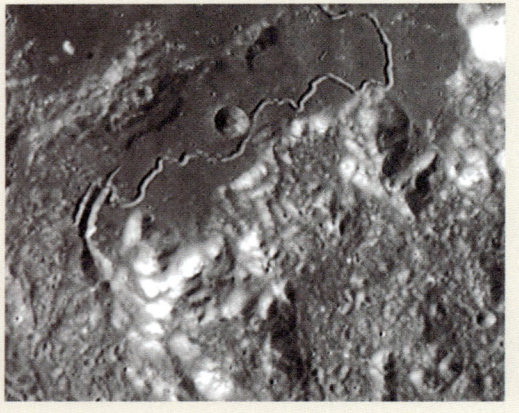

Hadley Rille, viewed by Apollo 15

Located at the southeastern edge of Mare Imbrium, Hadley Rille is a steep-walled valley that snakes across Palus Putredinis (the Marsh of Decay) at the foot of the Montes Apenninus (see pp.232–233). It is possibly the most famous rille on the Moon, being both visually striking and one of the sites explored by Apollo 15's David Scott and John Irwin (see pp.176–177). Although originally thought to have been created by water, samples proved that it is volcanic in origin and is most likely the remains of a lava tube (see pp.124–125). Like many such rilles, it has a crater at its source—in this case, the elongated Béla—which may have been a volcanic vent. Photographs taken by Scott and Irwin, who explored the area by rover, also show that it is layered in a way that is typical of a lava bed. The region is named after John Hadley, an 18th-century British mathematician and meteorologist who invented both the reflecting octant (a navigational aid) and the parabolic Newtonian telescope.

PLATO CRATER

Hemisphere	Age
Northern	3.8 billion years
Type	**Diameter**
Impact crater	68 miles (109 km)

Located at the northeastern edge of Mare Imbrium, Plato crater is about 3.8 billion years old. Isolated from the surrounding mare by its 6,600-ft- (2,000-m-) high rim, it is filled with lava that must have welled up from beneath, rising through cracks that opened when the crater was excavated. Although foreshortening makes it appear elliptical, it is almost perfectly circular, and its basalt floor makes it one of the darkest points on the Moon. Its surface is also remarkably smooth, which inspired 17th-century Polish-Lithuanian astronomer Johannes Hevelius to name it the Greater Black Lake. A prominent feature of its western rim is a 14-mile- (23-km-) wide slump block—a roughly triangular piece of rim that broke free as the crater subsided. One mystery about Plato is that most craters of its size have a central peak (see pp.106–107); maybe Plato's is buried in its core.

An LRO photograph of Plato at sunrise

TAURUS-LITTROW VALLEY

Hemisphere	Age
Northern	3.9 billion years
Type	**Depth**
Valley	6,562 ft (2,000 m)

Famous for being the landing site of Apollo 17, the Taurus-Littrow Valley gave scientists the final major pieces of the Moon's geological puzzle (see pp.180–181). It was vital that the final Apollo mission visited a highland area, since the highlands account for 85 percent of the lunar surface and previous missions had focused on maria. The astronauts found three types of rock: 4.5-billion-year-old magnesium olivine, dating to soon after the Moon's formation; 3.9-billion-year-old basalts, dating to the Late Heavy Bombardment; and equally old breccias that had been ejected from a nearby crater. The samples showed that the Moon was 40 million years older than scientists had previously estimated.

A close-up view of the valley taken by the LRO

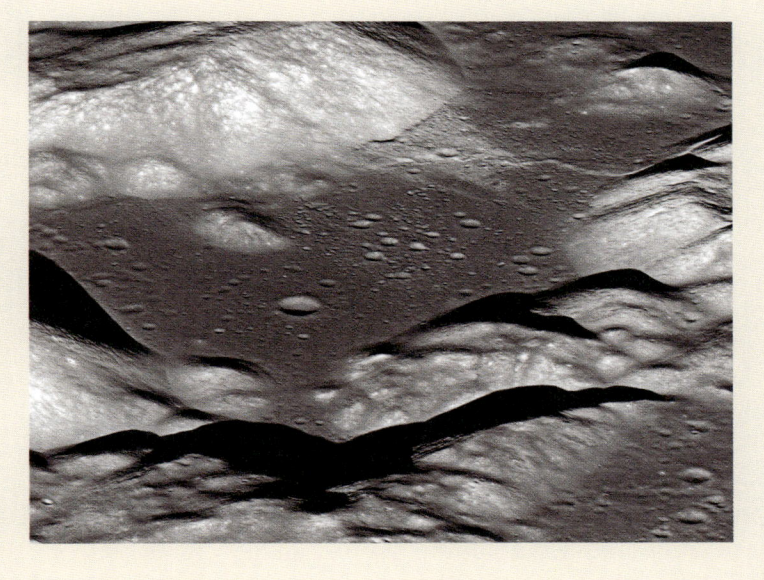

Alphonsus, with its rilles and curious dark-halo craters

ALPHONSUS CRATER

Hemisphere	Southern
Type	Impact crater
Age	4 billion years
Diameter	67 miles (108 km)

An ancient crater dating to the Pre-Nectarian period (see pp.14–15), Alphonsus stands in stark contrast to the younger craters, such as Aristarchus (see pp.232–233). Erosion has smoothed its rim; erased its rays; and, by chemically altering it, darkened its surface. It is 8,800 ft (2,700 m) deep and has a central, pyramidal, anorthosite peak that rises 4,920 ft (1,500 m) above its surface. Its floor is scored by a number of rilles and dotted with craters, some of which are surrounded by halos of darker material. These halos may be evidence of explosive pyroclastic eruptions—small-scale but explosive ejections of molten material caused by gases building up beneath the lunar surface. Located at the eastern edge of Mare Nubium (the Sea of Clouds), Alphonsus slightly overlaps two other craters: Ptolemaeus to the north and the smaller Alpetragius to the southwest. The crater was closely surveyed in 1965 by Ranger 9 (see pp.142–143), which intentionally impacted on its surface, and was shortlisted as an Apollo landing site.

RUPES ALTAI

Hemisphere
Southern

Type
Escarpment

Age
4.2 billion years

Length
300 miles (480 km)

The jagged curve of the Rupes Altai escarpment

Best seen a few days after a new Moon, when the Sun is low in the Moon's western sky, Rupes Altai (the Altai Cliff) is the Moon's largest escarpment—a 300-mile (480-km) cliff that marks the southwestern edge of the Nectaris basin and rises to a height of 13,000 ft (4,000 m). It is the most visible part of the basin, which, although 500 miles (800 km) in diameter, has had 4 billion years in which to erode. It stretches south from the western side of Catharina crater, and then southeast to Piccolomini, a 3.2–3.8-billion-year-old crater that brings the escarpment to a sudden end.

The unique Catena Abulfeda (Chain of Abulfeda)—a line of tiny craters—stretches westward from the northern end of the escarpment. It extends for over 140 miles (230 km) across the surface, comprising mainly craters that are less than 6.2 miles (10 km) in diameter.

HUMBOLDT CRATER

Hemisphere
Southern

Type
Impact crater

Age
3.8 billion years

Diameter
128 miles (207 km)

An oblique view of Humboldt, taken by Apollo 15

Situated on the Moon's southeastern edge, Humboldt is one of those features that comes and goes with the Moon's libration (see pp.116–117). At 128 miles (207 km) in diameter (its depth is currently unknown), it is one of the largest craters of the Late Imbrian epoch (see pp.14–15). It has various distinguishing characteristics: a perfect "bull's-eye" double crater on its eastern side; instead of a central peak, it has a range of hills that swirls northeast from its center to its rim; a complex network of rilles radiates from its center, forming spokes and concentric arcs; and its floor is stained with dark patches, which may be where lava reached the surface during the Moon's volcanic stage. The crater is named after German philosopher and linguist Wilhelm von Humboldt (1767–1835).

TYCHO CRATER

Hemisphere
Southern

Type
Impact crater

Age
108 million years

Diameter
53 miles (85 km)

Named after Danish astronomer Tycho Brahe (1546–1601), Tycho is one of the Moon's most prominent features. At 53 miles (85 km) wide and 15,420 ft (4,700 m) deep, it is by no means the largest crater, but at only 108 million years old, it is relatively young, bright, and uneroded. Its ejecta rays are one of the marvels of the night sky, visibly stretching some 930 miles (1,500 km) across the lunar surface.

Bright Tycho beneath the dark Mare Nubium

GRIMALDI CRATER

Hemisphere Southern	**Age** 3.9 billion years
Type Impact crater	**Diameter** 146 miles (235 km)

Located on the Moon's western edge, to the southwest of Oceanus Procellarum, Grimaldi is another feature that comes and goes with the Moon's libration (see pp.116–117). At 3.9 billion years old, it belongs to the Pre-Nectarian period (see pp.14–15), and so is severely eroded; its rim has been reduced so much by subsequent meteor impacts and space weathering that today it resembles a smooth ring of hills. Its other notable feature is its dark mare floor, which makes it easy to spot on the lunar surface.

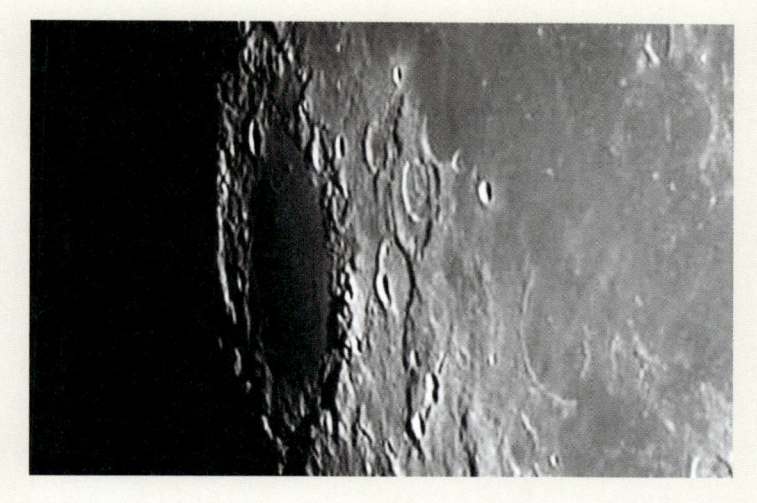

Grimaldi, on the Moon's western edge, viewed from Earth

GASSENDI CRATER

Hemisphere Southern
Type Impact crater
Age 4 billion years
Diameter 68 miles (110 km)

Gassendi crater was so inundated with lava from its formation that only the tops of its rim and central peaks lie above its central floor. It is streaked with fractures—some of which are tens of thousands of feet wide—which may have formed as the crater cooled and settled. Gassendi A, a smaller crater on its north side, gives the feature its "diamond ring" appearance.

Double-ringed Gassendi on the border of Mare Humorum

RUPES RECTA

Hemisphere Southern
Type Escarpment
Age 3.2–3.9 billion years
Length 110 miles (177 km)

An escarpment scarring the southeastern region of Mare Nubium, Rupes Recta (the Straight Cliff) is 110 miles (177 km) long and on average 8,200 ft (2,500 m) wide. It is 790–985 ft (240–300 m) high, and, although it appears to be vertical, it is in fact gently sloped. It is an example of a linear fault—the result of subterranean pressures causing the western side of the area to drop relative to the eastern side.

CLAVIUS CRATER

Hemisphere Southern
Type Impact crater
Age 3.8 billion years
Diameter 144 miles (231 km)

At 144 miles (231 km) wide, 11,480 ft (3,500 m) deep, and more than 3.85 billion years old, Clavius is one of the largest and oldest craters on the Moon. It is distinguished by a chain of craters—each of diminishing size—that arcs across its floor, and for the craters on its rim. In 1968, it was immortalized as the site of humankind's encounter with an alien artifact in the film *2001: A Space Odyssey*. More recently, it became the first sunlit surface of the Moon to be known to harbor water.

Ancient Clavius, with its arc of inner craters

The straight Rupes Recta cutting across Mare Nubium

FEATURES OF THE FAR SIDE

PASCAL CRATER

Hemisphere
Northern

Type
Impact crater

Age
4.1 billion years

Diameter
67 miles (108 km)

Named after Blaise Pascal (1623–1662), the French mathematician who invented the mechanical calculator, Pascal crater lies to the west of the lunar north pole. Although the Moon's libration brings it partially into view each month (see pp.116–117), the crater was only first photographed in detail by the European Space Agency's Smart-1 lunar satellite in 2005. This provided a clear picture of Pascal and its neighboring Brianchon crater—an image that was later improved by NASA's LRO in 2017. These pictures show a heavily eroded crater with a smoothed outer rim and worn-down terraces. Its lava-filled floor is also smooth, although peppered by tiny craters and breached by a central ridge. A line of craterlets crosses its northern edge, while its northwestern, southwestern, and southeastern sections are interrupted by three larger craters—Pascals F, A, and G, with Pascal A being notably more eroded.

LRO's mosaic of Pascal, made in 2017

TSIOLKOVSKY CRATER

Hemisphere	**Age**
Southern	4.2 billion years
Type	**Diameter**
Impact crater	115 miles (185 km)

Named after Konstantin Tsiolkovsky, the pioneering Russian rocket scientist (see pp.128–129), Tsiolkovsky crater is one the largest features of the Moon's far side. It is at least 16,400 ft (5,000 m) deep and has a central peak that rises over 10,000 ft (3,200 m). It is also distinctively dark, being one of the few locations on the far side where basalt has welled up and spilled across the surface. (Most of the Moon's maria are on its near side—see pp.118–119.) Its peak is roughly 12 miles (20 km) wide and is particularly interesting geologically; it was wrenched approximately 18 miles (30 km) from the Moon's liquid mantle—all in a matter of seconds, when the impactor struck—and remains a snapshot of the Moon's interior structure. For this reason, the crater was suggested as an Apollo 17 landing site (see pp.180–181).

Tsiolkovsky, photographed by Apollo 15 in 1971

KOROLEV BASIN

Hemisphere
Southern

Type
Ringed impact crater

Age
3.7 billion years

Diameter
272 miles (437 km)

Lying across the lunar equator, the Korolev basin was first photographed in detail by NASA's Lunar Orbiter I in 1966. The photograph showed a highly eroded double-ringed basin that was later dated to the Nectarian period. Its inner ring is half the diameter of its outer ring and contains a small key-hole-shaped crater that Apollo 8 used as a reference point both for photographing and later mapping the Moon (see pp.148–149). In 1970, the basin was formally named after Soviet engineer and spacecraft designer Sergei Korolev (1909–1966).

A vertical view of Korolev taken by Lunar Orbiter 1 in 1966

VAN DE GRAAFF CRATER

Hemisphere
Southern

Type
Impact crater

Age
3.6 billion years

Diameter
155 miles (250 km)

Van de Graaff, photographed by Lunar Orbiter 2 in 1966

Situated to the north of the South Pole-Aitken basin, the figure-eight Van de Graaff crater is a conjoined pair of craters that has no separating central wall. Only its southern half includes a central peak, its northern half having a relatively smooth basalt floor. It is one of the largest craters of the Nectarian period (see pp.14–15) and lies in a region that was marred by the shockwaves that passed through the Moon when the Imbrium basin was excavated on the opposite side (see pp.232–233). Perhaps relatedly, the crater has two anomalies that intrigue scientists; it has its own magnetic field (unlike Earth, the Moon lacks a global magnetic field) and is rich in elements such as thorium that are usually found on the near side of the Moon.

MARE ORIENTALE

Hemisphere
Southern

Type
Impact basin

Age
3.8 billion years

Diameter
578 miles (930 km)

Half-lit Orientale, shot by Lunar Orbiter 4 in 1967

Located on the Moon's southeastern edge, Mare Orientale (the Eastern Sea) is a multi-ring impact basin that can be seen from Earth when the Moon's libration permits (see pp.116–117). At only 3.8 billion years old, it is the Moon's most recent basin. Roughly 460 miles (900 km) wide, it is unusually well defined due the minimal amount of lava that welled up when it formed. Its central mare may be less than 3,280 ft (1,000 m) deep, which is considerably less than the average near-side maria. Scientists do not agree on why this is so, but it may be that its impactor struck at a shallow angle. Its center is surrounded by the inner and outer Montes Rook mountains, which in turn are encircled by the Montes Cordillare, which together give the basin its "bull's-eye" appearance.

South Pole-Aitken—the largest, oldest, and deepest lunar basin

SOUTH POLE-AITKEN BASIN

Hemisphere
Southern

Type
Impact basin

Age
4.2–4.3 billion years

Diameter
1,550 miles (2,500 km)

The largest impact structure in the solar system, the South Pole-Aitken basin is approximately 4.3 miles (7 km) deep. It was excavated roughly 4.2–4.3 billion years ago, during the Pre-Nectarian period, which makes it the Moon's oldest basin. Discovered in the 1960s, it was visited by China's Chang'e 4 in 2019 and by Chang'e 6 in 2024. The latter returned samples of its surface to Earth—the first samples to be taken from the far side of the Moon.

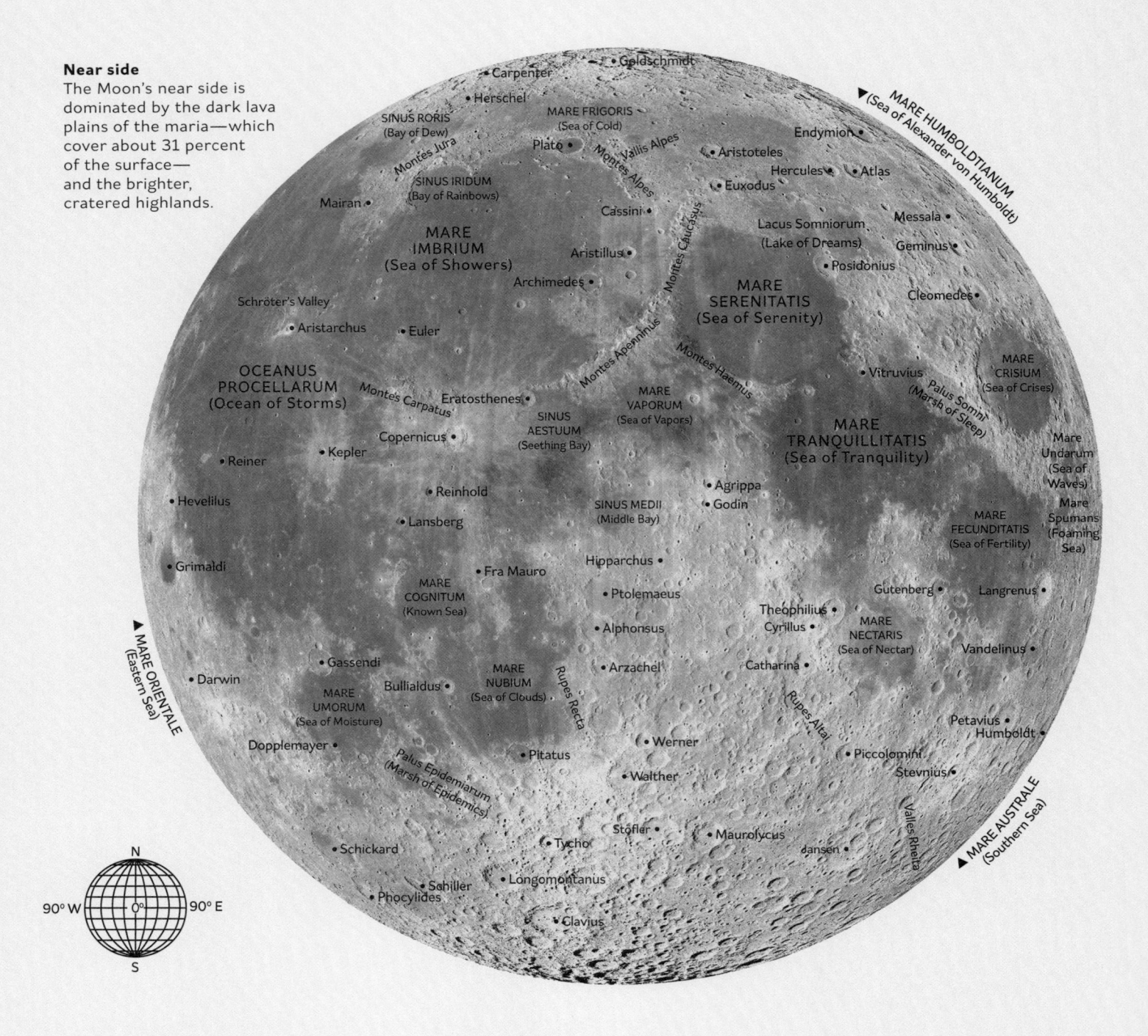

Near side
The Moon's near side is dominated by the dark lava plains of the maria—which cover about 31 percent of the surface—and the brighter, cratered highlands.

Carpenter • Geldschmidt

Herschel •

SINUS RORIS
(Bay of Dew)

MARE FRIGORIS
(Sea of Cold)

Plato •

Montes Jura

Vallis Alpes

Montes Alpes

Aristoteles •

Endymion •

► (Sea of Alexander von Humboldt)
MARE HUMBOLDTIANUM

Hercules • • Atlas

Mairan •

SINUS IRIDUM
(Bay of Rainbows)

Cassini •

• Euxodus

Messala •

Lacus Somniorum
(Lake of Dreams)

Geminus •

MARE
IMBRIUM
(Sea of Showers)

Aristillus •

Montes Caucasus

• Posidonius

Cleomedes •

Archimedes •

MARE
SERENITATIS
(Sea of Serenity)

Schröter's Valley

• Aristarchus

• Euler

Montes Apenninus

MARE
CRISIUM
(Sea of Crises)

OCEANUS
PROCELLARUM
(Ocean of Storms)

Montes Carpatus

Eratosthenes •

Montes Haemus

• Vitruvius

Palus Somni
(Marsh of Sleep)

MARE
VAPORUM
(Sea of Vapors)

Copernicus •

SINUS
AESTUUM
(Seething Bay)

MARE
TRANQUILLITATIS
(Sea of Tranquility)

Mare
Undarum
(Sea of
Waves)

Reiner •

• Kepler

• Agrippa
• Godin

SINUS MEDII
(Middle Bay)

Mare
Spumans
(Foaming
Sea)

• Hevelius

• Reinhold

• Lansberg

Hipparchus •

MARE
FECUNDITATIS
(Sea of Fertility)

• Grimaldi

MARE
COGNITUM
(Known Sea)

• Fra Mauro

• Ptolemaeus

Gutenberg •

Langrenus •

Theophilius •

Cyrillus •

MARE
NECTARIS
(Sea of Nectar)

Vandelinus •

• Alphonsus

• Gassendi

Bullialdus •

MARE
NUBIUM
(Sea of Clouds)

• Arzachel

Rupes Recta

Catharina •

Rupes Altai

• Darwin

MARE
UMORUM
(Sea of Moisture)

Petavius •
Humboldt •

Dopplemayer •

Palus Epidemiarum
(Marsh of Epidemics)

• Pitatus

• Werner

• Piccolomini

Stevnius •

• Walther

Valles Rheita

► MARE AUSTRALE
(Southern Sea)

• Schickard

Stöfler •

• Maurolycus

Jansen •

► MARE ORIENTALE
(Eastern Sea)

• Schiller

• Longomontanus

• Tycho

• Phocylides

• Clavius

N
90° W   0°   90° E
S

MAPS OF THE MOON

These detailed pictures of the Moon's near and far sides were made using the Wide Area Camera on NASA's Lunar Reconnaissance Orbiter (LRO) over a 15-month period during 2009–2010. More than 15,000 images were assembled into a mosaic to show the Moon's surface illuminated at low Sun angles, ranging from 21° to 8° above the horizon. This highlights the details of the Moon's topography at the expense of surface brightness variations, and so looks quite different to a photograph of the full Moon from Earth. But such maps are invaluable when reconnoitring potential landing sites for future missions.

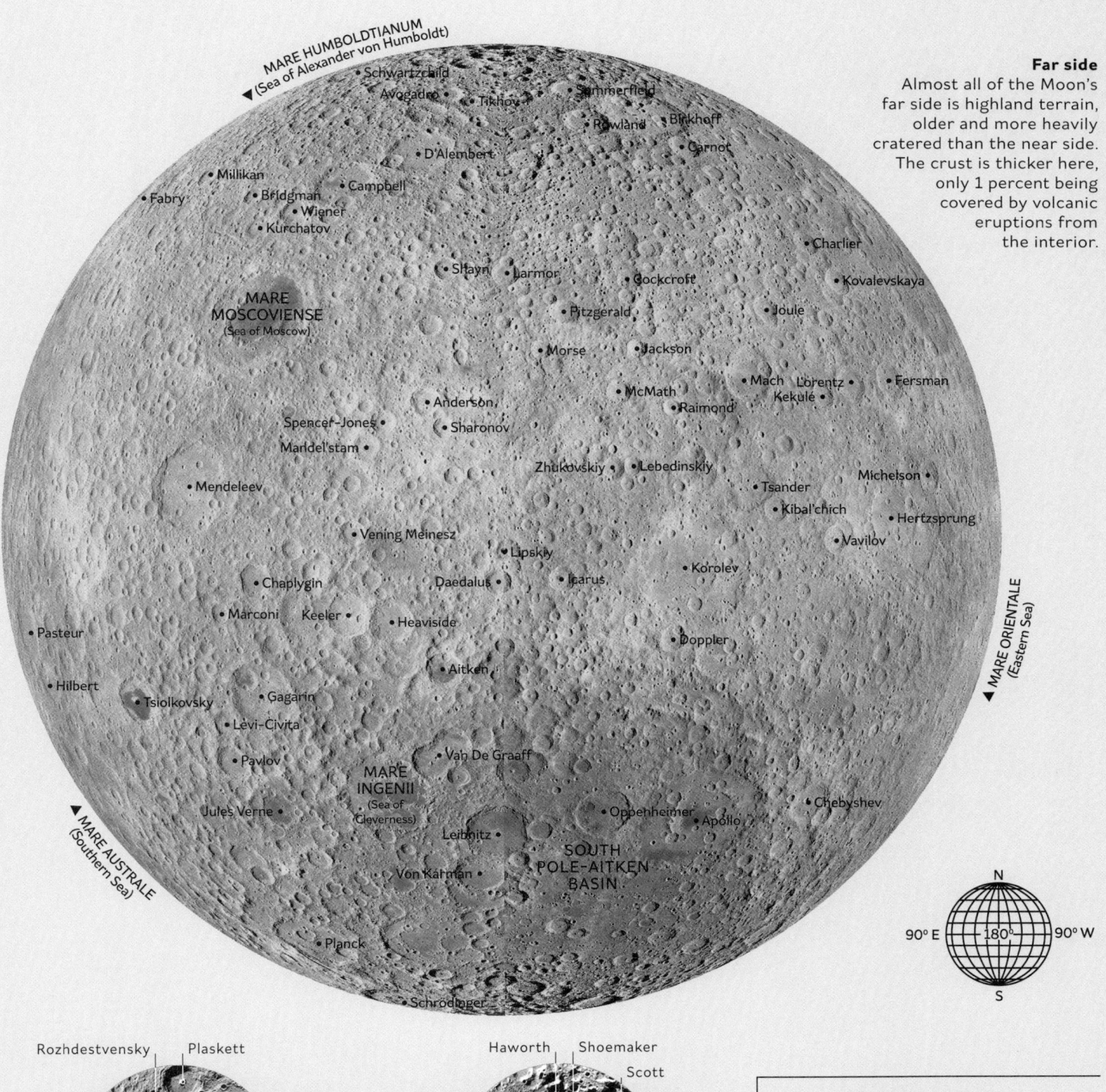

- Schwartzchild
- Avogadro
- Tikhov
- Summerfield
- Rowland
- Birkhoff
- Carnot
- D'Alembert
- Millikan
- Fabry
- Bridgman
- Campbell
- Wiener
- Kurchatov
- Charlier
- Kovalevskaya
- Shayn
- Larmor
- Cockcroft

MARE MOSCOVIENSE
(Sea of Moscow)

- Fitzgerald
- Joule
- Morse
- Jackson
- McMath
- Mach
- Lorentz
- Fersman
- Kekulé
- Anderson
- Raimond
- Spencer-Jones
- Sharonov
- Mandel'stam
- Zhukovskiy
- Lebedinskiy
- Tsander
- Michelson
- Mendeleev
- Kibal'chich
- Hertzsprung
- Vening Meinesz
- Vavilov
- Lipskiy
- Chaplygin
- Daedalus
- Icarus
- Korolev
- Marconi
- Keeler
- Heaviside
- Pasteur
- Doppler
- Hilbert
- Aitken
- Tsiolkovsky
- Gagarin
- Levi-Civita
- Pavlov
- Van De Graaff

MARE INGENII
(Sea of Cleverness)

- Jules Verne
- Oppenheimer
- Apollo
- Chebyshev
- Leibnitz
- Von Kármán

SOUTH POLE-AITKEN BASIN

- Planck
- Schrödinger

Far side
Almost all of the Moon's far side is highland terrain, older and more heavily cratered than the near side. The crust is thicker here, only 1 percent being covered by volcanic eruptions from the interior.

90° E   180°   90° W

North Pole
- Rozhdestvensky
- Plaskett
- Haskin
- Lovelace
- Nansen
- Sylvester
- Peary
- Hermite
- Byrd
- Florey

Libration (see pp.116–117) enables us to glimpse the lunar polar regions, but they remain shadowed due to the small tilt of the Moon's rotational axis in relation to the Sun.

South Pole
- Haworth
- Shoemaker
- Scott
- Drygalski
- Nobile
- Amundsen
- Hédervári
- Sverdrup
- Shackleton
- Idel'son
- Kocher
- Wiechert

The deep craters here possibly harbor ice deposits in permanent shadow, making this pole a prime target for future crewed Moon landings or a permanent base.

NAMING CONVENTIONS

Lunar craters are mostly named after prominent scientists, engineers, and explorers. The maria are given Latin names describing weather conditions, mental states, or other concepts. Some mountain regions are named after terrestrial ranges, others after scientists or nearby craters. Russian names dominate the far side, which Russia's Luna 3 first photographed in 1959. Aptly, many features at the Moon's poles are named after polar scientists and explorers.

OBSERVING THE MOON

With an ever-changing appearance, the Moon is a rewarding subject for the amateur astronomer. Observation aids reveal many more details of the lunar landscape than are visible to the unaided eye.

Latitude
The orientation of the Moon's features relative to the local horizon changes with latitude, as well as between moonrise and moonset. The Moon's appearance at moonrise during its eight phases is shown here for the equator, the poles, and midlatitudes.

While dark maria, bright highlands, and the largest craters can be seen with the unaided eye, binoculars or a telescope give a much clearer view of the lunar surface. Both instruments use an aperture wider than the pupil of the human eye (which is about 0.27 in, or 7 mm, across in darkness) to capture more light and enable dimmer objects to be seen, then magnify the image with an eyepiece. Binoculars offer moderate magnification and a wide field of view. A large pair of binoculars might have a front lens diameter of 2 in (50 mm) and a magnification of 7× or 10×, described as 7×50 or 10×50.

Telescopes offer greater magnification but cover a smaller part of the sky, and thus of the Moon's surface. Refracting telescopes use a glass lens to collect light; reflecting telescopes use mirrors. Even a small telescope, of either type, needs a good mount—ideally an equatorial mount that can be aligned to Earth's rotation.

Full Moon is not the best time for lunar observation—shadows almost disappear, making features difficult to locate. Viewing is better at crescent, half, or gibbous phase, particularly near the edge of the Sun's illumination, where the shadows lengthen and highlight the topography.

	New Moon	Waxing crescent	First quarter	Waxing gibbous	Full Moon	Waning gibbous	Third quarter	Waning crescent
North Pole 90°N								
London 51.5°N								
Equator 0°								
Sydney 33.9°S								
South Pole 90°S								

LIGHT

The photos below, taken from a low lunar orbit with Apollo 15's Metric Camera in 1971, demonstrate how the Sun's changing illumination dramatically affects the appearance of the Moon's surface. The images show the same area in Mare Imbrium (the Sea of Showers), a lava plain peppered with craters of various sizes. With the Sun at a high angle, the brightness of the surface dominates and the dark, undisturbed mare contrasts with the bright ejecta around the craters. When the Sun's angle drops, shadows lengthen and the surface topography becomes more apparent, revealing crater rims, wrinkle ridges, and shallow rilles (valleys).

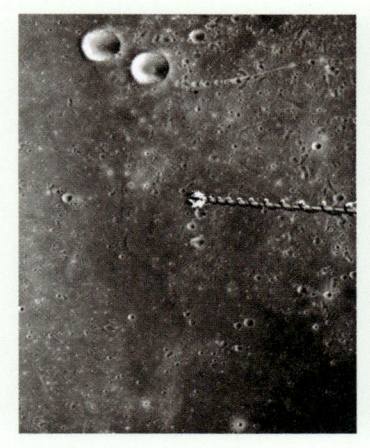
Sun at 42° above the horizon

Sun at 17° above the horizon

Sun at 7° above the horizon

Sun at 2° above the horizon

MAGNIFICATION

A pair of 10x50 binoculars will magnify 10x, giving a good view of the entire lunar disc. A digital camera fitted with a 15.7 in-(400 mm-) telephoto lens will do a little better at 20x magnification, but to clearly see details on the Moon's surface, a telescope is required. A small telescope will magnify 50–200x depending on its focal length and the eyepiece used. Focal length is the distance from a lens or mirror to the point where it brings light rays together (into focus). Telescopes are also measured by the diameter of their aperture.

Camera with telephoto lens
The full Moon on December 8, 2022, during an occultation of Mars (red dot, lower right), photographed from London, UK, with a digital camera fitted with a 15.7 in-(400mm-) lens.

Telescope view
Small craters are clear with a 4 in-(102 mm-) diameter reflecting telescope. A focal length of 51.2 in-(1300 mm) and a 0.63 in-(16 mm-) eyepiece provide 81x magnification (1300 ÷ 16).

> "**Astronomy** compels **the soul** to **look upward**, and leads us **from this world** to another."
>
> Plato, Greek philosopher, *The Republic* (375 BCE)

OBSERVING ECLIPSES

To ancient cultures, eclipses could be terrifying and were sometimes interpreted as ill omens. Now that their causes and duration are understood, they are appreciated for the spectacle they provide.

Total solar eclipses (see pp. 38–41) are not particularly infrequent. Each one can only be seen from a small section of Earth's surface, however—and then only under clear skies—so for an individual, they remain a rare event. They are also entirely predictable, so it is possible for observers to ensure they are in the right place at the right time to enjoy the experience. Even a partial solar or a lunar eclipse can inspire awe among onlookers.

Spectators must take care when observing a solar eclipse, however. Except for the brief phase of totality during a total eclipse, it is not safe to look directly at the Sun. Severe eye injury will result if the bright Sun is observed through a camera lens, a pair of binoculars, or a telescope. Effective safeguard methods include using special-purpose solar filters or utilizing a telescope to project an image of the Sun onto white card. Spectators can align the telescope without having to look at the Sun by tilting it to minimize the size of its shadow on the ground.

SOLAR ECLIPSES FROM 2025 TO 2035

Date	Type	Location
2025, March 29	Partial	N. W. Africa, Europe, N. Russia
2025, September 21	Partial	S. Pacific, New Zealand, Antarctica
2026, February 17	Annular	Antarctica (partial: S. Argentina, S. Chile, S. Africa, Antarctica)
2026, August 12	Total	Arctic, Greenland, Iceland, Spain (partial: N. North America, W. Africa, Europe)
2027, February 6	Annular	Chile, Argentina, Atlantic (partial: S. America, Antarctica, W. and S. Africa)
2027, August 2	Total	Morocco, Spain, Algeria, Libya, Egypt, Saudi Arabia, Yemen, Somalia (partial: Africa, Europe, SW. Asia)
2028, January 26	Annular	Ecuador, Peru, Brazil, Suriname, Spain, Portugal (partial: N., C., and S. America; W. Europe; N. W. Africa)
2028, July 22	Total	Australia, New Zealand (partial: S. E. Asia, E. Indies, Australia, New Zealand)
2029, January 14	Partial	N. America, C. America
2029, June 12	Partial	Arctic, Scandinavia, Alaska, N. Asia, N. Canada
2029, July 11	Partial	S. Chile, S. Argentina
2029, December 5	Partial	S. Argentina, S. Chile, Antarctica
2030, June 1	Annular	Algeria, Tunisia, Greece, Turkey, Russia, N. China, Japan (partial: Europe, N. Africa, Asia, Arctic, Alaska)
2030, November 25	Total	Botswana, S. Africa, Australia (partial: S. Africa, S. Indian Ocean, East Indies, Australia, Antarctica)
2031, May 21	Annular	Angola, Democratic Republic of the Congo, Tanzania, Zambia, S. India, Malaysia, Indonesia (partial: Africa, S. Asia, E. Indies, Australia)
2031, November 14	Hybrid	Pacific, Panama (partial: Pacific, S. North America, C. America, N. W. South America)
2032, May 9	Annular	S. Atlantic (partial: S. South America, S. Africa)
2032, November 3	Partial	Asia
2033, March 30	Total	E. Russia, Alaska (partial: N. America)
2033, September 23	Partial	S. South America, Antarctica
2034, March 20	Total	Nigeria, Cameroon, Chad, Sudan, Egypt, Saudi Arabia, Iran, Afghanistan, Pakistan, India, China (partial: Africa, Europe, W. Asia)
2034, September 12	Annular	Chile, Bolivia, Argentina, Paraguay, Brazil (partial: C. and S. America)
2035, March 9–10	Annular	New Zealand, Pacific (partial: Australia, New Zealand, S. Pacific, Mexico, Antarctica)
2035, September 1–2	Total	China, Korea, Japan, Pacific (partial: E. Asia, Pacific)

Solar eclipse time-lapse
Pictured here is a sequential breakdown of a solar eclipse in July 2019, as viewed from a site in Chile. Each shot was taken 5 minutes apart with a digital camera.

On average, any **location on Earth** experiences a **total solar eclipse** only once every **375 years**.

Lunar eclipse time-lapse
A blood Moon (see pp. 38–39) appeared during the total lunar eclipse of July 2018. At almost 43 minutes, it was the longest lunar eclipse of the 21st century.

LUNAR ECLIPSES FROM 2025 TO 2035

Date	Type	Location
2025, March 13–14	Total	Pacific, Americas, W. Europe, W. Africa
2025, September 7–8	Total	Europe, Africa, Asia, Australia
2026, March 2–3	Total	E. Asia, Australia, Pacific, Americas
2026, August 27–28	Partial	E. Pacific, Americas, Europe, Africa
2027, February 20–21	Penumbral	Americas, Europe, Africa, Asia
2027, August 16–17	Penumbral	Pacific, Americas
2028, January 11–12	Partial	Americas, Europe, Africa
2028, July 6–7	Partial	Europe, Africa, Asia, Australia
2028/2029, December 31–January 1	Total	Europe, Africa, Asia, Australia, Pacific
2029, June 25–26	Total	Americas, Europe, Africa, SW. Asia
2029, December 20–21	Total	Americas, Europe, Africa, Asia
2030, June 15–16	Partial	Europe, Africa, Asia, Australia
2030, December 9–10	Penumbral	Americas, Europe, Africa, Asia
2031, May 6–7	Penumbral	Americas, Europe, Africa
2031, June 5	Penumbral	East Indies, Australia, Pacific
2031, October 29–30	Penumbral	Americas
2032, April 25–26	Total	E. Africa, Asia, Australia, Pacific
2032, October 18–19	Total	Africa, Europe, Asia, Australia
2033, April 14–15	Total	Europe, Africa, Asia, Australia
2033, October 7–8	Total	Asia, Australia, Pacific, Americas
2034, April 2–3	Penumbral	Europe, Africa, Asia, Australia
2034, September 27–28	Partial	Americas, Europe, Africa
2035, February 21–22	Penumbral	E. Asia, Pacific, Americas
2035, August 18–19	Partial	Americas, Europe, Africa, SW. Asia

VIEWING AN ECLIPSE SAFELY

Observers can safely watch a solar eclipse through eclipse glasses or handheld solar viewers. Fitting the front lenses of binoculars or telescopes with stronger solar filters also offers protection. Another safe approach is to use an indirect viewing method, such as projecting an image of the Sun onto the ground or into a pinhole camera.

Using eclipse glasses
At least 1,000 times darker than ordinary sunglasses, eclipse glasses are fitted with special solar filters that reduce the sunlight's intensity and filter out damaging ultraviolet wavelengths.

Watching the shadow of an eclipse
During a partial solar eclipse or the early stages of a total eclipse, the dappled sunlight on the ground beneath trees takes on the crescent shape of the partially obscured Sun.

GLOSSARY

A

albedo The amount of light reflected by an astronomical body.

anomalistic month The roughly 27.5 days it takes for the Moon to travel from perigee to perigee.

anorthosite The oldest type of rock on the Moon, composed almost entirely of plagioclase feldspar.

annular eclipse An eclipse that takes place when the Moon passes between Earth and the Sun but is too distant to obscure the latter completely.

aphelion The point in an object's solar orbit at which it is farthest from the Sun.

apocynthion The point in an object's lunar orbit at which it is farthest from the Moon's center.

apogee The point in an object's orbit at which it is farthest from the body it is orbiting.

asteroid A small, rocky object that orbits the Sun.

B

Baily's beads Bright spots visible around the Moon's shadow just before and just after an annular or total solar eclipse.

barycenter The center of mass (and the central point) around which two objects orbit each other. The Earth–Moon barycenter lies within Earth itself.

basalt A dark, iron-rich type of volcanic rock. The Moon's maria are largely composed of basalts.

basin An impact crater larger than 186 miles (300 km) in diameter.

blood moon The red Moon of a total lunar eclipse. It is caused by the blue wavelengths of sunlight being scattered by Earth's atmosphere and its red wavelengths being refracted toward the Moon.

breccia A type of rock made up of fragments of rocks that were fused together by the force of meteor impacts.

C

Callippic cycle The 76-year period at the end of which the Moon's phases occur on the same date and at the same time.

celestial equator Earth's equator projected into space, cutting a plane through the celestial sphere.

celestial poles The two points in the sky where Earth's axis (extended north and south) intersects with the celestial sphere.

celestial sphere An imaginary globe, centered on Earth, around which the Moon, the Sun, and other celestial objects appear to be positioned. The sky above is the half of the sphere seen from Earth.

circumpolar At, or around, a celestial body's poles. Circumpolar stars are those that never set at a given latitude on Earth.

Coordinated Universal Time (UTC) The standard global time system—Greenwich Mean Time.

Copernican The geological period from roughly 1.1 billion years ago to the present day.

crust The outermost layer of a planetary body.

D

declination The angle at which a celestial object lies either above or below the celestial equator.

diamond ring A thin band of sunlight with only one Baily's bead visible (the ring's "diamond"), which appears around the Moon's shadow just before and just after an annular or total solar eclipse.

draconic month The 27.2 days it takes for the Moon to return to the same node (ascending or descending) as it crosses the ecliptic.

E

Earthlight Light reflected from Earth that dimly illuminates the darkened portion of a crescent Moon. It is best seen at dawn and dusk.

eccentricity A measure of how elliptical an orbit is, or how much it varies from a perfect circle.

eclipse A event in which one celestial body moves into the shadow of another. Seen from Earth, this causes the shadowed body (either the Sun or the Moon) to become partly or wholly obscured.

ecliptic An imaginary plane representing Earth's orbit around the Sun.

ejecta Material thrown up by an explosive event, for example, when one celestial body violently strikes another (often forming a crater, as on the Moon) or during a volcanic eruption.

epicycle A circle made by a planet in its orbit to account for its apparent retrograde motion. The idea was popularized by Ptolemy (c.100–170), who believed that the planets (including the Moon) orbited in circles around Earth.

equinox One of two times a year (in March and September) at which Earth's axis is angled neither toward nor away from the Sun. This gives the northern and southern hemispheres almost equal periods of night- and daytime.

Eratosthenian The geological period from 3.16 to about 0.8 billion years ago.

escape velocity The minimum speed required for an object to break free from the gravitational pull of a celestial body.

exeligmos cycle A cycle of eclipses over a period of 54 years and 33 days. Astronomers use it to predict the location and features of future eclipses.

extra-vehicular activity (EVA) Tasks performed by an astronaut outside a spacecraft.

G

general relativity A theory published by Albert Einstein in 1915 proposing that the effect we feel as gravity is caused by the curvature of space and time (collectively, "spacetime"), which are distorted by massive bodies such as planets.

geocentric Having Earth as a central focal point, or relating to the center of Earth.

gibbous The Moon when it is more than half full but not yet completely full. A "waxing gibbous" Moon is one that becomes fuller each night.

graben A trench caused by the Moon's surface contracting, forcing surrounding areas to slump.

gravitational lensing A visual effect that occurs when a massive object, such as a galaxy, bends light traveling from a distant source toward an observer.

H

HALO orbit A periodic three-dimensional orbit involving the Sun, Earth, and a spacecraft or artificial satellite centered on a Lagrange point.

heliacal rising The annual rising of a star or planet just before dawn on the eastern horizon.

heliocentric Having the Sun as a central focus, or related to the Sun's center.

High Earth Orbit An orbit of Earth at around 22,370 miles (36,000 km) from its surface or 26,199 miles (42,164 km) from its center.

I

Imbrian The period between the formation of the Imbrium Basin, 3.8 million years ago, and the beginning of the Eratosthenian period, approximately 3.2 billion years ago.

isotope Atoms with the same number of protons but a different number of neutrons.

K

Kepler's laws Three laws of planetary motion, published by astronomer Johannes Kepler (1571–1630), which describe the elliptical orbits of the Moon and planets.

KREEP An acronym for K (the atomic symbol for potassium), REE (rare-earth elements), and P (for phosphorus)—a geochemical component of certain lunar basalts and breccias that may have formed when Theia collided with Earth.

L

Lagrange points The points of gravitational equilibrium between two orbiting bodies, such as Earth and the Moon. Lagrange points enable spacecraft to maintain stable positions in space with the minimum of fuel consumption.

Late Heavy Bombardment The period around 4 billion years ago when the evolving orbits of the solar system's giant outer planets disrupted the paths of comets and asteroids, causing them to bombard the inner planets and their moons.

latitude The distance north or south of a celestial body's equator, expressed in degrees.

libration A visual effect that creates an apparent shift in the Moon's position in the sky, exposing areas of the lunar surface that are not otherwise visible. It is caused by a misalignment between the Moon's orbital and rotational planes.

lobate scarp Long, clifflike structures caused by compressions or extensions of the lunar surface.

longitude The easterly or westerly position of an object on a celestial body relative to a meridian line (an imaginary circle running through both poles).

Low Earth Orbit An orbit relatively close to Earth, defined by NASA as up to 1,200 miles (2,000 km).

lunar calendar A calendar based on the monthly cycle of the Moon's phases.

lunar eclipse An eclipse that occurs when Earth comes between the Moon and the Sun, causing Earth's shadow to fall on the Moon.

lunar parallax The apparent change in the Moon's position as seen from two different points on Earth.

Lunar Reconnaissance Orbiter (LRO) A NASA-built robotic spacecraft launched in 2009, currently still in orbit around the Moon.

lunar standstill The most northerly or southerly points that the Moon reaches in a month (in other words, when it is farthest above or below the equator). Also known as a "lunistice."

lunisolar calendar A calendar that indicates both the Moon's phase and the date in the solar year.

M

mantle The layer within a planetary body between its outer crust and its inner core.

maria Latin for "seas" (singular: "mare")—the large, relatively flat expanses of the lunar surface that appear dark when seen from Earth.

mascon A "mass concentration" of dense material below the surface of a planetary body. It creates a stronger gravitational pull in that area that can affect orbiting spacecraft.

meteorite A meteoroid that has passed through a planet's atmosphere and landed on its surface.

meteoroid A small, mobile celestial body, varying in size from a dust grain to an asteroid. If it enters a planet's atmosphere, it is known as a "meteor."

Metonic cycle The period of almost 19 years it takes for the Moon to return to exactly the same point in the sky at the same date on the calendar.

N

Nectarian The period between the formations of the Nectaris Basin, 3.9 billion years ago, and the Imbrium Basin, 3.8 billion years ago.

nodal precession The gradual alteration in the orbital plane of a satellite (such as the Moon) around the rotational axis of a planetary body.

nodes The two points at which the Moon crosses the ecliptic plane in its monthly orbit of Earth: the ascending (south-to-north) node and the descending (north-to-south) node.

O

orbit The regular, repeated path of a planet around another celestial body. All planetary bodies in the solar system have elliptical (oval-shaped) orbits.

outgassing The release of gas held within material.

P

parallax The apparent change in the position of an object in the sky when seen from different points of view on Earth—a vital tool in calculating the distances of very remote stars.

payload The cargo and/or people carried by a spacecraft, necessary for the successful completion of a mission's main objectives.

penumbra The outer, lighter part of a cone-shaped shadow cast by a celestial body during an eclipse.

pericynthion The point in an object's lunar orbit at which it is closest to the Moon's center.

perigee The point in an object's orbit at which it is closest to the body it is orbiting.

perihelion The point in an object's solar orbit at which it is closest to the Sun.

phase cycle The series of different shapes that the Moon appears to assume during the course of a lunar month (about 29.5 days). This ranges from an illuminated circle (full Moon) to an entirely dark form (new Moon).

pre-Nectarian The period between the formation of the Moon's crust, approximately 4.5 billion years ago, and the formation of the Nectaris Basin, 3.9 billion years ago.

precession The gradual shift in a celestial body's orbital path or the direction of its rotational axis. Apsidal precession affects a celestial body's orbit; axial precession affects its axis of rotation.

protoplanet A planetary embryo, consisting of a revolving mass within a cloud of dust and gas, in orbit around a sun.

R

reflector telescope A telescope with curved mirrors that reflect light to form images.

refractor telescope A telescope with lenses that bend light to a focal point to form images.

regolith Loose material, ranging from dust to rocks, lying above the bedrock of a celestial body.

right ascension The angle at which a celestial object lies on the celestial equator measured eastward from the point of the March equinox.

rille A long, narrow channel on the lunar surface, usually the remains of a lava stream.

S

saros cycle The 223 synodic months (roughly 18 years and 11 days) it takes for the Sun, the Moon, and Earth to return to the same relative positions.

satellite A machine, moon, or planet in orbit around a star or planet.

Saturn V The super heavy-lift launch vehicle used by NASA from 1967 to 1973 to send the crewed Apollo missions to the Moon.

sidereal month The roughly 27.3 days it takes for the Moon to return to the same position in the sky relative to the background of stars.

slingshot The technique of using the Moon's gravity to alter the velocity or direction of a spacecraft. Also known as a "gravity assist maneuver."

soft landing A landing in which a spacecraft uses retrorockets to slow its descent.

solar calendar A calendar based on the roughly 365.4 days it takes for Earth to orbit the Sun.

solar wind The continuous flow of charged particles spreading across the solar system from the Sun's corona (outermost atmosphere).

solstice The time when the Sun reaches its highest or lowest point in the sky. This happens twice a year, on June 21 (the summer solstice—the longest day) and December 21 (the winter solstice—the shortest day).

Space Launch System (SLS) NASA's latest super heavy-lift launch vehicle designed to take astronauts to the Moon and beyond.

space race The rival space programs of the US and USSR during the Cold War period. It is often dated from 1957 (when the USSR launched the first functional satellite, Sputnik 1) to 1969 (when the US Apollo 11 astronauts set foot on the Moon).

spacetime A conceptual model that combines the three dimensions of space with the one dimension of time, creating a four-dimensional whole, or "spacetime continuum."

special relativity A theory presented by Albert Einstein in 1905 that describes how space, time, mass, and speed are related. It is associated with the equation $e = mc^2$, in which energy (e) is equal to mass (m) times the speed of light (c) squared.

spectrometer An instrument that splits light into various wavelengths.

synodic month An alternative term for "lunar month"—the time it takes for the Moon to complete a single orbit of Earth (roughly 29.5 days).

T

terminator The transition zone between the day (illuminated) and night (unilluminated) sides of a planet or moon; as the body rotates, this line moves around it.

terrae The highland areas of the Moon that rise above the darker, younger *maria*.

Theia A theoretical protoplanet of the early solar system. According to the giant-impact hypothesis, it collided with Earth roughly 4.5 billion years ago, creating the debris that formed the Moon.

tidal locking The phenomenon of a celestial body's rotation around its own axis taking the same amount of time as its orbit around another body. The Moon is tidally locked to Earth, so we only ever see one of its sides.

U

umbra The inner, darkest part of a cone-shaped shadow cast by a celestial body during an eclipse.

INDEX

Page numbers in **bold** denote main topics.

synodic months 28, 29
Synthetic Aperture Radar (SAR) 193
syzygy 68

T

Tabwa people 97
Taqī ad-Dīn 78
tarot cards 83
Taurus-Littrow Valley 161, 180–181, **235**
tectonic activity 10, 106, **170–171**
telescopes 110, 115, 116, 226
    invention of **102–103**
    magnification by 243
terminator line 52, 103
terrain
    highlands 74, **118–119**, 125, 136, 160, 180
    maria 74, 109, 118–119, 124, 136, 171, 184
    mountains 112
    see also rocks
The Regolith Ice Drill for Exploring New Terrain
    (TRIDENT) 225
Theia 12–13, 14
theory of relativity **132–133**
thorium 224–225
Thoth 50–51
tidal locking 17, 30, 31, 116
tides 16, **18–19**, **20–21**, 31, 93, 115
tilt, orbital 16–17, 33, 34
TLPs (Transient Lunar Phenomena) 191
Todd, Mabel Loomis 40
tourism 218, 219
toys 140
Tracy's Rock 180–181
Tranquillitatis, Mare (the Sea of Tranquility)
    124, 154, 156, 184, 186, 205, **233**
Transient Lunar Phenomena (TLPs) 191
TRIDENT (The Regolith Ice Drill for Exploring
    New Terrain) 225
tropical months 28
Tsiolkovsky (crater) 136, **238**
Tsiolkovsky, Konstantin 128, 238
Tsukuyomi 48
turtles 92, 93, 95
Tycho (crater) 107, 108, 116, 194, 234, **236**
Tycho Brahe 83, 100, 101, 236

U

umbra 39, 41
unicorns 91
United Arab Emirates (UAE) 195
United States of America (USA) 218–219, 204–205
    indigenous Americans 63, 72, **94–95**
    see also NASA; "space race"
Unurgunite 58
Ur-Nammu, King 45

V

V-2 missiles 128, 129, 135
valleys 124, 176–177
    Taurus-Littrow Valley 161, 180–181, **235**
    Vallis Alpes (Alpine Valley) 234
    see also rilles

Vallis Alpes (Alpine Valley) **234**
Van Allen, James 135
Van de Graaff (crater) **239**
Vanguard satellite program 135, 146
Vari-ma-te-takere 48
Venus (planet) 46
Venus (Roman goddess) 32–33
Verne, Jules 76, 77, 128, 130
Vikram lander 195
Virgin Galactic 219
volcanic activity 14–15, 119, **124–125**, 136,
    183, 184
    early theories of 106, 115, 180
    tectonic features 170
von Braun, Wernher 134, 135, 146, 147, 178
Von Kármán (crater) 200–201
Vostok 1 139

W

Wadjet 50
Wan Hu 77
Wang Yingming 69
waning crescent Moon 29, 242
waning gibbous Moon 29, 242
water carriers 74–75
water sources 12, 17, **196–197**, 212, 224, 225
    on Moon bases 220, 221
waxing crescent Moon 28, 47, 72, 242
waxing gibbous Moon 28, 242
Wells, H. G. 76, 130
werewolves 90, 91
wheels, rover 179
Whipple, John 122, 123
White, Edward 149
winter solstice 26, 33, 34
Wiseman, Reid 209
witchcraft 90, 91
women 87, 89, 95, 97, 147
    menstrual cycles and pregnancy 32, 87, 89
woodcutter 74, 75
Worden, Alfred M. "Al" 177
World War II 128, 129, 135, 140
wrinkle ridges 143, 171, 234

X

Xerxes I, King 37

Y

Yanktonais Nakota people 94
Yin and Yang 66
Yolngu people 58
Yoshitoshi, Tsukioka 86
Young, John 141, 149, 180
Yutu (Moon rabbit) 66
Yutu rover 194, 200, 205

Z

Zhang Heng 68
Zhang Sengyou 66–67
Zhu Chongzhi 68
Zhuravlov, Vasili 130

Ziggurat of Ur 44–45
zodiac 56, 57, 82, 88
Zond space program 153, 204–205
Zulu people 73

ACKNOWLEDGMENTS

Dorling Kindersley would like to thank Christine Stroyan for her managerial read; Diana Vowles for proofreading; Tim Topper for his authenticity read; Elizabeth Wise for the index; Noor Ali for design support; Priyanka Lamichhane for additional fact-checking.

The publisher would like to thank the following for their kind permission to reproduce their photographs:

(Key: a-above; b-below/bottom; c-centre; f-far; l-left; r-right; t-top)

2-3 NASA. 4 Science Photo Library: NASA. **6-7 NASA. 8-9 123RF.com:** pockygallery. **10-11 Rich Addis. 11 NASA:** Goddard Space Flight Center / Arizona State University (bc). **14-15 NASA:** Goddard Space Flight Center. **18 Michael Marten Photography. 19 CNES:** (b). **20 Don Dixon/www.cosmographica.com. 22-23 Getty Images:** Moment / Javier Zayas Photography. **24 Getty Images:** Francois Ducasse (b). **Wikipedia:** Joey Kentin (cl). **25 Peabody Museum of Archaeology & Ethnology Harvard University:** Gift of Elaine F. Marshack, 2005 (2005.16.2.318.38). **26-27 Landesamt für Denkmalpflege und Archäologie Sachsen-Anhalt. 27 John Davis / www.bucksnortobservatory.com:** (tr). **28-29 Science Photo Library:** NASA's Scientific Visualization Studio (b). **31 Alamy Stock Photo:** Visual Arts Resource. **32 Alamy Stock Photo:** Japhotos. **32-33 Getty Images:** David Clapp (b). **33 Alistair Coombs / alistaircoombs.com:** (t). **34 Alamy Stock Photo:** Charles Bowman (l). **35 Bridgeman Images:** From the British Library archive (bl). **36-37 Anthony Murphy / mythicalireland. 37 Alamy Stock Photo:** Universal Art Archive (t). **© The Trustees of the British Museum. All rights reserved:** 1876,1117.1961 (b). **39 NASA:** Aubrey Gemignani (cr); Bill Ingalls (cb); Hinode / XRT (crb). **Science Photo Library:** Miguel Claro (tr); Juan Carlos Casado (Starryearth.com) (ca, cra). **41 Science Photo Library:** Juan Carlos Casado (Starryearth.com) (t); NASA Earth Observatory (cr). **42-43 Greg Meyer / gregmeyerphotos.com. 44 Alamy Stock Photo:** Hoberman Publishing. **44-45 Dreamstime.com:** Sergey Mayorov. **45 © The Trustees of the British Museum. All rights reserved:** 1907,1014.1 (t). **48 Alamy Stock Photo:** Universal Art Archive. **Wikipedia. 49 Alamy Stock Photo:** Visual Arts Resource (tr). **CONSTANTINE:** (l). **The Metropolitan Museum of Art:** Rogers Fund, 1985 (cr). **50-51 Science Photo Library:** Juan Carlos Casado (Starryearth.com). **50 Alamy Stock Photo:** funkyfood London - Paul Williams (b). **51 The Metropolitan Museum of Art:** Rogers Fund, 1948 (b). **Wikipedia:** Diego Delso (tr). **52 Alamy Stock Photo:** Pavel Dudek (b). **Science Photo Library:** Eckhard Slawik. **53 Ángel M. Felicísimo. 54 Science Photo Library:** Science Source. **56 Images First Ltd / Tony Freeth PhD:** (bc). **Shutterstock.com:** Cardiff University / EPA (t). **57 Getty Images:** SSPL (br). **Images First Ltd / Tony Freeth PhD. 58 Dreamstime.com:** Kontrymphoto (l). **Stuart McKay:** (tr). **59 National Museum of Australia:** Buku-Larrnggay Mulka Centre (b). **60 Los Angeles County Museum of Art:** Gift of Paul F. Walter (M.83.219.2). **61 British Library:** (t). **The Cleveland Museum of Art:** Howard Agriesti (b). **62 SLUB Dresden. 63 Alamy Stock Photo:** Janzig / USA (b). **Princeton University Art Museum:** Bequest of Gillett G. Griffin (t). **64-65 Science Photo Library:** Miguel Claro. **66-67 Wikipedia. 67 The Metropolitan Museum of Art:** Fletcher Fund, 1928 (br); Gift of Douglas Dillon, 1981 (bl). **68 Alamy Stock Photo:** Imago. **69 Alamy Stock Photo:** Granger - Historical Picture Archive (c); Robert Burch (t). **Library of Congress, Washington, D.C.:** (br). **70-71 Alamy Stock Photo:** Universal Art Archive. **72 Alamy Stock Photo:** Xinhua (br). **Getty Images:** John Theodor (bl); muratekmen (c); Narinder Nanu (cr). **73 Alamy Stock Photo:** Neil Cooper (bl). **Getty Images:** Kevin Frayer (t); maodesign (cl); NurPhoto (cr); Sowetan (br). **74 Alamy Stock Photo:** Colin Underhill. **75 akg-images:** Roland and Sabrina Michaud (ca). **Coco Obsidain:** (tc). **Shutterstock.com:** Jaya Kesavan (c). **76 Alamy Stock Photo:** Makota Sakurai (r). **Roland Smithies / luped.com:** Ruth Cobb (l). **77 Bridgeman Images:** North Wind Pictures (br); Universal History Archive / UIG (bl). **Science Photo Library:** Sheila Terry (tr). **Wikipedia:** United States Civil Air Patrol (tl). **78 Alamy Stock Photo:** robertharding. **79 Alamy Stock Photo:** Aliaksandr Mazurkevich (b). **Bridgeman Images:** Photo Josse (t). **80 Getty Images:** Alan Dyer / Stocktrek Images (l); Seth Joel (r). **81 Alamy Stock Photo:** eye35.pix (cr). **NASA. 82 Alamy Stock Photo:** Universal Art Archive (l, r). **83 Bridgeman Images:** Photo Josse (l). **Wellcome Collection:** MS.8932 (r). **84-85 AF Fotografie. 86 Alamy Stock Photo:** Makota Sakurai. **87 AF Fotografie:** (br). **John Aster Archive:** (t). **The Metropolitan Museum of Art:** Rogers Fund, 1921 (l). **88-89 Bridgeman Images:** From the British Library archive. **89 Alamy Stock Photo:** Album (tr). **Bavarian State Library, Munich:** BSB Clm 14456 (br). **Wikipedia:** DeFacto (l). **90 Alamy Stock Photo:** Alexeyev Filippov (t). **Bridgeman Images:** Leonard de Selva (br); UCL Art Museum, University College London (bl). **91 Alamy Stock Photo:** Universal Art Archive (tr). **Bridgeman Images:** Stefano Bianchetti (br). **Getty Images:** Photo Josse / Leemage (l). **92 Getty Images:** Arterra (t). **Juan Ma Contortrix:** (b). **93 Alamy Stock Photo:** Blue Planet Archive (b). **naturepl.com:** David Tipling (t). **94 Alamy Stock Photo:** Science History Images (t). **Buffalo Bill Center of the West:** (b). **95 Alamy Stock Photo:** Lee Rentz (t). **Wikipedia:** Pierre5018 (b). **96 Getty Images:** Anthony Pappone. **97 The Metropolitan Museum of Art:** Purchase, Mrs. Howard J. Barnet Gift, 2015 (l). **Photo Scala, Florence:** The Metropolitan Museum of Art / Art Resource (r). **98-99 Barry Lawrence Ruderman Antique Maps Inc. 100 The Metropolitan Museum of Art:** The Elisha Whittelsey Collection, The Elisha Whittelsey Fund, 1959. **101 Getty Images:** Heritage Images (tr). **Science Photo Library:** Paul D Stewart (br). **102 Getty Images:** Eric Vandeville / Gamma-Rapho. **103 Science Photo Library:** Max Alexander / Lord Egremont (cl). **SuperStock:** Iberfoto Archivo (cr). **104 SMA - Sistema Museale di Ateneo, Alma Mater Studiorum - Università di Bologna. 105 SMA - Sistema Museale di Ateneo, Alma Mater Studiorum - Università di Bologna. 106 NASA:** Goddard / Arizona State University. **107 NASA:** GSFC / Arizona State University (cla); Lunar Reconnaissance Orbiter Camera (bl). **108 NASA:** GSFC / Arizona State University (bc); GSFC / Arizona State University (br). **Science Photo Library:** Babak Tafreshi (tl). **109 NASA:** Johnson Space Center (t, b). **110 Alamy Stock Photo:** Visual Arts Resource (bl). **Science Photo Library:** Max Alexander / Lord Egremont (cl). **Wikipedia:** (cr). **111 AF Fotografie:** (br). **Bridgeman Images:** British Library Archive (tr). **Getty Images:** Sepia Times / Universal Images Group (bl). **Science Photo Library:** Royal Astronomical Society (tl). **112 NASA:** NASA's Scientific Visualization Studio (cla). **113 Shutterstock.com:** elRoce. **114 Bridgeman Images:** Giancarlo Costa. **115 Science Photo Library:** Library of Congress, Rare Book And Special Collections Division (crb). **116 NASA:** NASA's Scientific Visualization Studio (tl, cla, clb, bl). **Science Photo Library:** Damian Peach (br). **118 NASA. 119 NASA:** GSFC / Arizona State University (t). **120-121 AF Fotografie. 122 The Metropolitan Museum of Art:** (br). **NYU Special Collections, New York University:** (bl). **The Royal Society:** (tr). **123 David Rumsey Map Collection, David Rumsey Map Center, Stanford Libraries:** (bl). **Image courtesy of the Hastings Historical Society, Westchester County, New York:** (tl). **Rich Addis. 124 NASA:** GSFC / Arizona State University (crb). **Science Photo Library:** NASA / GSFC / Arizona State University (tr). **125 NASA:** JSC. **126-127 NASA:** JSC. **128 Alamy Stock Photo:** Alexeyev Filippov (ca). **129 Bridgeman Images:** Stefano Bianchetti (bl). **Science Photo Library:** Detlev Van Ravenswaay (r). **130 Alamy Stock Photo:** SilverScreen (r). **Shutterstock.com:** Melies / Kobal (bl). **131 Alamy Stock Photo:** Cinematic / Universal (br). **BFA**; MGM (bl). **Shutterstock.com:** Everett (tr). **132 Science Photo Library:** European Southern Observatory (c). **133 Alamy Stock Photo:** Archivio GBB (tr). **NASA:** ESA / Hubble (br). **Science Photo Library:** Claus Lunau (bl). **134 Dan Beaumont:** Collier's. **135 Getty Images:** Keystone-France / Gamma-Keystone (bl). **NASA:** NASA's Scientific Visualization Studio (br). **Science Photo Library:** Detlev Van Ravenswaay (ca). **136 NASA:** Caltech-JPL / MIT / SRS (cra); JPL-Caltech / IPGP (bl). **136-137 NASA:** GSFC / Arizona State University. **137 NASA:** GSFC (crb). **138 John F. Kennedy Library Foundation. 139 Alamy Stock Photo:** Alexeyev Filippov (tr). **140 Alamy Stock Photo:** AF Fotografie. **Garpenhus Auktioner AB:** (bl). **141 1stdibs, Inc:** (tr). **Alamy Stock Photo:** Futuras Fotos (br); Victoria Lipov (cl). **National Air and Space Museum Archives, Smithsonian:** Norman Rockwell Licensing, Niles, IL (bl). **Science Photo Library:** Detlev Van Ravenswaay (cr). **Roland Smithies / luped.com:** (cra). **142 NASA:** JPL (tl, tc, tr). **Science Photo Library:** Detlev Van Ravenswaay (bc). **143 NASA:** JPL (tl, tc, tr). **National Air and Space Museum Archives, Smithsonian:** (ftr). **U.S. Geological Survey:** (bc). **144 Science Photo Library:** Carlos Clarivan (bl); Carlos Clarivan (r). **145 Science Photo Library:** Carlos Clarivan (t); Carlos Clarivan (b). **146 NASA. 147 Alamy Stock Photo:** Alexeyev Filippov (tl); Futuras Fotos (tc). **NASA. 148 NASA. 149 NASA:** JSC (t); KSC (bc). **150-151 Alamy Stock Photo:** William Anders / NASA / Futuras Fotos. **152 National Reconnaissance Office. 153 Anatoly Zak / RussianSpaceWeb.com:** (br). **Getty Images:** Keystone-France / Gamma-Keystone (tr). **Wikipedia:** USSR Post (bl). **154 NASA. 155 Alamy Stock Photo:** Futuras Fotos (cr). **NASA. 156 NASA:** JSC (t); JSC (bl). **157 NASA:** JSC (tr); JSC (bl). **158-159 NASA:** Goddard Space Flight Center / DLR / ASU. **160 Library of Congress, Washington, D.C.:** Geography and Map Division Washington, D.C. **161 Alamy Stock Photo:** Futuras Fotos (Badges); **NASA:** Lunar Reconnaissance Orbiter Camera / Science Operations Center (NASA LRO Images). **National Air and Space Museum Archives, Smithsonian. Roland Smithies / luped.com:** Roy G. Scarfo (br). **162-163 National Air and Space Museum Archives, Smithsonian. 162 National Air and Space Museum Archives, Smithsonian. 163 NASA. National Air and Space Museum Archives, Smithsonian. 164 NASA. 165 NASA. 166-167 Getty Images:** Bettmann. **168 Alamy Stock Photo:** Futuras Fotos (b). **NASA. 169 Alamy Stock Photo:** Futuras Fotos (t). **170 NASA:** Dr. Mark Robinson, Arizona State University. **171 NASA. 172-173 NASA. 174 Alamy Stock Photo:** Futuras Fotos (t). **NASA:** (b). **175 NASA. 176 Alamy Stock Photo:** Futuras Fotos. **176-177 Andy Saunders - ApolloRemastered.com:** NASA / JSC / ASU (b). **177 NASA. 179 NASA. 180-181 NASA. 180 NASA. 181 Smithsonian National Air and Space Museum:** Daniel Soñé (t). **182 Alamy Stock Photo:** Matteo Chinellato (cr). **NASA. Science Photo Library:** NASA / JSC. **183 NASA. Science Photo Library:** Detlev Van Ravenswaay (b). **184-185 Alamy Stock Photo:** Science Photo Library. **185 NASA. Southwest Research Institute:** Robin M. Canup (tr). **186 Alamy Stock Photo:** Universal Art Archive (cr). **NASA. 187 Alamy Stock Photo:** RGB Ventures / SuperStock (br). **NASA. 188-189 NASA:** JSC. **190 NASA:** JPL / USGS. **191 Judit Slíz-Balogh, András Barta, Gábor Horváth:** (cr). **NASA. Society for Popular Astronomy / www.popastro.com:** (b). **192 NASA:** JPL-Caltech / CSM (b). **193 NASA:** JPL-Caltech (t). **Science Photo Library:** NASA (bl). **194 Alamy Stock Photo:** Xinhua (b). **NASA:** GSFC / Arizona State University (t). **195 ISRO. SpaceX:** NASA TV (t). **196 ISRO. 197 NASA. 200-201 CNSA. 200 CNSA. 202-203 Alamy Stock Photo:** Panoramic Images. **204-205 NASA. 205 NASA:** LRO (bc/background). **Shutterstock.com:** bodrumsurf (bc/Soviet Flag); Kasman Vector (bc/India Flag); tayyab900 (bc/US Flag). **206 ISRO. 207 CNSA:** (bc). **NASA:** (cr). **State Space Corporation Roscosmos:** (br). **208-209 Alamy Stock Photo:** Futuras Fotos. **208 NASA. 209 NASA. 210-211 Science Photo Library:** David Ducros. **210 NASA. 211 NASA. 212 Blue Origin:** NASA (bc). **SpaceX:** (b). **214 NASA. 215 NASA. 216-217 Wikipedia:** Jinzhu Ji, Dijun Guo, Jianzhong Liu, et al. (2022). **218 SpaceX. 219 NASA:** Intuitive Machines, LLC (t). **220 ESA:** Pierre Carril (b). **222-223 NASA. 224-225 NASA:** Lunar Prospector (b). **226 NASA:** Concept art by Volodymyr Vustyansky. © 2024 Jet propulsion Laboratory, California Institute of Technology. Government sponsorship acknowledged (c). **227 NASA:** DOE / Brookhaven National Laboratory (t); WMAP Science Team (b). **228-229 NASA:** JPL. **228 NASA:** JPL-Caltech (c). **229 Alamy Stock Photo:** Stocktrek Images, Inc. (t). **230-231 Science Photo Library:** Royal Astronomical Society. **232 NASA. 233 NASA. 234 ESA:** SpaceX, Space Exploration Institute (cr). **NASA. 235 NASA. 236 NASA. Science Photo Library:** Babak Tafreshi (b). **237 NASA:** LRO; LRO (tr); LRO (cl, br). **238 NASA. 239 NASA. 240-241 NASA:** GSFC / Arizona State University. **242 Science Photo Library:** NASA's Scientific Visualization Studio (b). **243 Philip Eales:** (bl). **Gerd Waloszek. NASA:** JSC / Arizona State University (t). **244 Wikipedia:** Callan Carpenter (bl). **245 Alamy Stock Photo:** Associated Press (br); Constantinos Iliopoulos (tl). **Getty Images:** Tuul & Bruno Morandi (cr)

Cover images: *Front:* **Alamy Stock Photo:** Chronicle l, NASA Image Collection cr; **Getty Images:** Moment / Javier Zayas Photography r; **Science Photo Library:** Miguel Claro cl; *Back:* **Adobe Stock:** Anastasia. *Endpapers* image: **Science Photo Library:** Eckhard Slawik

DK LONDON
Project Editor Daniel Byrne
Project Art Editor Daksheeta Pattni
Editors Robert Dimery, Elizabeth Dowsett,
Steven Setford, Alison Sturgeon, Andrew Szudek
US Senior Editor Kayla Dugger
US Executive Editor Lori Cates Hand
Designers Emma Forge, Tom Forge
Illustrator Jason Solo
Managing Editor Gareth Jones
Managing Art Editor Luke Griffin
Production Editor Robert Dunn
Production Controller Nancy-Jane Maun
Picture Researcher Sarah Smithies
Senior Jacket Designer Surabhi Wadhwa-Gandhi
Senior Jackets Art Editor Suhita Dharamjit
DTP Designer Vikram Singh
Senior Jackets Coordinator Priyanka Sharma
Jacket Designer Rhea Menon
Publishing Director Georgina Dee
Managing Director Liz Gough
Art Director Maxine Pedliham
Design Director Phil Ormerod

First American Edition, 2025
Published in the United States by DK Publishing,
a division of Penguin Random House LLC
1745 Broadway, 20th Floor, New York, NY 10019

Copyright © 2025 Dorling Kindersley Limited
21 22 23 24 25 10 9 8 7 6 5 4 3 2 1
001–341848–May/2025

A catalog record for this book
is available from the Library of Congress.
ISBN 978-0-5939-6670-9

DK books are available at special discounts when purchased
in bulk for sales promotions, premiums, fund-raising,
or educational use.
For details, contact: DK Publishing Special Markets,
1745 Broadway, 20th Floor, New York, NY 10019
SpecialSales@dk.com

Printed and bound in China

www.dk.com

This book was made with Forest
Stewardship Council™ certified
paper—one small step in DK's
commitment to a sustainable future.
Learn more at www.dk.com/uk/
information/sustainability

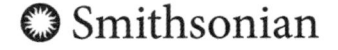

FOR SMITHSONIAN ENTERPRISES

Licensing Coordinator Avery Naughton
Editorial Lead Paige Towler
Senior Director, Licensed Publishing Jill Corcoran
Vice President of New Business and Licensing Brigid Ferraro
President Carol LeBlanc

FOR NATIONAL AIR AND SPACE MUSEUM

Curator of Rocketry Colleen Anderson
Curator of Aeronautics Russ Lee
Curator of International Space Programs and Spacesuits
Cathleen Lewis
Curator of Contemporary Spaceflight Emily Margolis
Curator of the Apollo Collection Teasel Muir-Harmony
Curator of Astronomy Samantha Thompson
Curator of Planetary Science and Exploration Matt Shindell

Established in 1846, the Smithsonian is the world's largest
museum and research complex, dedicated to public education,
national service, and scholarship in the arts, sciences, and history.
It includes 21 museums and galleries and the National Zoological
Park. The total number of artifacts, works of art, and specimens in
the Smithsonian's collection is estimated at 155.5 million.